Book Promotion:

The book, *"It Never Happened," Volume 1*, is a riveting and amazing account of U.S. Air Force cover-up of the "UFO question." It is authored by retired Air Force Captain David Schindele who was a former Minuteman ICBM launch control officer stationed at Minot Air Force Base in North Dakota. It was there, in 1966, when he was involved with a spectacular incident where a flying object managed to take down all ten of his nuclear-tipped missiles, and render them unlaunchable! This very grave situation involved the strategic U.S. nuclear deterrent force, which was America's first line of defense in the Cold War. It resulted in the Air Force "instructing" Captain Schindele to never speak of the incident again, and telling him "As far as you are concerned, It Never Happened."

Captain Schindele maintained his silence for nearly forty years until other former missileers began to speak out about their incredible unearthly incidents. It was then that he attempted to learn the long history of the Air Force cover-up. He needed to find out why the Air Force publicly insisted that flying objects didn't exist, and why it was stated that there was "nothing to investigate." He also found he needed to confirm to the world the awesome Truth that, "We are not alone in the Universe."

The book reveals the enormous attention given to Unearthly Flying Objects by the Air Force, which was documented in its own investigative efforts code-named Project Sign and Project Grudge. Those projects are examined in detail, which outlines a trail of convoluted and contorted investigative efforts to determine modus operandi of the objects and their potential threat to national security. Those efforts, however, were impacted by crucial and essential efforts by the Air Force to protect Truth of the UFO question, and the closely held, magnificent, and Monumental Secret of Roswell and other recoveries.

Book Reviews:

Diane English (Vashon, Washington)–

I was someone who had heard about UFO's my entire life, but never took the information seriously. This book, however, has changed my outlook 100 percent. I do not see how it can be read without raising serious doubts among those who do not believe. Each incident gave me pause, forced me to reread and even raised goosebumps on occasion. The book is like a dam breaking for the author. A very interesting and, yes, fascinating read.

Bill DeGroodt (Mill Creek, Washington)–

As a former Combat Crew Commander on Minuteman I and III, at Minot Air Force Base (1967-1972), I read the book "It Never Happened" with great interest. While I did not experience any incidents at Minot AFB, I clearly remember others talking about sightings and situations. I particularly enjoyed the depth of research, and learning about those who were withholding the Truth. We were trained to such a high level of readiness that it is unconscionable, on the part of the USAF, that we did not receive instructions on how to react to situations such as those mentioned in the book. To say the least, the book was extremely enlightening, especially after now obtaining a good picture of how much of the Truth is truly being covered up!

Tom Waterman (Bothell, Washington)–

A very interesting book, and it makes one think that maybe there really are UFO's out there! It captured my imagination, especially the chapter about Roswell, New Mexico. There is vivid detail about UFO sightings, and things our government does to cover it up. Overall, it is very well written and thought provoking, and I found it hard to put down once I really got into it. It is an amazing and genuinely alarming book to read.

Book Endorsement:

Jim Klotz (Researcher, Co-author Faded Giant)–
Captain David Schindele, USAF retired, is a fully credible source which is rare in the field of Unidentified Flying Objects (UFOs). In his book, "It Never Happened"–(an instruction to keep a UFO experience secret)–Capt. Schindele has not only provided an account of extraordinary National Security level events which affected a USAF Minuteman nuclear missile facility to which he was assigned–strikingly similar to events which have occurred at other Minuteman facilities–but also presents a meticulously researched history of US Air Force UFO projects which are often, rightly in my opinion, characterized as cover-ups. This history is presented in a concise, straight-forward, thorough fashion which is a rare thing in itself.

If you think there are no credible reports of UFO events, think there never has been anything to UFOs–that certainly UFOs couldn't affect National Security–or don't believe in a long-standing UFO cover-up, I invite you to read "It Never Happened" and think again!

"It Never Happened"
Volume 1

"It Never Happened"

U.S. Air Force

UFO Cover-up

Revealed

Volume 1

Capt. David D. Schindele

USAF Retired

Send inquiries to the author at:
DavidSchindele@EdgarRockPublishing.com

Ask for this book at your favorite book store,
or order it at: EdgarRockPublishing.com

ISBN: 978-0-9986890-4-3 (softcover)

First Printing: May 2017

Printed in the United States of America.

EdgarRock Publishing, LLC
181 Fontaine Lane
Naches, WA 98937

EdgarRockPublishing.com

CONTENTS

CONTENTS

Foreword

By Robert L. Hastings

The reality of UFO incursions at American nuclear weapons facilities has been convincingly established. Hundreds of U.S. military veterans now openly discuss these ominous incidents, and thousands of declassified government documents expose a longstanding, widespread, and ongoing situation.

Indeed, it is now known that in January 1945, seven months prior to the atomic bombings in Japan, U.S. Navy pilots unsuccessfully attempted to intercept a "ball of fire" that had hovered over the Hanford plutonium processing plant in Washington State on three different nights, all of the intrusions confirmed by radar. And, as recently as October 2010, U.S. Air Force nuclear missile technicians reported observing a huge, cigar-shaped craft maneuvering over portions of the F.E. Warren AFB nuclear missile field, in western Nebraska, just as the base inexplicably lost the ability to communicate with five of its underground launch control capsules.

During the Cold War era, incidents such as these were, from the Pentagon's point of view, distressingly frequent. Over the past four decades, I have interviewed more than 150 Air Force veterans–ICBM launch, targeting, and maintenance personnel, as well as missile security guards–who report their involvement in one case or another. Without exception, those individuals state that they had been told by their superiors, or by Air Force intelligence officers, that the UFO activity they witnessed was highly classified and, therefore, never to be discussed, even among themselves. Often they were told, "This never happened."

Given these disclosures, it becomes clear that the UFO-Nukes Connection is highly significant and perhaps even the key reason these mysterious aerial craft have appeared in our skies over the past seven decades.

Perhaps the most stunning accounts are provided by former U.S. Air Force missile launch officers who have described multiple ICBMs mysteriously malfunctioning moments after a disc-shaped craft was observed hovering near their underground launch silos or launch control capsule. In this book, retired Captain David D. Schindele reveals what occurred at Minot AFB, North Dakota in 1966, when he and his missile commander were suddenly drawn into one of those incidents upon relieving the two launch officers who were directly involved in it.

Schindele, having been warned that the ICBM-shutdown event affected national security, kept his silence about it for more than 40 years, believing that to do otherwise would be unpatriotic. Nevertheless, over time, he began to sense that a larger issue was involved–the American public's right to know–and, after much soul-searching, decided that he must speak out about this amazing, still-classified case.

I stumbled upon Schindele's blog account in 2010, and interviewed him shortly thereafter. It was immediately apparent that he was determined to report his experience as accurately as possible and to provide as much detail as he could remember, given the passage of four decades.

At the same time, he was clearly conflicted–concerned that his candor might result in untoward consequences for himself and his family. I assured Schindele that not one of my U.S. veteran sources had ever been harassed by the Air Force or any other governmental organization, following their interviews. Whether that is due to the unwanted publicity that might result–from the government's perspective–is unknown. Regardless, none of them have been so much as warned since revealing what they know about these incidents. To date, Schindele has not suffered any repercussions as a result of his actions.

Fortunately for the reader, David Schindele approaches the history of the UFO Cover-up in a scholarly manner, using a wide range of declassified documents to provide a comprehensive summary of the covert activities and secret policy implemented by the U.S. government, all designed to keep the American people in the dark. Year after year, public relations personnel at the Pentagon routinely dismissed UFO reports as being due to the misidentification of manmade aircraft, meteorological and astronomical phenomena, or hoaxes. Meanwhile, behind the scenes, the military's internal, classified communications reveal a deep, ongoing concern over the presence of unknown, highly-advanced aerial craft in American airspace, often near nuclear weapons installations.

For example, the same year that Captain Schindele and his colleagues experienced the mass-missile-shutdown incident at Minot's November Flight, disc-shaped UFOs were sighted at a nearby group of ICBMs, designated Mike Flight. According to U.S. Air Force documents, on August 24, 1966, a missile officer in Mike's underground launch control facility discovered that communications on his two-way radio were being disrupted by static. At the same time, Air Force security guards at ground level reported a UFO maneuvering in the vicinity. At one point it appeared to land, at which time the missile site control dispatched a security "strike-team" to investigate. At about ten miles from the landing site, the team's radio was also disrupted by static. Shortly thereafter, the UFO became airborne and passed beneath a second, identical object which was higher up in the sky. The report stated that Air Force radar had tracked both of the objects.

Were these mysterious aerial craft Soviet secret weapons, sent to disrupt American strategic missile systems? Unlikely, given that Soviet Army and KGB documents smuggled out of Russia in the early 1990s reveal multiple UFO incursions at Soviet ICBM sites during the Cold War era. One incident, in October 1982, involved the temporary, unexplained activation of several missiles just as a huge disc was sighted hovering over the base. According to the documents, after 15 terrifying seconds the unauthorized launch count-down ceased and all systems returned to standby status.

In other words, those operating the UFOs—whoever they are—appeared to be playing an even-handed game during the decades-long standoff between the U.S. and U.S.S.R., monitoring and occasionally tampering with both superpowers' deadliest weapons.

To say that these revelations are momentous is an understatement. While many questions remain unanswered at this time, humankind appears to be confronted with an unprecedented challenge, from unknown entities, and one that seemingly continues today. Recent reports from civilians living within Malmstrom AFB's missile field, in Montana, suggest an ongoing pattern of UFO activity near the ICBM sites.

Consequently, David Schindele's efforts to enlighten and educate the public regarding the UFO reality and its link to nuclear weapons is an important and necessary development. Other Air Force veterans who have preceded him in coming forward with the facts, including former Minuteman missile launch officer Robert Salas, welcome his contributions to the revelation of Truth. Indeed, each of us owes these individuals a debt of gratitude for their courage and candor.

Prologue

Truth Revealed? Yes, that is what this book proclaims and establishes in regard to the U.S. Air Force cover-up of the UFO question. But what is Truth? That is a question philosophical thinkers have debated and thought about for eons. For many, the concept of Truth is absolute and self-evident, and it encompasses a very specific reality that cannot be denied. But Truth is also a concept embedded in the mantra of various religions and ideologies, and supported by strong "faith," which is antithetical to an agnostic view of what is materially knowable. Parenthetically, if you are someone who supports a particular religion, you may also understand and proclaim that there is no conflict between your God and the UFO question, because you know that God rules over Heaven, Earth, and the entire Universe. But that is another subject, which is not addressed in this book.

This book examines and identifies the specific "reality" of the UFO question, a reality that the U.S. Air Force has been vitally concerned with. It has nothing to do with "faith," or in "believing," but it is supported by sightings and encounters made by highly-reliable, reputable, and competent people. And it is something the Air Force has investigated and repeatedly documented in secret and top secret documents. It is also a reality supported by tangible recovered evidence, which has been carefully stashed away and protected by those in-the-know.

Many people, upon an introduction to something like this, would be inclined to put this book down, because it is obvious that any kind of proclamation about reality of flying objects is a myth promulgated by individuals who have nothing better to do. If that is what you believe, you are one more individual the Air Force need not be concerned about, because a major goal of the Air Force, and also a certain "controlling" agency of the government, is to steer public attention away from the UFO question, which helps protect the magnificent secret that they have maintained and managed for decades.

While serving in the military as an Air Force officer in the 1960s, I was one of a dozen individuals involved in an incredible incident. We stumbled upon the Truth, and we were caught totally by surprise. It was a momentous and unforgettable event in my life, and it was also a very serious one that took place in a format of pure science fiction, but also within earthly reality. Although others were with me at the time, I later found that our experience was not unique, and I discovered that many others in the Air Force were also instructed to remain silent about their incidents, and the magnificent secret.

Truth comes to light in this book because of major efforts by the Air Force to investigate a multitude of mysterious sightings and encounters reported by highly-reliable, reputable, and competent people. Although Air Force investigative action began in years prior to initial recovery of evidence, serious investigative activity by the Air Force began afterward in a series of specific projects to find answers. It became necessary to determine motivation and intentions of the flying objects, which often demonstrated harassing and threatening behavior around nuclear facilities and other highly-important military areas. Those actions by the objects warned of serious potential impact to national security.

The history of Air Force involvement with the UFO question, and the contorted investigative actions it has taken while protecting the magnificent secret, reveal Air Force duplicity and the Truth behind it all.

"When all is said and done, the axiom of Truth is that it cannot be denied"

David D. Schindele

CHAPTER ONE

Background

A New Perspective of Reality

The words, "It Never Happened," still echo loudly in my ears, even after fifty years. And it was not just to my ears those words were spoken. There were a dozen of us involved in the same incident who were all told the same thing. It was a verbal instruction (an order), which came through quite distinctly: "**As far as you are concerned, It Never Happened**." But we were not alone in hearing those particular words, which are words that have become a trademark symbol in U.S. Air Force parlance referring to something that will not be revealed or spoken of again. That "something" is more or less embraced as a subject the Air Force has managed to cover-up for decades. It involves a magnificent and monumental secret this book will reveal in startling detail. Some know it as the "UFO conspiracy."

At the time, people I worked closely with in the Air Force were some of the finest, most capable, and most competent individuals one could find anywhere. This is not to imply the Air Force and people with me were unique in that regard, but those with whom I worked were highly educated, psychologically profiled, given high security clearances, and specifically trained for their extremely critical and very technical jobs. They, like me, were entrusted with control, operation, and security of nuclear-tipped Intercontinental Ballistic Missile (ICBM) systems that were poised at a moment's notice to execute America's first line of defense in the Cold War. We were the inconspicuous warriors in a very serious business of nuclear deterrence. And the Air Force had complete confidence in us to perform our duties as required, and as directed through appropriate order.

We were under constant review and training to insure that our technical skills and ability to follow complex procedures met high, stringent requirements set before us. Total perfection in performance was mandatory, and no tolerance was allowed for error. That was the way the Air Force maintained ultimate confidence in us, and we served proudly in that regard. But as years went by, especially after leaving and retiring from the Air Force, those haunting and unforgettable words remained in the back of my mind. They served to remind me over and over again of the incident I once experienced. But I was not the only one for whom those words were left echoing in one's mind.

Truth of the incident, however, wasn't easy for many of us to incorporate into reality, or reconcile. It was an experience we were not prepared for. It was foreign to us, we had not been trained for it, and we were caught totally off guard that such an incident would happen.

Afterward, when told to forget the whole thing as if it never occurred, there was no further verification or confirmation available in talking about it with others. It was forbidden, and not allowed. We didn't talk with one another about it, with our wives, or with anyone else. The experience was put on the back shelf, and out of mind. For me, it would be relegated to a haunting mysterious incident from long ago, but it would reappear from time to time in recesses of the mind, which would then jog consciousness back to recall details impossible to forget.

This continued until several years ago when a shocking reinforcement brought the old experience back to a new life, and Truth became startlingly clear and more than obvious. It was a revelation for me, and my mind was set free with a new perspective of reality. It was then that the urge to find the conviction and courage to speak out began to take hold, for I knew I didn't have a moral right to further contain my silence.

Hard to Believe

I am quite aware that few people in this world would pause to thoroughly contemplate what I and those with me experienced. This is because it was such an incomprehensible enigma, and so very difficult to incorporate into reality. It was so far beyond normal human experience that any attempt at comprehension of what took place quickly creates doubt, suspicion, and skepticism.

There have been many others before me who have attempted to reveal Truth about their experience, and I'm referring to many reputable, credible, respected, and reliable people with high credentials who have been left by the wayside in their earnest and solemn efforts to speak out. None of them received backing by high authority, either within government, within the scientific community, or within high-profile media, where it would be hoped that Truth might be corroborated in an effort to properly inform the public about something so immensely important to all of humankind.

Yes, I am quite serious about this, but indulge me as I toss forth a couple of scenarios that come to mind.

Imagine that major news networks suddenly come out with an announcement from high national authority that radio signals originating from some distant star system, or galactic planet, have revealed the existence of other intelligent life. It might be recalled that astrophysicists have discovered thousands of planets circling stars in the heavens, some orbiting within zones that could potentially harbor life. It might also be recalled that scientists involved with the Search for Extraterrestrial Intelligence (SETI), a private nonprofit organization, have been searching distant radio signals in the cosmos to find such evidence.

More recently, in July 2015, a massive project was announced called Breakthrough Initiatives, which intends to enlist the world's leading experts and scientists, using a grant of $100 million, to search for life-indicating signals. Famous British physicist, Stephen Hawking, stated the following in regard to the project:

> *...I believe the Breakthrough Initiatives are critically important. To understand the universe, you must know about atoms, about the forces that bind them, the contours of space and time, the birth and death of stars, the dance of galaxies, the secrets of black holes, but that is not enough. These ideas cannot explain everything. They can explain the light of stars, but not the lights that shine from planet Earth. To understand these lights, you must know about life, about minds. We believe that life arose spontaneously on Earth. So, in an infinite universe, there must be other occurrences of life. Somewhere in the cosmos, perhaps, intelligent life may be watching these lights of ours, aware of what they mean. Or do our lights wander a lifeless cosmos - unseen beacons, announcing that here, on one rock, the universe discovered its existence? Either way, there is no bigger question. It's time to commit to finding the answer, to search for life beyond Earth. The Breakthrough Initiatives are making that commitment. We are life. We are intelligent. We must know.*

Imagine hearing such an announcement that life-indicating signals were found. How might people react to such news? Would the reaction be, "Hmmm, that's interesting," and then give no more thought to the subject? Possibly. But would some people be inclined to let their imagination proceed a bit further with thoughts of flying object stories once heard about, and then wonder about a possible connection?

Well, it's likely most people would not care to consider this scenario, or the possibility of such an announcement. It would be an exercise in fanciful dreaming, and there are better things to dream about. Evidently, it's the same with many leading scientists, some who may be convinced about life elsewhere, and even believe that the odds are overwhelmingly in favor. For many of them, with exceptions such as Stephan Hawking, a proclamation regarding the undeniable possibility of life elsewhere would risk a hard-earned reputation.

How about another scenario? Imagine that major news networks suddenly broadcast a sensational announcement from high national authority stating and confirming that "extraterrestrial" flying objects under "intelligent" control have regularly occupied air space above our heads, have done so for many years, and evidence is being released to confirm it. On hearing this, what would go through one's mind now? With surprised amazement, perhaps there would be expressions of, "Wow!"; "Oh my God!"; "Show me the evidence!" With keen interest, people might be on edge to learn more.

It is also possible, however, that most people would not care to give a second thought to this scenario either. For the general public, it stretches current reality to the extent that it's doubtful any sort of casual attention, or consideration, would be given. The majority of people would not care to contemplate either of the above "fairytale" scenarios. The subject is too quirky, and not worth the time to mull over such science fiction. That is the normal, rational, and understandable response to something like this.

 On the other hand, if one were to pause and actually contemplate for more than a moment, would it be possible that some "high authority" in those scenarios might actually force serious consideration, in comparison to the subject being addressed in this book?

If one were to stop and really contemplate the later scenario, one's initial reaction might depend on how the announcement was presented in regard to tone and character. Perhaps it would depend on whether it was presented in a factual manner, or tinged with emotional concern suggesting serious impact on continued progression with everyday life. Would sudden interest or thoughts about the announcement produce trepidation, foreboding, or other emotion? Would there be panic? Or would people just go on with daily life and simply absorb the new information?

It would undoubtedly come to attention, at some point, that the words "extraterrestrial" and "intelligent" suggest that an alien power arrived on planet Earth with far greater technical achievement and knowledge than currently possessed by humankind. Also, there might be realization that the arrival may have occurred long ago, especially when recalling flying object incidents and stories from the past. Perhaps there would be a reminder of previous jokes, speculation about flying objects, and realization that incidents such as "Roswell" were possibly true after all.

With the above, I'm simply pointing out that I have no illusions about what I'm up against with this book. I'm in the same boat as many others before me who had no agenda other than to draw serious attention to the fact that government agencies, including the Air Force, have been hiding a monumental secret for decades. It involves a cover-up of immense and unfathomable proportion.

With nothing to gain in speaking out, it is quite likely I'll simply be another of those added to a long list of lost voices. But I will have fulfilled a heartfelt moral obligation to inform others, and my mind will find peace. When you have experienced a mind-shaking and eye-opening unearthly experience, and have been warned by the Air Force to maintain a lie, an immense burden is suddenly thrust upon you. But it goes against all sense of what is morally right, or morally right toward all of humanity. It is therefore necessary to voice the Truth, which the Air Force itself has avoided and will continue to refute.

Credible People vs "Kooks"

There are a number of very credible and reputable individuals, including ranking Air Force and a few government people, who have come out in recent years to spread the Truth. But skeptics, debunkers, and other government officials have continued with their repeated attempts to bat it down. Those attempting to spread Truth understand that debunkers are basically of one mind, or have an "institutional" motivation to debunk. Debunkers, and especially skeptics, see it as their duty to discredit the "kooks," and they look upon their efforts as sort of a mental hobby. This continuing standoff has taken place quietly in the background, while the public continues to proceed with reality of daily life, and care less about such fantasy. Altogether, it is a subject nearly impossible to consider realistically, or rationally without some confirmation by high authority. But such apathy and disinterest by the public is what imbues the Air Force and other government agencies with continued confidence to remain quiet on the UFO question, and maintain the cover-up.

A great many respected, responsible, and reputable people with high credentials know the Truth. They know it because they have experienced it. For many years, they have waited patiently for the Air Force, or other high government authority, to make an announcement and confirm to the public that extraterrestrial objects under intelligent control do exist. They are acutely aware that the Air Force has known about this for decades, and they have also experienced firsthand how the

Air Force has covered it up. When you have been part of the government itself, and have been a pawn in the cover-up because of being instructed to never speak of what you witnessed, you then know the Truth, and you know the government does too.

Attracting attention to reality and Truth of the UFO question is difficult. No one cares to be distracted from everyday life and bothered by something that seems a bit preposterous, or very unlikely. For me, I continue to live a normal life, even though I have experienced the fact that the human race is not alone in this world, or the universe. I realize it is extremely difficult, or impossible, to relate to this idea without a confirming experience. That is why I encourage an open mind, at least when initially reading this book. In the discovery of what is detailed between its covers, it can become a confirming experience for the reader when finished. It will be seen that an extensive cover-up has taken place, and managed over many decades by certain agencies of the government, and especially the U.S. Air Force and Central Intelligence Agency (CIA). Answers to the UFO question are revealed in previously classified government memos, letters, and documents–all confirmed by affidavits and statements by credible witnesses.

Fact and Fiction

One might be amazed at the great amount of evidence and number of people involved in the cover-up that has taken place for many decades, yet the Air Force has been able to keep mum on the subject. How the Air Force has been able to do so is documented here in detail, and it reveals extraordinary duplicity. The duplicity, however, has a strange twist to it, because some units and personnel within the Air Force, who are not in-the-know, are delegated with responsibility for investigating the mysterious flying objects. This situation arises because of continued contact with the objects, which has been an on-going problem for the general military and certain government agencies, especially for those not "in-the-know" and needing to find answers. When Air Force investigative efforts begin to close in on the Truth, and military personnel begin to express extreme concern about reality of the objects, the Air Force hunkers down with denial, deception, diversionary tactics, debunking, and subterfuge. Investigators who get too close to the Truth are then transferred, removed from the scene, and instructed to keep silent.

When I began to work on this book, I knew I would be tasked with performing my own investigation, and to discover exactly how the Air Force has been involved. I did not know where it would lead, but like anyone considering the UFO question, I heard old stories and rumors of mysterious sightings, alien contacts, abductions, crop circles, and hoaxes, including all the myriad fanciful experiences from those considered to be lunatics, kooks, and mentally deficient oddballs. I wanted to bypass all the fanciful stuff, if possible, and concentrate instead on the mass of information available on Air Force and government involvement, which is available through many sources. The information I found required very careful selection and sorting, with relevant facts needing verification and confirmation. Repeated connections between people and events were noted, and information was arranged according to subject area and time frame. After further examination and filtration of data, it all needed to be laid out in a concise format, which would ultimately reveal the construct, scenario, and history of the cover-up.

Through use of the Internet, and utilizing a multitude of books written on the UFO question, I found a vast amount of information available, although a great percentage could be categorized as trash or pulp fiction. I worked with a great degree of caution to make sure that information I obtained was verified, checked, and then rechecked against many sources. Contradictions needed to be resolved, and associations connected. Inferential logic was sometimes necessary, especially when Freedom of Information Act (FOIA) documents pointed in certain directions. Those particular documents, some buried in government files, and some previously classified secret or top secret, provided valuable guidance to a proper path of discovery, which allowed other information to fall into place. Along the way, previous familiarity and insight regarding the Air Force environment, how it operates and functions, provided an ability to make reasonable assessments regarding motivation and policy. All of that, together with logic and analytical skills developed during a thirty-two year career as a systems analyst, guided me to the following presentation. I was acutely aware and mindful that "garbage in equals garbage out," and I was on guard to avoid pitfalls of falling into such a trap.

In the task of finalizing information, a few situations were noted where statements or conclusions did not match with that by other researchers. In the effort to determine why research paths diverged, it was revealed that critical information was sometimes missed, or misperceptions or illogical assumptions were made by others. As time proceeds, and when more data becomes available in the form of newly released documents and other information, it is entirely possible that new revelations will come to light requiring updates to conclusions, speculation, and conjecture.

Book Content

In chapters that follow, various documents are referred to, and people quoted, including Dr. Allen Hynek, who was Chief Scientific Investigator and consultant for the U.S. Air Force on Project SIGN, Project GRUDGE, and Project BLUE BOOK. All three Air Force projects were a continuation of the same effort in the investigation of flying objects. Projects SIGN and GRUDGE became very contentious, and demonstrated much Air Force duplicity, but Project BLUE BOOK became better known, especially during the administration of Captain Edward Ruppelt, its first leader.

Active Air Force investigation into the UFO question, continuous since the late 1940s, primarily involved the Directorate of Intelligence (DI) at Air Force Headquarters at the Pentagon, and Air Material Command (AMC) at Wright-Patterson Air Force Base (AFB). The Joint Research and Development Board (JRDB), and later the Research and Development Board (RDB) of the Defense Department, became a major controlling influence until the Office of Scientific Investigation of the CIA (CIA-OSI) became involved. It has continued to maintain a clandestine effort as primary controller of the UFO question.

Some of us who were in the Air Force during the 1960s have come to realize that Dr. Hynek was kept in the dark about many of our experiences. During my time in the Air Force, he and others at Project BLUE BOOK were purposely kept away from incidents where the Air Force did not want his investigative involvement. Hynek was not aware how much was kept from him, but he later came to understand that the primary role of Project BLUE BOOK, especially after January 1953, was to debunk the UFO question. In his 1972 book, *The UFO Experience: A Scientific Inquiry,* he stated:

> *The entire Blue Book operation was a foul-up based on the categorical premise that the incredible things reported could not possibly have any basis in fact.*

The term "UFO" is commonly used as an acronym for "Unidentified Flying Object." This and other names, including "flying saucer" and "flying disk" (or "disc"), are labels used for lack of observable, or reliable clues, on what the objects are. Under normal public usage and connotation, the terms have acquired a sense of fantasy, which tends to catch attention. They are often the subject of tabloids, and used in association with sightings of flying objects by crackpots, kooks, or rational people fooled by circumstance. To Air Force investigators, the objects are often attributed to natural phenomena, hoaxes, or misidentification of something man-made, such as planes or balloons of various types. This book attempts to avoid using the UFO acronym, except when referring to the "UFO question."

"Flying objects," which is the generalized term used in this book, are objects usually defined as having a shaped structure, but not normally recognized in relation to natural or man-made objects. They might be disc-shaped, cigar-shaped, triangle-shaped, V-wing shaped, or other configurations. They are often observed to hover in place, with the ability to speed silently and instantly away, and they are sometimes detected at extremely high altitude. They exhibit terrific speeds, with ability to make sharp right-angle turns, and they can instantly blink out of sight, disappear, or suddenly materialize. They do not necessarily appear as a point of light, except in distant views, and they may exhibit strange white, yellow, orange, or red "glowing" colors. They may also present an extremely bright surface that is difficult to look at, or have a silvery color, or they may incorporate strange and unusual flashing features, or pulsating lights. Those are the signature traits of flying objects that have captured notable interest by competent and reliable observers. They are traits not commonly found in our worldly setting and not indicative of recognizable natural phenomena, or man-made objects that fly or float through the air. They can be more accurately labeled as "Unearthly Flying Objects," especially if reported by reliable, reputable and credible people.

This book, Volume 1, consists of five parts. Part One begins with an introduction to the UFO question and a review of public perception of the subject. The stage is then set for the amazing and surprising flying object incidents observed and encountered at Minot Air Force Base (AFB), North Dakota, in 1966. It documents the revelation and confirmation of a new life perspective as I experienced it, which was probably the same for many others who were there with me at the time.

Particular attention is given in this book to a period of time from mid-1947 to the middle of March 1952, with certain world events noted to provide perspective. It begins in Part Two with introduction to initial public awareness of the UFO question, which was also the time when the Air Force became greatly involved. But the Air Force then became critically involved upon discovery of the magnificent secret.

The magnificent secret pertains to the Roswell Incident, which initiated the Air Force's monumental cover-up. A unique historical overview is provided of the incident with a comprehensive and detailed sequence of events that took place. It includes a review of circumstantial but incriminating empirical evidence, including evidence never previously offered by other researchers. The entire incident was confirmed by many hundreds of witnesses, with many offering signed affidavits.

A review is then provided, in Part Three, of the aftermath of Roswell, and critical government efforts to protect the secret, while others in the Air Force, who were not in-the-know, were still independently investigating what they considered to be "real" flying objects. It was a time of very active government involvement.

A historical review then follows of two investigative efforts by the Air Force called Project SIGN (in Part Four) and Project GRUDGE (in Part Five). The two projects demonstrate Air Force attempts to interfere with or manipulate flying object investigators. They provide overwhelming justification to confirm and condemn extensive duplicity by the Air Force and other government agencies.

Chapter headings identify sequential periods of time covered, and the combined narrative can be considered a chronological history of U.S. Air Force investigation into flying objects, which reveals a massive cover-up of the UFO question.

A follow-on book, Volume 2, continues with a comprehensive review of Project BLUE BOOK, which eventually took the Air Force out of the business of "overt" flying object investigation, which went deep underground in 1969.

The purpose of this book is to strongly convey, through documented actions of the Air Force and other government agencies, the duplicity that has taken place against the civilian, scientific, and world communities. It will be seen that the Air Force, contrary to its proclaimed disinterest, has always possessed extreme concern about flying objects and their potential for impacting U.S. national security. That in itself is enough to show blame and extreme duplicity by the Air Force in hiding the Truth.

So that the reader might truly comprehend, this is not a book about my experience per se, but it is about a grand deception of Air Force cover-up and duplicity regarding their secret. More than that, however, it is about the experiences and participation of many people in government who have been involved with the UFO question. Some have been pawns in the Air Force cover-up, with many knowingly accepting the requirement to keep the "secret." Some have passively objected, but have also been careful to preserve their jobs and their futures. Others, only after retiring from government service, have spoken out. And others have held back with great trepidation. This is a book about all of them, including the experiencers, the investigators, and the secret keepers. It is also recognized, however, that many people not involved with government service have also been affected, and have experienced Truth. Those who have previously spoken to the Truth, but were relegated to "oddball" status, are casualties to Air Force and government duplicity, and they will be owed deep respect–eventually.

❧

The incident I experienced provides more than enough incentive for me to look into the UFO question, and determine why the Air Force sternly instructed me and others that "It Never Happened." While totally unskilled in writing a book, it's been a huge effort for me to document what I found in my research, but it's been important to put it down in my own words rather than have someone else do it for me. My only reward for this is the release of a moral obligation to humankind, and revelation of Truth.

After uncovering a great amount of material, I found answers, and I learned the Truth, but only a small part of it is documented here. In this process, I also discovered that a huge amount of exceedingly important information remains, which has not been released, reported, or included in what may someday become official U.S. or world history. It remains in secured vaults, with only a smidgen of information released through the FOIA. Information available, however, highlights the fact that the Air Force, CIA, and other government agencies have been seriously involved in a massive cover-up for a very long time. This book presents only a small portion of the cover-up. A much larger story remains to be told of a cover-up that has continued to this present day. It is a cover-up that will remain unchecked unless disclosure of Truth is revealed by officials in "high authority," but only when they acquire courage enough to do so and take action.

Who might that be? Well, the highest authority in the land in a democratic nation is, by definition, an elected politician—at least in the United States. That person is tasked with making very difficult decisions on a daily basis, often in the face of great adversity, since they are also influenced by very strong political winds and corrupting pressures. Because they must cater to political winds, they are prevented from being "in-the-know" on many things, especially in matters that other career officials in high positions of government, or industry, are sworn to protect. That leaves elected high officials, such as the president of the United States, powerless to take action unless they summon enough courage to unearth and expose those responsible for protecting the secret. The highest authority in the land, however, must ultimately be held responsible and accountable for revealing the Truth on the matter.

On 30 April 2013, I appeared at the National Press Club in Washington, D.C. before a mock "congressional hearing," which was organized by the Citizens Hearing on Disclosure, where I provided six former members of Congress with details of my incident. There was no way that current congressional members would ever consider the subject. Although my voice was briefly heard, I was certain, even before my testimony, that it would take much more than my speaking out to make a difference. It will take no less than strong demand by the American public, or perhaps a worldwide clamorous demand on social media, to send a message to high authority. Will that happen?

PART ONE

Truth Revealed

❧

Never in my wildest dreams would I have ever considered that such science fiction might be experienced in an earthly reality. I was always interested in the physical sciences, but science fiction or any kind of fiction was just that, someone's fanciful imagination put into storybook form. I was only interested in proven results, and the reality of pure science, which actually put me in the same domain occupied by skeptics and those dwelling in selective objective thought.

I became an Air Force officer, not because I chose that as a career, but because I considered it the best option for me when faced with a military obligation that all young men were presented with in the early 1960s. It did, however, prove to be a very interesting time in my life, and it provided an unbelievable but memorable incident that resulted in a life-changing experience. It was an experience that challenged my perception of earthly reality, but it took place in a dramatic setting of conscious reality, and within a format of pure science fiction.

Truth was discovered, but then it was suddenly reconfirmed after the passing of about thirty-five years. The confirmation was almost as startling as the initial experience, but then it was repeatedly confirmed in unexpected ways when others who were with me in the Air Force began to reveal their experiences. It then became obvious that the Air Force, and other government agencies, had managed an extensive cover-up of massive proportion.

Truth and a cover-up was revealed to me, but they continue to remain hidden from the public, while also leaving many from the Air Force with a lie to maintain. The decision to voice Truth can be fraught with trepidation, to say the least, but those before me who voiced Truth were brave enough to challenge and confront Air Force deception, and they did so with a moral sense of obligation to humankind. I have chosen to join them—by voicing Truth as it was revealed to me.

CHAPTER TWO

Public Perception

Paradigm Challenge

Dr. Allen Hynek, the former Chief Scientific Investigator for the Air Force on flying objects stated:

> *The question of UFO's has developed into a battle of faiths. One side, which is dedicated to the Air Force position and backed up by the "scientific establishment," knows that UFO's do not exist; the other side knows that UFO's represent something completely new in human experience. And then we have the rest of the world, the great majority of people who, if they think about the subject at all, don't know what to think.*

For a good part of my life, I was strongly bonded with that "scientific establishment" mentioned by Hynek, which was due to my education and working associations in the missile and space industry. I have an understanding of the "scientific method" of research, and the necessity for collecting and validating appropriate data. I recognize this process as a valid method for discovering Truth. Unfortunately, that bond also positioned me in the same realm where many skeptics and debunkers reside.

My appreciation of the scientific method, however, has taken on a somewhat different perspective because of an enlightening experience I became involved with. I now understand there are Truths out there that cannot be resolved, or reconciled, because our current database of knowledge has not sufficiently matured, but those Truths do exist, nonetheless.

With age comes wisdom. That often stated remark is intuitive, but I mention it because it has taken me many years, and a lot of learning, to come to a true understanding of how profound it really is. Sometimes I wonder why it took so long to come to a deeper understanding, especially when I relate it to the attitude of skepticism that previously occupied much of my life. I'm referring to situations where I debunked ideas or statements of others, which were not in agreement with my educated interpretation of the natural and physical world, or with its associated laws. It is those laws, and associated truths, that were put together by human intelligence, and built into a database of knowledge through the ages. When professing skepticism, I took great pride in exposing the flaws of others, and deriding the intelligence of those who obviously did not have knowledge, or a correct understanding.

I've always been enthralled with the sciences, and I've enjoyed learning the latest in technological advances, latest astronomical discoveries, and latest theories in physics and laws of nature that are continually revealed and revised. I educated myself in those areas, and received a college degree in physical science. I supported myself with technical jobs in aerospace engineering, project management, systems analysis, and computer programming. I became centered on what was real and possible using current technology, and I understood that the scientific method of investigation was paramount to revealing Truth. To me, "fantasies of the unknown" were irrelevant, especially if they could not be proven, deciphered, observed, and rationally explained with current scientific laws, or our database of knowledge.

I now take solace in the fact that I have not been alone in the paradigm I once found myself in. I understand that ignoring something considered "irrelevant," and then exhibiting skepticism, is a natural human trait. It is displayed by many people, including leading scientists and researchers. Their goal is to gain knowledge and Truth through the use of inherent logical processes, which leads to repeatable and verifiable results. The scientific method, and the use of "objective reality" reasoning, helps produce new truths through a selective construct that builds upon previous truths from knowledge gained. This systematic process makes it easy to be skeptical of most anything not fitting into today's puzzle of objective reality. If something is not logical, easily observable, provable, and fitting within today's database of knowledge, it must then be fiction, fantasy, or a hoax. It becomes easy to say "not possible," or "show me" (seeing is believing), or "prove it," or any number of comments sometimes tainted in a derisive tone. Expressing such derision becomes a heady and egotistical approach in showing disdain and a know-it-all attitude.

Blinders of Skepticism

Yes, the pursuit of knowledge is truly divine, but could it be possible that blinders of skepticism actually hinders progress in finding Truth? My realization of this occurred over the last few years while thinking back on technological progress made over the last couple hundred years or so, especially in the last few decades.

From my perspective, there has been an exponential increase in scientific and technological progress since the time my parents were born. We cracked open the atom and demonstrated the horrendous power of fission, and then fusion. Our science and technology has put us into space and on the moon, and we have looked at photons originating from outer reaches of the universe. There is no limit to what the human mind and technology can accomplish, and no end to progress in that regard.

With every accomplishment, new areas are brought forward for additional investigation, with mysteries to solve. Scientists now conclude that dark matter and dark energy (whatever that is) are the major components of the universe, some ninety-seven percent, which cannot be seen. This dramatic statement provides something to think about. Everything we are aware of, and have ever known, including everything observed with our most powerful telescopes, has been reduced to something significantly insignificant. This is an enigma, or paradox, that the scientific community is obligated to consider, especially after previously dealing with only the other three percent of the universe. Of course, this presents further avenues to explore and mysteries to solve, resulting in new proposed theories and conclusions. But skeptics need not be part of this. Our technology will continue to grow, but it is not currently known what new laws of nature and physics will be discovered beyond this particular moment in time.

Everyone lives in their own particular moment of time, just as in ages past, with little idea of what the future might bring. Skeptics in the past strongly professed that the world was flat, and that the earth was the center of the universe. Were they right or wrong? How might they have altered their thinking, and adjusted their views in that regard?

Imagine what would go through the minds of persons living in the Stone Age, or even a couple hundred years ago, if they were suddenly placed in the middle of a large modern city or modern home. How would they process what they saw? How would they describe what they saw to others when they returned? What would their contemporaries later think when told about it? If their current database of knowledge and current laws of nature and physics were used to explain what was observed, what would they come up with? How would skeptics of that era respond when told?

It is important to recognize that someone from the past would likely exhibit utter amazement when brought forward to our current time. Very quickly, however, it would turn into a "seeing-is-believing" confirmation of Truth, and it would take hold with stark realization of something very real, although still beyond comprehension. Upon return to the past, the reaction of others to a description of what was seen would likely be skepticism, or outright disbelief. One might compare this to flying object sightings made today by credible people. It's possible that things in existence today, or in the future to come, will not be explained with present day science and technology. But Truth will still exist, even though proof may not be possible within our current base of knowledge.

There have been skeptics since the time humans have been able to think logically and rationally. But skeptics live in a certain paradigm that prevents them from considering that future science may provide answers leading to Truth. It is easy to fall into a trap of denying something that, although not currently scientifically provable, may actually prove to be true. Sometime in the future, there is great potential of being proven wrong, especially if one is susceptible to being stuck in the paradigm of ignoring potential possibilities, or something considered "irrelevant."

When I come upon something mysterious and unexplainable, I now prefer to keep an open mind, while believing all things are possible. It then becomes a matter of conducting more research, and pursuing more knowledge. It is a process that feels much more comfortable. This is especially true when encountering reasonable, respectable, and very reputable people who profess to have some experience, or knowledge of something that is beyond my understanding. It then becomes a challenge to investigate, research, and obtain additional information in order to attach some sense of reality and meaning to it all.

To continue this a bit further and looking at the extent of human history today, how does it compare with the long history of the universe? The age of the universe is thought to be about 13.75 billion years, and the age of our solar system about 4.5 billion years. Dinosaurs roamed the earth for about 165 million years, dying out about 65.5 million years ago. And from my understanding, depending on measurement criteria, modern humans evolved about two hundred thousand years ago, and human tool making started about fifty thousand years ago. Evolution and expansion of technology, along with philosophical and scientific thinking, has only been within the last few thousand years. When comparing the age of the universe to other measurements given above, the time of human habitation, and subsequent sophistication in today's world, is historically infinitesimal, and exceedingly recent.

Astronomers tell us there are 100 to 400 billion stars in our Milky Way galaxy, although some scientists put that figure at a trillion or more. Obviously, the estimation process on this is not an exact science. But it's also said that each star has at least one or more planets, and many of the planets exist in the "goldilocks zone" where liquid water allows support of life. Astrobiologists tell us there are at least 100 billion planets in just our galaxy alone capable of sustaining life of some sort. If that is difficult to contemplate, just think about the 100 billion (or many more) additional galaxies in the universe, and the planets they might contain. When considering that there are infinitely more stars in the universe than grains of sand on the beaches and deserts of Earth, the mathematical probability does not allow our planet to be unique, or alone, in supporting life. With that in mind, we wonder about the long history and technical sophistication of supposed life originating on other planets eons ago.

As a scientist and astronomer, Allen Hynek was acutely aware of overbalanced statistics that support the idea of life elsewhere. He once stated:

> *We all suffer from a cosmic provincialism—the notion that we on this earth are somehow unique. Why should our sun be the only star in the universe to support life, when the number of stars is a 1 followed by 20 zeros? Suppose that only one star in 10 is circled by a planetary system that has life; that means that the number of life-supporting stars in the universe would be 1 followed by 19 zeros.*

Many scientists, including the great theoretical physicist and cosmologist Stephen Hawking, agree with the premise that there must be a great many life forms much more advanced than us humans. If that is the case, how advanced might technology be for intelligent "life forms" originating a mere billion years ago, which is fairly young in the scheme of things? Or, maybe just a few hundred million years ago, or say only a million years ago, or even much less? What advanced scientific and technical abilities would it possess compared to us? Would it still be true that "nothing" could exceed the speed of photons or electromagnetic waveforms in its framework of knowledge? How might it connect space, time, gravity, and other universal forces in its knowledge of the universe? Would it be possible that its technology might be able to create huge amounts of energy on demand? Would it have mastered technology involving positrons, anti-matter, anti-gravity, dark energy, or magnetic forces? Would it have mastered technology allowing travel of great distances in an instant? Would it have found a way to enter other dimensions, or go back and forth in time? Would it have unlocked paranormal or psychic technology for use in communications and object control? Could it have already visited planet Earth, perhaps eons ago?

One might presume that such ideas or questions are overreaching, and not reasonable. But when considering the short history of human existence, and the potential for intelligent life elsewhere, no one has the insight or ability to absolutely deny such possibilities, which may be in the realm of Truth in some other world.

There exists many situations where observations and circumstances have given us clues to something truly mysterious and unexplainable, such as dark matter and dark energy, but problems often develop in connecting clues with knowledge currently available. It is the pursuit and investigation of what is mysterious, while discovering clues leading to furthering of one's knowledge that serve to increase human advancement. Those who write off something mysterious and unexplainable, or casually label it with a nebulous conclusion, or resort to debunking and ridicule, are simply exhibiting their unenlightened skeptical nature. That is why I now refuse to accept the label and paradigm that occupied my way of thinking for many years, which I now look at as a mental lapse of wasted time. Instead, I prefer to take on a new perspective, and a mind-set of searching for new possibilities.

The real question here is not whether skeptics or debunkers would want to contemplate the UFO question—they wouldn't. It is whether there are any scientists or researchers out there who might be open-minded enough, brave enough, and dedicated enough to contemplate and recognize they may be living in a paradigm that prevents them from investigating and pursuing knowledge outside their current realm of reality, outside of their concepts and constructs of visualizing Truth, and outside their comfort zone. It is a process that goes beyond the empirical, and explores possibilities they currently refuse to consider. When they do recognize their imposed limitations by taking on a new perspective, and then pursue knowledge of what is mysterious, unknown, and seemingly unreal, the doors to Truth and scientific advancement will truly be open. For others, the process of recognizing that they are currently living in a particular moment of time, and within an expanding evolution of technological change, this should be enough to alert them to keep an open mind for new, magnificent, and unimaginable Truths to behold.

The Scientists

In regard to scientists, and others who openly disregard the UFO question, there are mitigating reasons for stubbornness. They know and understand that nothing in our universe can travel faster than the speed of light. It is an established fact associated with Albert Einstein's Theory of Relativity; it's plain and simple. With other celestial bodies being light-years away from us, extraterrestrial entities cannot travel here in their lifetimes, or expend the enormous amount of energy necessary to make it remotely possible. For a vast majority of scientists, the case is open and closed. End of discussion.

As far as the general public goes, elementary Newtonian physics is as far as most people get in education and knowledge of celestial mechanics. Few people know anything about "relativity," which exists within the fabric of space and time. That is the realm where scientists work to understand cosmology and wonders of the universe.

Those who know something of Newtonian physics, are aware that an object of certain mass remains at rest unless influenced by an external force that imparts acceleration (F = M x A). It then follows that for two or more objects located in space, and moving away or toward each other due to an imparted acceleration, each would perceive (in its specific frame of reference) that it was the one at rest, and the other moving. The calculation of speed between the two is dependent on the specific frame of reference of an observer, and the time it takes an object to travel a measured distance. With an object experiencing additive increases of acceleration, resulting in even greater speed, it would seem that an object might increase its speed an infinite amount in relation to the observer.

Relativity now enters the picture with consideration of the speed of light. This is represented in the equation stating that cumulative energy of an object is proportional to its mass, modified by the constant speed of light (E = M x C²). This is what throws a confusing factor into the situation, and separates most of the public from scientists. The confusing part is that the speed of light, in a space/time environment, is a determinant factor when considering frames of reference.

Speed of light is a constant, which is supported in propagation of its electromagnetic waves by the "ether." Some would refer to that as "nothingness," and some might relate it to "dark matter." Seen by an observer, its speed is about 186,280 miles per second in any frame of reference the observer is placed. On Earth, it takes about eight and one-third minutes for sunlight to reach us. Light can also exert a force, be bent by a transparent medium, and be affected by gravity.

In effect, the speed of light has its own frame of reference, and everything else is located within it. If you were to somehow speed along with a leashed pet photon, and another photon was heading almost directly toward you, that photon would theoretically be observed passing by at the same speed of 186,280 miles per second. Not twice the speed, as one might expect, which is strange indeed! The thing that is even stranger and limits the above analogy is that the faster an object travels, the smaller it gets, the more massive it gets, and the slower time passes. The speed of light becomes a speed limit, where it becomes impossible to impart enough energy to an object to reach that speed. As more energy is applied, the "apparent" mass of the object becomes infinitely heavier, while it also "grows" infinitely smaller, and time nearly comes to a stop.

In comparison to intuitive Newtonian logic outlined previously, in regard to distance, time, speed, and frames of reference, Einstein's theory of relativity becomes twisted into a great monster of complexity. The theory of relativity, however, and its associated laws have been proven over and over again, and a great many scientific minds of yesterday and today (physicists, astrophysicists, cosmologists, and astronomers) are in total consensus with the theory. They undoubtedly have it perfectly right, especially when it involves the UFO question, and they really cannot be so embedded in their own paradigm of complexity, and proven mathematics, that they have not seen the light!

Astronomers acknowledge that the universe is expanding away from us with increasing speed. Celestial objects furthest away have greater speed, as evidenced by a large "red shift" to their spectrum of emitted light. One then deduces that increased speeding "visible" spectrums of those objects, at outer reaches of the universe, virtually vanish out of view from our frame of referenced observation. Outer reaches of the universe are so highly red-shifted out of view that those speeding spectrums are not visible to our eyes or instruments. The wavelength of light becomes so stretched out, that it becomes virtually undetectable. It becomes impossible to see beyond that light spectrum boundary at outer reaches unless we take on extra speed to move closer toward that boundary. In that Newtonian sense, the universe does extend infinitely further on, and those far reaches are just "dark" and invisible from where we are positioned. To the lesser educated, it might seem that a majority of the unseen "dark" part of the universe, beyond that spectrum boundary, would be composed of the same sort of matter and energy as the visible part, and it could be very much larger and very much older. And this does not discount the fact that there may be some areas within the visible universe that may also be "dark" due to localized speeding spectrums, especially when one considers all the "black holes" now recognized.

Of course, the forgoing does not consider quantum mechanics, quantum field theory, and gravity curved space-time of general relativity, where the speed of light could possibly be exceeded, or the situation where the edge of the expanding universe may exceed the speed of light, which is sometimes referred to as the "Hubble Sphere." This is the area where today's physics is in limbo, which is an area similar to the "event horizon" surrounding the surface of a black hole, a boundary where nothing makes sense beyond it. But then again, there really is no center of the universe. Every point in space is expanding as if it were the center, which means that there is no edge of the universe at its "outer reaches." All of this taken together is where even the educated become confused in their enlightened perspective. It generates thoughts of multiple universes, duplicate realities, and many other abstractions.

Despite the limits to our knowledge database, the theory of relativity takes precedence when considering cosmological matters, and relegates most Newtonian ideas to basic elementary concepts, although it would be interesting to contemplate all the additional billions upon billions of galaxies, and additional planets with intelligent life, residing in the rest of that "dark" universe. Likewise, in a relative sense, it would be interesting to contemplate the undefined boundary of the ever expanding and speeding mass of the universe at its outer limits, where it "apparently" becomes infinitely heavy and infinitely small with time at a standstill. It would be like a never ending treadmill to catch up with and locate in a Newtonian sense, which also makes one consider, if so inclined, how scientists might then reinterpret microwave background noise and the Big Bang Theory.

At this point in time, all scientists are mostly in the dark about dark matter and dark energy, although they are busy studying it and constantly submitting ideas on the subject. They claim we are imbedded in it and it surrounds us, although it cannot be seen. It is dark energy that contributes to the acceleration of space, and the expanding universe. And it is the gravity of dark matter that keeps galaxies from flinging apart.

Scientists are also in the dark about the UFO question, without paying notable attention to it, at least not publicly. Hynek, in a presentation to a body of the United Nations, on 27 November 1978, stated:

> *There is a surprisingly large number of individual scientists who have expressed to me, privately and personally, their involved concern with the challenge of the UFO phenomenon... These scientists are in many cases associated with large and prestigious scientific organizations, both government and private, which as organizations are silent or even officially derisive about the UFO phenomenon. The individuals within these organizations who have intimate knowledge of the UFO phenomenon are restrained by organizational policy to remain officially silent about their interest and in private work with UFO matters.*

This amply explains why most scientists refrain from becoming publicly involved with the UFO question. They fear it could destroy their careers. Similarly, it explains why government officials and politicians keep silent and don't get involved. Their careers would be immediately destroyed by skeptics, debunkers, and detractors.

In later years, Hynek became concerned when he realized he was running across very puzzling cases, where he personally viewed startling physical evidence at sighting locations. He also found radar data that matched visual observational data. More important, he came upon very credible people, with impeccable credentials, who were involved with the sightings.

In 1966, during the time of my experience, Hynek stated:

> *Of the 15,000 cases that have come to my attention, several hundred
> are puzzling, and some of the puzzling incidents, perhaps one in 25,
> are bewildering.*

In reported sightings of some flying objects observed at ground level, Hynek
stated:

> *These are the cases in which it is reported that concomitant and
> simultaneously with the occurrence of a UFO event, there appears
> physical evidence of the immediate presence of the UFO. This can take
> the form of immediate physical effects on either animate or inanimate
> matter, or on both. Thus, physiological effects on humans and animals
> and plants have been very reliably reported, as have the interference
> with electrical systems in the immediate vicinity and the appearance of
> disturbed regions on the ground also in the immediate vicinity of the
> reported UFO event. Now over thirteen hundred physical trace cases
> are on record.*

Hynek saw overwhelming evidence in those "bewildering" cases, and realized
that more scientific investigation was definitely called for. He may have also sensed,
from evidence he saw, that the science of today might be incapable of providing
rational answers. It would be something akin to how scientists now theorize about
dark matter or dark energy, while having no idea of the science behind it all. The
current realm of science is not capable of making it rational to us. Hynek made
inference to this when he stated:

> *The UFO phenomenon…bespeaks a higher reality not yet recognized
> by science.*

In all that has been mentioned above, it is not my purpose to degrade great
thinkers, accomplished scientists, and researchers who have contributed greatly
to Truth and knowledge existing in this world. It is disappointing, however, to
recognize that my previous attitude was associated with an inadequate paradigm
that occupied much of my life. That paradigm, however, continues to reside with
others.

As a scientist, Dr. Hynek expressed his frustration with the scientific community
many times. He once said:

*Nearly all of my scientific colleagues, I regret to say, have scoffed at
the reports of UFO's as so much balderdash, although this was a most
unscientific reaction since virtually none of them had ever studied the
evidence. My friends were obviously mystified as to how I, a scientist,
could have gotten mixed up with "flying saucers" in the first place...
The general view of scientists was that UFO's couldn't exist, therefore
they didn't exist, therefore let's laugh off the idea. This, of course, is a
violation of scientific principles.*

In utter frustration with scientists who put down flying objects as hoaxes or
hallucinations, Hynek commented:

*This, of course, is the view that a number of my scientific colleagues
have taken. I think that enough evidence has piled up to shift the
burden of proof to the critics who cry fraud.*

This statement by Hynek shows the great frustration he built up toward the
scientific community, especially after he became very much involved with the UFO
question and the many puzzling cases he investigated. He came to the conclusion
that there was valid reason to believe flying objects were the real thing, but he didn't
have scientific data or evidence to prove it.

It is understandable how the scientific community, in general, is linked to the
stance the Air Force has taken on the UFO question. "UFOs and flying saucers do
not exist" is the authoritative but deceptive cry of the Air Force, and the scientific
community does not raise an eyebrow to it. The scientific establishment is then
seen by the public as backing the Air Force. Behind the scenes, however, and
unknown to the public, the Air Force is critically concerned about maintaining
secrecy of flying object incidents that are continually and frequently experienced
by its airmen. I had firsthand experience with that.

They Do Exist

And then, as Hynek once said, there are those who absolutely know "UFOs
represent something completely new in human experience." This includes a good
number of respected, responsible, and reputable people with high credentials who
have experienced Truth. They understand, with absolute certainty, that exotic
objects not resembling or performing like anything man-made, and guided and
controlled by some sort of unknown intelligence, do exist. Many have collectively,
not just individually, experienced the same situations at the same time. That is the
case with many highly-educated Air Force officers who were handpicked for their

high integrity, passed rigorous security checks, obtained high security clearances, passed psychological profiling, and were entrusted with control, operation, and security of nuclear weapon systems.

It must be noted, however, that certain factors enter the equation in regard to particular airmen who have experienced Truth with unearthly flying objects. Some are reluctant to speak out, and they prefer to keep their secret. Some sincerely believe they are morally responsible to the U.S. government, and must maintain their secret as instructed. To some, it's also a matter of personal integrity to keep the secret, and they cannot be faulted for that. Others are fearful or terrified of potential consequences that might come to them, or to those close to them, and they feel constrained in speaking out. I've run into instances of this with people I've associated with in the Air Force, particularly those I've come into contact with after retiring. Not long ago, one former missileer stated to me that he had been forced to sign a document that committed him to silence, and he could not tell me anything further about his "very serious" unearthly incident.

Just few years ago, when I confided to my old Air Force buddy Paul Johnson about my incident–I'm a godparent to one of his children–he then confided to me about a couple of incidents he was involved with. Only after I told him of my experience, did my good friend open up and tell me about his two disturbing experiences, which was some forty-five years previous. When I indicated that I intended to publicly tell the Truth of my incident, he immediately exclaimed, "Oh no, I would never do that!" I suspect he was fearful of losing his Air Force retirement pension, which was a precious asset needed in his advanced age. Then, following an unfortunate bout of ill health, his obituary mentioned the following:

> ...He entered the Air Force and the newlyweds moved to Minot, North Dakota. Paul was a missile combat crew member and discovered a UFO. When he sent an Airman out to investigate, the airman asked Paul if he should shoot at it. Paul said he didn't think that was a good idea and the UFO flew away...

It is significant that my friend Paul would have his experience mentioned very briefly in his obituary, along with other major events of his life. There were so many other things in his productive life that could have been considered more important. In the end, however, he finally felt freedom to have his experience made known.

Like me, other airmen I'm familiar with also experienced situations where flying objects seriously compromised our nuclear ICBM forces. Their stories will be reviewed in Chapters 5 and 6. The Air Force simply instructed all of us to keep quiet, to never say another word, and pretend "It Never Happened." The Air

Force covered up those situations, and left us with awe-struck open mouths in a conspiracy of silence. It was a silence that would penetrate deep into our psyches, and with stark memories hauntingly and periodically recalled with passage of time.

Public Opinion

So how does the average person look at the UFO question? Hynek indicated that a majority of people do not know what to think. This is understandable, because none of it is in the realm of reality for the average person, unless personally involved with a sighting incident or encounter. There is no reason to think about flying objects, let alone contemplate their existence. Of course, the stigma of being labeled an idiot for considering the subject may have something to do with this. Someday, this may provide good reason for a scientific study on how human thought, reasoning, and social interaction have played out in regard to the UFO question. But there is also another factor regarding media impact upon the public, and its influence on public attitude toward the "question," which will be discussed shortly.

It is possible that a positive movement on the UFO question, within the public sector, is turning toward the side of Truth. People sometimes remark to me that "there must be some form of life in other areas of the Universe." And now there are often TV programs related to flying object incidents, and the potential for alien life elsewhere, although much is still presented in tabloid format. Recent focus on the subject, however, makes one consider that a possible government shift is taking place to slowly prepare the public, and to condition everyone about the possibility of life elsewhere in the universe.

In the last few decades, opinion polls from many sources have shown that 50 to 75 percent of Americans believe that an extraterrestrial presence has engaged our world. More than 80 percent believe the government is not telling the truth. Percentages are even greater for educated people holding college degrees. What does this tell us? It could mean the public is ready to hear the Truth, that they have great mistrust in government, and that they believe the government knows far more than Air Force denials indicate

When taking a careful look at outward public apathy, and the apparent contradiction in public opinion polls, there is a reason for this to be connected. People have more important things on their daily agenda than to ponder something not affecting them personally. But when seriously confronted with the UFO question, people will speak out about what they believe. It's not uncommon to run into someone claiming to have seen a strange flying object in the sky, or of hearing second or third-hand reports of others witnessing sightings, or hearing about encounters. With increased exposure and curiosity about the UFO question, people are being conditioned to have personal opinions, as well as having doubt about truthfulness of government. On the other hand, the public is also exposed to debunking by skeptics, and the antics of "misfits" taking pictures of automobile hubcaps thrown in the air, which manage to effectively dispel the UFO question for some. The public, however, needs to hear much more from reluctant, but respectable, reliable, reputable, and credible people. And also those who are still on the fence about speaking out.

The Media

Several factors enter into the equation regarding media orientation toward the UFO question. Major TV networks, famous for their "tongue in cheek attitude" on the subject, are primarily involved with entertainment value of their programming, and they cannot afford to air something that doesn't increase viewer ratings. It is a matter of making money, and they understand that their programming must attract serious public attention. Therefore, they concentrate on presentation of soap operas, sitcoms, episodes of dark mystery and violence, sporting attractions, and news programs presented in bad news or negative opinionated format.

Another important factor, however, has little relation to economics. Media reporters have been conditioned to be flippant on the subject, but it can be more insidious than that, with government involvement using influence and interference. Certain agencies of the government understand, quite well, how the general public can be covertly manipulated through what they refer to as "psychological warfare," or what the commercial advertising business refers to as "applied psychological principles." It will later be learned how government agency penetration of media, and other organizations, have diverted public opinion away from Truth.

In the last few years, television networks and social media have taken the opportunity to provide the public with information on the UFO question, but there are also situations with this that tend to mislead the public. Tabloid and reality TV formats dominate the cable industry with programs that frequently offer little redeeming value. Some programs depict hunters of flying objects who search through deserts and forests for clues, but come up empty-handed. Those programs tend to ignore former Air Force personnel, or government officials with meritorious credentials, who have been witness to incidents and have spoken out on the subject.

Also, the Internet provides a great number of websites, with a huge percentage of them making undocumented fanciful claims about saucers, aliens, and various encounters. Within the Internet, however, one can also view released FOIA documents, Project BLUE BOOK files, and other reliable information. But there is also much trash that camouflages it all, which includes document forgeries and much disinformation provided by certain government agencies.

Public Officials

When it comes to public officials and politicians, they purposely keep their lips zipped on the UFO question to maintain electability. When in office, they are purposely kept clear of highly sensitive information on the UFO question. They cannot be totally trusted, and they are considered at high risk for potentially leaking or releasing such highly classified information. Politicians do not remain in their jobs forever. Sooner or later, they might chance to speak out or leak the sensitive information for personal gain, but only if they are willing to risk their "glowing" reputations, and threats by certain agencies of government.

On 4 November 2011, the White House, and President Obama's administration, came out with the following statement after considerable pressure was applied by those pressing for disclosure of the UFO question:

> *...The U.S. government has no evidence that any life exists outside our planet, or that an extraterrestrial presence has contacted or engaged any member of the human race. In addition, there is no credible information to suggest that any evidence is being hidden from the public's eye...The fact is, we have no credible evidence of extraterrestrial presence here on Earth.*

This blanket statement by the White House demonstrates that it did not do its homework, or was actively contributing to public deception, or did not have courage enough to reveal the Truth. Later chapters in this book will confirm that Truth was avoided, and that active deception underlies the very thing this book addresses. Could it be that a specialized high level "Control Group," or a "government within the government," is effectively managing this situation and the White House? This may seem outrageously preposterous, but this statement originated from the executive office of a previous administration.

❦

In regard to public perception of the UFO question, we are left with a public that is somewhat apathetic, although a majority would definitely like the government to come forth with Truth. There are also scientists who have no reason to get involved. It doesn't fit their paradigm of thinking. More importantly, however, it would destroy their careers. Politicians do not discuss the subject, or initiate investigations for the same reason. Also, the paranoid media cannot escape controlling influences. This leaves secretive government agencies, and also the Air

Force, with free rein to do as they please to divert public attention from Truth.

The Air Force now remains quiet, and does not comment on sightings made. Its stated position is that "flying objects have previously been investigated with no proof they exist," and that flying objects "constitute no threat to national security." This, however, is definitely not what certain elements within high levels of the Air Force know and understand. Instead, they are critically concerned about flying objects and their harassing behavior, which they constantly have to deal with.

The skeptic in us says, "Seeing is believing," but there are those who have seen the Truth and know that flying objects are real. In between, are people who have no idea what to believe. Some will tell you they do not believe in "that stuff," and would rather not listen to any of it. Others will listen, and then shrug their shoulders, or maybe say "wow," with no further interest. And some will listen intently, and want to hear more. Statistics indicate that a majority of people have intense curiosity and want to believe, but they need to be motivated and convinced, either by some authority, or some "happening" that makes it real. But the Air Force exercises official authority with denial. In this way, and with an unknowing public, the Air Force has a great advantage in maintaining the status quo, and the cover-up.

Could it be possible that lonely voices out there might actually be heard someday, or make a difference at some point in time? Many former Air Force personnel and others from government with reliable and credible standing have tried, and continue to stand up for Truth. So far, the UFO question has been effectively squelched, and it will remain that way until an interested public becomes restless, agitated, and united enough to demand Truth. But even then, official authority (as recognized by the public) must first be convinced, committed, and courageous enough to expose the Truth. Will that happen?

CHAPTER THREE

Preparing the Way

I Was Involved

Having participated as a pawn, or perhaps better expressed as being held "hostage" to a colossal cover-up by the Air Force, I'd rather think this book has nothing to do with me. It's really a historical narrative about a grand deception that has continued for nearly seventy years (at the time of this publication). I inadvertently became part of an Air Force deception, participated in it, and lived it silently for a good part of my life. It certainly was not something I anticipated, expected, or wanted, but I was caught within the government's decision to withhold information on the UFO question. I just happened to be in the "right place" at the "right time" to discover Truth, and then I was "instructed" to keep silent.

In that respect, there is obligation on my part to provide an account of what brought me into this situation. The following is certainly not a biography, but it is provided as a backdrop for a better perspective on me and the Air Force environment in which I participated.

Early Interest in Astronomy

Throughout my years, I was always interested in astronomy. I learned early on in my childhood how to spot the Big Dipper and Orion constellations, and I also began to learn names and locations of certain stars such as Sirius, Aldebaran, and Polaris (the North Star). On summer nights, I camped out in the backyard and spent hours watching stars roll by, and I often saw a shooting star before falling to sleep. At my parents' summer beach cabin, located far from city lights, I sometimes saw the northern lights with their hanging ribbons of faintly glowing light.

In my early teens, in the mid-1950s, I would follow the astronomy section in the local newspaper, which listed the positions of planets and other celestial happenings. This helped me identify the planets of Jupiter, Venus, and Mars. I also followed fanciful articles about flying objects, which were sometimes reported on the front page of the paper. They, however, were just curiosities to snicker at. Much more interesting were articles on rocket science, future possibility of manned space flight, and expected government participation in the 1958 Geophysical Year, which would include planned launching of earth satellites. I kept a scrapbook of articles that caught my interest, and I read them many times over.

In contemplating the idea that satellites would one day be whizzing endlessly around earth, I pondered how it would work. Curiosity grabbed hold of me one day while throwing rocks into the waters from our local beach. By throwing a rock far out, I imagined the ability to throw it further, perhaps a mile out, and then further and further. Each time, I visualized that gravity would eventually pull it down. But if thrown really hard, I might have to duck when the rock came all the way around from behind, perhaps landing just beyond, or maybe continuing around earth again, causing me to duck once more. With that idea, the concept of an orbiting satellite seemed not so farfetched, but it would take a lot of muscle power to make that happen. But then it did happen, in October 1957, when the Russians launched Sputnik. The whole world watched while it sped across the sky reflecting the sun in early evening or morning hours.

Just before that, however, my interest in space really took hold when I noticed a large telescope in the driveway of a neighbor down the street. It was late afternoon when I was delivering newspapers, and I stopped to have conversation with the man who was making adjustment to his apparatus. He indicated he built his telescope in his spare time, and he showed me another larger version in his garage. He then invited me to come back later in the evening to view what could be seen through his telescope. That experience really opened my eyes. I had no idea an ordinary person could actually construct a telescope and bring wonders of the heavens so up close and personal. This was vastly different from astronomical pictures in magazines or books. It was the real thing. I was viewing the heavens live before my very eyes.

My newfound friend showed me the moon, and I was amazed that his telescope could focus on an area that showed a few large craters near the illuminated day/night terminator, where crater rims were casting long shadows on the pocked landscape. He also showed me many other moon craters distinguished by central peaks in the middle, and he showed other features such as long sinewy rills and gullies. It was all so fascinating. Then he showed me the distant planet Jupiter, the largest planet in the solar system. Centered in the field of view was the small orb of the planet, with four of its moons aligned with it. Ever since, I have often found myself gazing upward at night to see where various planets are located.

In college, I took a course in astronomy, and I began a project to hand-grind a lens for my own reflecting telescope. That unfinished adventure was abandoned by my impatience when I bought a commercially built telescope. It opened up wonders of the universe during opportunities of clear night viewing, and I was able to look whenever I wanted at the moon, the moons of Jupiter, the rings of Saturn, the nebula of Orion, and many other wonders of our solar system, and the universe.

Conscious Thoughts Confined

When I consider the night sky now, I marvel at the immensity of the universe, and I contemplate our existence on our small planet called Earth. I think of astronauts who walked on earth's moon, from 1969 to 1972, and I ponder their unique experience in escaping from Earth's grasp. I speculate on what they must have thought when viewing Earth positioned above them in the darkness of space. From that viewpoint, it would not be difficult to imagine standing on some other celestial body, looking upward, and viewing the sun positioned among countless other stars in the void. Those stars, billions upon billions of them, are enveloped in mostly dark empty space. And the sun and its orbiting earth are merely specks of insignificant proportion. But it is the Earth where we humans reside, and where our conscious thoughts are mostly confined.

In regard to those confined thoughts, however, which is woven into my Air Force experience, there has been previous hesitation, and great reluctance, to discuss or expound further on that subject with others. I suspect it's the same with other credible and reputable people who previously held, or currently hold, sensitive government jobs and had an enlightening experience such as mine. They fear being labeled a "kook," or being slightly "off base," or fear retribution in not keeping silent. When someone with government authority over you, or a security branch of the government makes you sign a commitment of silence, or tells you "from now on you will not talk about this again," or "from now on, It Never Happened," it sends a powerful message.

Physical Science Education

As reflected by my early interest in astronomy, I was always interested in the physical sciences. In fact, I graduated from college with a Bachelor of Science degree in physical science. Some of my courses included physics, radiation chemistry, geology, microbiology, and astronomy. After obtaining my degree, I regretted that additional courses related to my specific areas of interest were not available, which included aeronautics, astronautics, rockets, and manned space flight. My attention was focused on the fact that the United States, and President John Kennedy, were committed to landing a man on the moon. NASA was engaged in that plan, and I wanted to be part of it, as well as connected to the aerospace industry in some way.

Military Duty

In the early 1960s, after graduating from college and facing mandatory military service, the Air Force was my choice. I attended Officer Training School (OTS), where I received a commission as a U.S. Air Force Officer. After many more months of training, I became a Launch Control Officer in the Strategic Air Command (SAC) with the Atlas E and Minuteman I ICBM systems. This was the job I envisioned upon entering the Air Force, and I felt very lucky to be selected, because it would be my connection to aerospace. Of course, I had to have technical abilities, fit within the proper psychological profile, and have the necessary security credentials.

I soon became intimately familiar with specific aspects of the nuclear capability entrusted to me. The job involved frequent review, training, and testing on procedures. It required stringent perfection in performance, and constant attention to regulations and procedures. Later, after attaining the rank of captain, and becoming a crew commander and instructor with the Minuteman I missile system, I was rated "highly qualified." I had arrived at the pinnacle of my active Air Force career, and I felt privileged to be involved in my area of interest while serving my country. Most of all, I gained great respect for the position in which I was placed. I became quite proud of my hard work, and satisfied with my accomplishment.

Dream Come True

After four and a half years in this position, and knowing I would soon be transferred to a remote assignment, I started to look for another way to extend my career in the missile and space business. I traveled to places such as Washington D.C. and Cape Canaveral for job interviews with various firms. I remember one interview where the job was so super-secret they could not tell me what it entailed, but they wanted to meet me to see if I might fit their qualifications. I soon received a job offer from Hamilton Standard in Connecticut, which was a division of United Technologies, and I accepted the job of Senior Experimental Engineer. I then left the Air Force.

When I arrived at Hamilton Standard, I was assigned to work on a project developing an "environmental system" for the Air Force Manned Orbiting Laboratory (MOL), which was being contracted by the Air Force. My top secret security clearance was an asset in obtaining the job. I sat at a desk working as a project engineer supervising the design, manufacture, and testing of components for MOL. This system included a Gemini spacecraft mated with a laboratory module that would enable the Air Force to have a presence in space. Several Air Force personnel were already picked to man this vehicle, and two of the nominees were launch officers from my missile wing at Minot AFB. After months of working on this, an announcement by President Richard Nixon cancelled the project on 10 June 1969.

I was subsequently tasked with making final adjustments to Portable Life Support Systems (PLSS, or astronaut backpacks) for NASA Apollo astronauts who were scheduled to walk on the moon. This work was truly exciting. Astronauts would come to our facility to test out Apollo space suits made by the company, and then exercise in a vacuum chamber with the backpacks. I examined the space suits fitted for Neil Armstrong and Buzz Aldrin, the first men on the moon. When they landed on the moon on 20 July 1969, then walked on the moon, and then left their backpacks on the moon surface with my fingerprints on them, it provided real satisfaction. My dream of participating in the national effort to put man on the moon came true, and I was pleased.

ICBM Missile Duty - Minot AFB

Now, I need to backtrack a bit, as I have actually jumped ahead of my real story. I need to go back and expound more on my military duty, because I need to set the stage regarding the Air Force experience I had at Minot AFB, North Dakota, and with the Minuteman Launch Control job I had there.

Prior to arriving at Minot, I previously served as a launch control officer with the Atlas E missile system at Fairchild AFB in Washington State. That is where nine Atlas ICBMs were phased out in favor of the more efficient and reliable Minuteman system. I was privileged to pull crew duty for the last Atlas missile when it was taken off "alert status" at Site Seven, which was Fairchild's Alternate Command Post (ACP) in the 567th Strategic Missile Squadron (SMS). I was then transferred to Minot AFB, and stationed there for almost three years, where I became qualified as a Minuteman ICBM Launch Control Officer.

Minot AFB is one of several bases in the United States that supports a wing of Minuteman ICBMs. When I was there, the base was part of the Strategic Air Command (SAC), and it supported one-hundred and fifty missiles of the 455th Strategic Missile Wing (SMW, now the 91st MW) within the 810th Aerospace Division. The missiles were designated Minuteman I, but they were subsequently upgraded to Minuteman III, beginning in 1969, and they are still "on-alert" today. These nuclear-tipped missiles are spread over an eight thousand five hundred square-mile area located in a clockwise semicircle from southeast of the town of

Minot, North Dakota, to northeast of the town, and they are assigned to the 740th, 741st, and 742nd Strategic Missile Squadrons.

Each squadron is divided into five Flights, for a total of fifteen flights in the wing. Each flight is named with a letter of the alphabet, beginning with Alpha Flight, in the southeast of the missile field, to Oscar Flight in the northeast. The 742nd Squadron to which I belonged, still consists of the Kilo, Lima, Mike, November, and Oscar Flights. I was normally assigned to Mike Flight, the Squadron Command Post (SCP), but I frequently performed duty at other flights in the squadron, and occasionally at flights in the other two squadrons. Each flight includes a Launch Control Facility (LCF) building above ground, and a hardened Launch Control Center (LCC) sixty feet below ground that is connected electronically to ten outlying Launch Facilities (LFs or "silos") located four to fourteen miles away in all directions. Each LF houses a single Minuteman missile.

The hardened underground capsule of the LCC contains a launch control console and associated equipment, where a two-man officer launch crew maintains security and readiness to launch their outlying ten missiles with an appropriate order. When I was there, the topside LCF housed two teams of three security personnel, with one airman from each team designated a Flight Security Controller (FSC), and the other two as part of a Strike Team assigned to investigate LF security violations. Other personnel at the LCF were a facilities manager, usually referred to as a "site manager," and a facility chef.

The Air Force gave special permission for my wife and I to live in the town of Minot, which was about thirteen and a half miles south of the base. My wife taught at the high school in town, and it was convenient for us to be located there, especially during severe winter months. Also, by living in town, we were experiencing both a civilian style of life, as well as the camaraderie of Air Force life when visiting the base exchange, commissary, officers club, or friends on base. This arrangement worked quite well for us, and I was also able to take advantage of graduate courses offered at Minot State University, which was located in the middle of town.

At this particular time, the furthering of one's education was strongly encouraged by the Air Force, especially for launch officers. It would have been foolish not to take advantage of this Air Force program, which paid the tab. I took courses in business law and graduate statistics. Many of the other officers in our missile wing were also taking university courses in their off time through a master's degree program specifically offered just to missile launch officers on base. When on twenty-four-hour "alert duty" underground, while monitoring missiles and communications, we were studying hard on courses we were taking.

Prior to assignment of duties underground with the Minuteman system, I spent many months in training before receiving my certificate as a Deputy Missile Combat Crew Commander (DMCCC). Much of this training was received at Chanute and Vandenberg Air Force bases. After that, I returned to Minot AFB for classroom and simulator training, including visits to LCCs to observe and learn real-time operations. For many long hours, my assigned crew commander and I were subjected to periodic and grueling scenarios in the LCC simulator on base, which tested our ability to operate various equipment and communication systems under emergency or unusual conditions. This was combined with simulated security violations at missile sites, and the receipt and handling of simulated launch orders. There was so much to learn, and so much that needed to be performed instinctively, and perfectly, without missing a step.

On 20 January 1966, more than six months after arriving at Minot AFB for missile duty, I was deemed qualified and ready for "alert duty" in an underground LCC. This is the term we used for our twenty four-hour duty. We would say we "pulled alert" or that we were "on-alert" that day. It was a very appropriate term, as it referred to exactly what we were doing. When below ground, we were constantly on-alert and monitoring the readiness and security of our missiles, and waiting for war orders to actually launch the missiles. This was during the Cold War, where things could get potentially very hot at a moment's notice, and we knew it.

Even after receiving my DMCCC credentials, which allowed me to perform alert duty at any one of the remote LCCs, my commander and I were often subjected to unannounced surprise visits by "Standboard" evaluation teams. They would appear on site for several hours to test our abilities and make sure we remained sharp. In failing of any critical part of the examination, there would soon be a replacement crew on site to replace us. Fortunately, we always did quite well on our exams, but pressure was constantly on us to insure we maintained perfection in performance.

Prior to going on duty at a LCC, all fifteen launch crews would gather at Wing Headquarters for a morning "Pre-departure Crew Briefing." As part of this briefing, we were always tested on Emergency War Order (EWO) procedures. If a crew member did not achieve 100 percent on the exam, he would be prevented from going on duty, told to report to the Wing Commander, or given a ticket for travel to SAC Headquarters in Omaha, Nebraska, to explain the failure. That was always the threat hanging over our heads. The threat never affected me or my commander, but it was rumored that some individuals suffered the consequences.

For some officers, there may have been moments of idle boredom underground, but that was not the case for me and many others because of demands on time. Just trying to keep up with stringent job requirements kept many of us on our toes, with constant review of operating procedures and regulations, which took priority over class studies. I suspect the situation is not the same today, in comparison to the stressful Cold War environment I was involved with. There is probably a different respect for mission, a more relaxed attitude, a different crew mix, and availability of social media, including TV, DVDs, and other ways to occupy time.

My crew partner and commander, Major Gordon Tollerud, and I became top-rated, and we worked as a close-knit team in pulling alert duty at all Launch Control Centers in the missile wing. I was a new first lieutenant at twenty-five years of age, and he was about fifteen years older and more than half again my age. He taught me much about Air Force life, and he became a great mentor and was someone I greatly respected and looked up to. In living twenty-four hours together during alert duty, we came to know each other very well, and it seemed as if we could communicate just by looking at each other. In a short time, we obtained an official rating of "highly qualified," and we became an elite crew. We were also designated as on-site instructors for other crews in training. We pulled about ten to twelve "alerts" together each month for about a year and a half, until 19 August 1967. Soon after that, I was promoted to captain, became a Missile Combat Crew Commander (MCCC), and had a crew of my own with a new partner.

A normal duty day in the LCC would consist of receiving and acknowledging messages from SAC headquarters over the Primary Alerting System (PAS), communicating with other flights in the squadron, and communicating with our security people above ground. Quite often, we were interrupted by security alarms on our launch control console, which was used to monitor our ten distant nuclear-tipped missiles. We had outer security sensors and inner security sensors at each missile site that would occasionally be set off for one reason or another. Often, the outer security alarm would be triggered by a mouse or jackrabbit investigating the surface area around a missile site, or by falling rain or a thunderstorm passing by. An inner security alarm could also be triggered by a heavy thunderstorm, or by a large truck passing in the vicinity. Whenever an alarm occurred, we first tried to reset it, but if it wouldn't reset, we would send a security strike team out to investigate the LF.

Usually, all our very reliable missiles would be on-alert and ready for launch. We seldom needed to call Job Control for any kind of maintenance. The missiles were, however, subjected to periodic scheduled maintenance, and their internal launch codes needed to be changed now and then. Sometimes, as extra duty, I would be called upon to transport top secret codes to a missile LF. I would travel with a maintenance crew and participate in the long procedure to gain entrance to the underground LF vaults and missile support equipment.

There were times when duty at a LCC was relatively hectic, such as when missiles were shut down for retargeting and realignment, when guidance and control system errors needed attention, or when new contractor upgrades to LFs or the LCC were under way. Sometimes we hosted VIP visitors or maintenance crews, or endured surprise visits from evaluation teams. All of that, combined with the usual security alarms, regular communication procedures, and frequent messages from SAC Headquarters over the PAS, would make for a day with a bit more work than usual.

In the course of our duty at various LCCs, there were only two particular times I can remember when things got a bit more hectic and varied from the normal routine. One was during the night of 12/13 October 1966, when an Air Force jet, an F-106A Delta Dart (registration #59-0001), came down in our area while we were pulling alert duty at Mike Flight. We were very busy communicating with base operations and with our security teams in trying to locate the plane, which was eventually found between us and the base, about thirteen miles away.

We had no idea what caused the plane to go down. Evidently the jet, piloted by Colonel John H. Fowler, Jr., who was the 5th Fighter Interceptor Squadron Commander on base, disappeared from radar as it was making its landing approach to Minot AFB. Later that morning, after my commander and I were replaced by our relief crew, we detoured to the crash site on our way back to base, and we noted that the plane had made a relatively soft wheels-up landing. It had created a long narrow furrow in a flat plowed field. The plane was sitting alone and intact, but with its nose cracked at the cockpit and drooping to the ground. No one else was around when we inspected the plane, but we later heard that Colonel Fowler did not survive. For us, it was a mystery why the plane went down, wheels up and virtually undamaged, and with no communication from the pilot indicating he had a problem.

This incident also brings to mind a very scary incident involving my commander and I when we were on a helicopter transporting us to alert duty at Kilo Flight. We were about ten miles from base when we felt a slight bump, which was followed by some frantic maneuvering by the pilot to control the chopper. He began communicating back to base requesting emergency clearance to land. As we made a long, wide turn toward base, the pilot decreased altitude while maintaining a minimum 90 mph airspeed. Before we knew it, we were screaming along about three feet above the tall prairie grass, and then over the runway. Then, the pilot suddenly dropped the craft onto the surface and we skidded "forever" to a final stop. We were immediately surrounded by fire trucks ready to spray us with foam, but it wasn't needed. Evidently, that small bump we earlier felt, was the helicopter losing control of its tail prop, or the mechanism keeping us on a straight path. The pilot could not slow down the chopper, or hover, without losing control, so he maintained enough air speed to keep the craft's tail behind us and crash land on the runway. After "landing," we drove to Kilo LCF in the standard Air Force blue station wagon. We were quite shaken by the incident, but we still needed to report to work at Kilo Flight.

🙠🙡

In my youth, little did I realize that my interests in the sciences, and especially astronomy and space, would prepare the way and lead me into a "situation" that would forever leave an indelible imprint on my mind. This, however, did not involve the crash of the F-106 Delta Dart, or of the controlled crash of our helicopter when heading to Kilo Flight, but it occurred on one other significant day prior to the above-described crash of the F-106. It truly left me with a huge new perspective on life itself, and of the universe we inhabit. It was so profound, and awesome, that nothing in my future life experience could possibly compare or even come close. This situation will be reviewed in the next chapter.

CHAPTER FOUR

A New Life Perspective

Strange Lights in Night Sky

The event that I and other airmen experienced at Minot AFB was truly life changing. It was something so far from normal reality that I would never have expected or imagined such a possibility, but it ultimately gave me deep respect for our place in the universe. It gave me reason to look outside our bubble of earthly conscious thoughts, and to view our world from afar as an insignificant speck of human habitation, and one that is not alone in the universe.

What on earth would give me this new perspective? Was it an Unearthly Flying Object I saw, or some alien creature from outer space? Well, not exactly, although it created the same effect. The "clincher," which verified the reality of my experience, was the Air Force telling me to never speak another word of the incident. The instruction, "As far as you are concerned, It Never Happened," was the Air Force confirming that my "out of this world experience" really did occur. It was all true and real, but this was aside from the fact that I was well aware that a dozen of us were involved. Official confirmation of this experience, however, after more than fifty years, still requires release of documents and an explanation from the Air Force.

At the time, only short news reports by local media at Minot, North Dakota, provided an indication that strange lights were sighted overnight by residents of a small town north of Minot. Witnesses to those lights were left with a mystery of their own, and they probably never realized that a bunch of airmen not far away were also impacted by the lights, which resulted in an amazing incident and an unearthly experience for them. Local town residents were never aware that such an incident took place, or that additional incidents occurred, which are likely documented within top secret files.

On this particular day, I awoke early to prepare for alert duty. During breakfast, while I was watching local TV news, the announcer mentioned that residents of Mohall, North Dakota, witnessed strange lights west of town during the night, which were attributed to a "UFO." This caught my attention, because I was scheduled for alert duty at November Flight, which was located about three miles west of Mohall.

After finishing breakfast and donning my distinctive white coveralls, a blue-grey ascot, and shiny brogan boots, which was the special missileer uniform, I drove my 1965 Pontiac Lemans to Minot AFB to report at 8:45 A.M. for the 9:00 A.M. crew pre-departure briefing. After receiving my customary salute at the entrance to the base, I passed through the main gate and headed to the far end of base to Wing Headquarters.

Main Gate, Minot AFB, North Dakota

The briefing room, located in an underground basement floor with other facilities of Wing Operations, was soon filled with arriving crews. This is where fifteen two-man crews would come before departing for their assigned LCCs. Each crew took their place seated at tables arranged in three columns by squadron, and in five rows facing the front of the room. The briefing officer gave the usual pre-departure information, which included necessary details on wing and squadron protocol and procedures, current world events with intelligence information, and current flight and missile status for the wing.

When missile status was addressed, we were told that the current crew at November Flight had experienced unusual circumstances overnight, where missiles went "off-alert." This general statement was made without detail, but there was an implication of an unexplained and mysterious situation involved. Normally, it was unusual to have one missile off-alert, but to have several missiles off-alert in one flight was unheard of. I immediately recalled the TV news report from earlier that morning, and I connected it to what I was hearing in the crew briefing.

At the end of the briefing, and upon being dismissed, several crew members approached me and commented about the morning news report, which mentioned strange lights seen west of Mohall. My crew commander had also heard the report, and a short conversation ensued with speculation on whether the report was linked to missiles now off-alert at November Flight. Our attention was now focused on this strange coincidence, and we began to suspect that we did not receive the whole story in our briefing.

Predeparture Crew Briefing Room, Minot AFB

Upon departing the briefing, my commander and I were given instructions on transportation arrangements to the LCF. Some crews were assigned to helicopters for transport, and the rest of us were assigned to the standard Air Force blue station wagon. Heading out by station wagon was the usual plan for November Flight, since travel distance was not as far as to other flights.

While traveling together to November LCF, which was about thirty-seven and a half miles from base, we were anxious and curious about what we would discover upon arriving. We talked about it, but all we could do was speculate, which seemed like a useless and fanciful endeavor. We were still on edge, however, about what seemed to be more than a coincidence regarding missiles going down. In driving through Mohall, a sleepy little town, we wondered who had observed those strange lights overnight. Perhaps a local police officer, or maybe others?

Truth Revealed

Upon arriving at the front gate of the LCF, I opened the phone-box at the gate and provided the necessary information to the FSC inside the LCF to allow entry, and the gate was opened. The normal procedure after parking our vehicle was to inspect outside surroundings of the facility to make sure things were in order, and then debrief topside personnel inside. My commander, however, immediately entered the main entrance of the building to talk with security guards.

After inspecting certain areas outside, I entered the south rear of the building where I encountered and greeted the site manager, a tech-sergeant. He immediately inquired whether I was aware of what transpired overnight. I told him I was not aware, but I said that I heard news about residents of Mohall sighting strange lights overnight. With that, he escorted me into the "dayroom" where he recounted the unique and bewildering incident that occurred in early morning hours. His account of what he observed was so startling, and so astounding, that I will never forget it.

The site manager took me to windows on the west side of the dayroom. While facing the windows with arms out-stretched from about the 11:00 to 1:00 o'clock position, he attempted to describe a large object with "bright flashing lights," which hovered just outside the fence of the facility about a hundred feet away, or maybe slightly further. I asked, "Was it a helicopter?" With a tone of voice indicating he expected the question, he said it was much larger, and it made no noise, or at least none heard inside the LCF. It was unlike the noisy choppers that would occasionally arrive for various reasons. He explained that the object was elongated, and looked like the silhouette of a disk.

I asked what the flashing lights looked like, but he was hard pressed to explain. He could not relate them to anything he had seen before. They were nothing recognizable, such as beacon lights of aircraft with red, green, and white blinking lights, but more continuous, undefined, and brightly flashing-pulsating-glowing. He was frustrated that he could not properly explain it to me, or give sufficient clarity and meaning. As I listened, I heard trembling in his voice, and there was no mistake that he experienced something from out of this world, and he knew it.

With several airmen gathered around as he briefed me, he said the object hovered in the same location for quite some time, while he and others in the darkened room stood and looked at the object through the windows. He indicated it was not a fun experience, which I could immediately tell by emotion in his voice and facial expressions. It was the same with the others, and it indicated to me that they all had a very frightening experience.

When I think about it, I can understand how close-up observation by an object exhibiting itself as something from out of this world, and totally separate from current reality, might easily generate fear. I can imagine it something like confronting a huge, strange, and fearsome beast in one's backyard staring intently at you. In its steady, motionless, uninvited, and menacing observation of you, and not knowing its intentions, it could generate intense emotion. This object was not an expected, invited, or recognizable visitor, and helicopters did not visit launch Control Facilities at night.

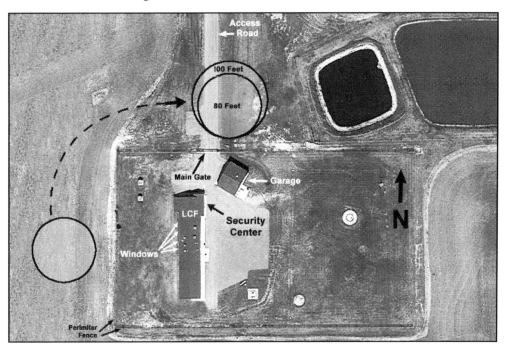

Overhead View, November Flight LCF, Minot AFB, ND.

The site manager indicated that after a certain amount of time, the object slowly moved clockwise around to the front of the LCF, and hovered just behind the main gate and slightly to the east. From this location, the object was easily viewed from windows of the security section of the facility, which faced north and east toward the gate. A portion of the object was hidden by a large garage located inside the fence to the right of the gate, and the object was positioned directly above and slightly north of the underground LCC, which was located partially under the garage and sixty feet below ground. The object was positioned for good viewing from the security center, probably no more than sixty to eighty feet away. No one, including security guards with weapons ready, cared to exit the building and confront it. It was far preferable to just let it be where it was with no observable harm taking place.

Based on their descriptions, I estimated the object to be eighty to one hundred feet in width. Admittedly, it's a rough guess based on how the site manager used his arms to indicate its width as viewed from west windows of the dayroom. But it must have been about this size to be partially obscured by the garage.

When I entered the security room, my commander was finished debriefing security personnel, who essentially confirmed the information I received from the site manager. The FSC then handed each of us a Smith & Wesson .38 caliber revolver. We each loaded our weapon with six bullets, and holstered it on our belt. On entering the adjoining service room, we stepped into a large elevator, and then descended sixty feet to where a massive hardened blast door awaited us. It opened into a vestibule with access on the left to a Launch Control Equipment Bay (LCEB), which was a hardened capsule providing air conditioning, emergency power and other services for the Launch Control Center. On the right was a smaller four-foot thick blast door leading to the LCC.

After entering the vestibule and closing the large massive door behind us, the deputy crew commander on duty opened the LCC blast door and welcomed us into their quarters. It was necessary to lower our heads upon entering through the small, confined entrance, and then we crossed a five-foot ramp into an enclosed lighted room that was suspended within the hardened capsule by massive chain links connected to large shock absorbing isolators at each corner of the room. This was the missile launch control nerve center within the hardened capsule, which would be our home during the next twenty-four hours.

After entering the LCC, our eyes were immediately transfixed on the launch control console's panel of lights at the opposite end of the room, which displayed status for all ten missiles of November Flight. All missiles were off-alert, and not launchable.

That caught us with great surprise. It was something we never expected to see. Never before had we heard of this kind of situation, or performed duty at a LCC where all missiles were off-alert. To have even one or two missiles down was an exception, except during times of scheduled maintenance.

Outgoing crew members were obviously quite ready and anxious to depart, but they were also prepared to give us an overview of the serious events that occurred during their long tour underground. I listened intently as they described the situation. They were on the phone with the FSC upstairs when lights on the launch control console lit up. This occurred at the same time the object approached the front main gate. With launch control console lights showing unexpected indications, the crew took available procedures to insure missiles would not launch, and the Wing

Command Post was contacted. Then, it was a matter of waiting for the object to leave, which it soon did. According to security airmen upstairs, it left with a quick burst of speed, and was instantly out of sight.

Of course, the crew was constantly on the phone with security upstairs, with the Squadron Command Post, and with personnel on base. Maintenance crews were dispatched to the missile sites by Job Control to bring missiles back on-alert, and additional security crews were deployed. From what I can remember, the primary malfunction remaining after missiles were taken down were indications of "guidance and control system failure," as indicated by the Voice Reporting Signal Assembly (VRSA), which was a voice reporting feature available from each missile.

While listening to outgoing crew members giving their briefing, I was struck by their demonstrated feelings of awe and wonder, their concerns about actions they took, and their expressed sense of helplessness regarding the situation. Each felt, however, that they performed to the best of their ability in taking all necessary actions and procedures. This was no different from topside airmen with whom I talked previously.

We briefly speculated on the possibility of an Electromotive Force (EMF), or Pulse (EMP) from the object, which may have created the situation. We knew the Air Force was concerned about possible electromagnetic effects from a potential nearby atomic detonation that could neutralize our missiles, which would make it imperative for early detection of incoming enemy missiles. But somehow, the object was able to send specific signals to our missiles without affecting or shutting down the LCC and its electrical power. The EMF idea was on our minds as a reason for our missiles being brought down by the object, but we began to speculate whether the Air Force was aware of something more specific that could generate a very special "EMF signal." We had no doubt that the ten outlying nuclear-tipped missiles of November Flight were compromised, tampered with, and put out of commission by the object that paid a visit.

My commander and I continued with normal procedures for "crew turnover" with the outgoing crew. This included opening the safe located on the floor just to the left of the launch control console, taking inventory of "go-code" documents and launch keys, and then relocking the safe with our individual locks. When all was completed, with signatures indicating turnover of responsibility, the outgoing crew departed, and the LCC blast door was closed and locked behind them.

Unfamiliar Reality

My commander and I settled into the routine of handling and responding to communications, waiting for maintenance crews to bring missiles back on line, and running various tests and calibration procedures as necessary. But it seemed a bit strange with no on-alert missiles to monitor for launch readiness.

We also contemplated all that we experienced that morning. It seemed so curious and surreal when thinking about it, especially in light of what the other crew and people topside had experienced. We just didn't know what to make of it all. It was so foreign, and apart from what we could wrap our minds around. We did not have the wild experience the previous crew did, but we were definitely looking at its results and ramifications, and we witnessed the very emotional concern and fear reflected in voices and faces. This was the reality, but it was a total unfamiliar reality.

One thing that needs to be mentioned is that most launch officers knew there would be attempts by headquarters to periodically test security preparedness at missile launch facilities. This was necessary in order to find any weakness in the response to security threats, and for determining potential vulnerability of our nuclear weapon systems. In no way, however, could we fathom that they would be using Unearthly Flying Objects to accomplish their mission, especially against our Launch Control Facility and underground LCC. What happened at November Flight was a major compromise of our "on-alert" missiles, which were deactivated and rendered unlaunchable by the mysterious object.

One of the key measurements used by headquarters to assess wing performance was "readiness" to launch, which was the percentage of time our wing's missiles maintained alert status. It was never possible to maintain 100 percent readiness, because of down time required for periodic maintenance, but keeping missile readiness above 95 percent was a major goal for the wing. There was no way SAC Headquarters, or Wing Headquarters, would ever jeopardize or dare compromise missile alert readiness of our strategic and deterrent missile force.

After much careful consideration about the situation, we came to understand that fate delivered something far beyond our mental comprehension. It was natural to wonder how on earth this situation could happen, but there were no reasonable alternatives to reflect on. This situation could not possibly happen without some unearthly interference being responsible. As mentioned previously, we could only come up with the idea that an electromagnetic signal had entered communication lines to our missiles. To contemplate that this was caused by some Unearthly Flying Object was hard to swallow, but we were absolutely unable to explain it with any other realistic scenario, especially with the facts that were presented to us by the crew, other people topside, and the reports of sightings overnight from some residents in the small nearby town of Mohall.

Incident Closed to Further Discussion

People who were involved with this incident, other than my commander and

I, were the two other officer crew members we replaced, six security personnel, the site manager, and facility chef. That made twelve of us who were left to review this incident over and over again in our minds.

I was also left with memory of that early morning news broadcast, which clung to the inside of my mind. It provided early notice that I might be in for a very interesting day. It was also confirmation that there really was some kind of reality to it, but it was so very hard to grasp in any rational sense.

In addition to the dozen people mentioned, there were many other people subsequently involved, including missile maintenance technicians, security police, and other personnel on base who were charged with managing the crisis and bringing missiles back on-alert. There must also have been others responsible for investigating the severe compromise that occurred, which exposed the supposedly secure nuclear-tipped missiles. It's possible that not all airmen involved with bringing missiles back to alert status were privy to what happened, but some Air Force officials were undoubtedly given high priority responsibility for finding answers to it all.

At this point, after so many years, I cannot remember how many missiles may have come back on-alert, if any, while we were on duty that day. It's almost as if the rest of that day was like all other tours we pulled at Launch Centers. Each has melded together in my mind except for special circumstances that made a particular tour unusual. My gut feeling is that a few missiles were brought back to alert status that day, but I am quite positive it took more than a day, or even several days, to bring all missiles back online to alert status. That information, and other details have departed from my mind, although I feel my memory may yet be rekindled as I further reconstruct, or have an unlikely return to the scene of the incident.

One thing I do remember, after being replaced by a follow-on crew the next day, is that I was anxious to talk again, eye-to-eye, with security guards topside. I had many more questions, and I wanted to verify and obtain more information and a further description of what was seen on the night of the incident. Upon returning topside, I attempted to question the FSC who was on duty the night previous, but he interrupted me by saying he was "instructed not to discuss the incident." And that is when my commander chimed in. He told me that he received a call during my rest period below ground, which instructed him that we were to "never speak of the incident with each other, or mention it to anyone." He was told that, "As far as you [we] are concerned, It Never Happened." I asked who told him that, and he replied, "It came from the OSI" (Air Force Office of Special Investigations). After hearing this, I knew the incident was closed to further discussion. There was no doubt about that, and I would just have to live with this great unbelievable mystery that was now imprinted on my mind.

Vulnerable Weapons

Now it's in the open. I have revealed something that has haunted me for more than fifty years. I was never informed why I was to keep silent, or what the security level of the incident was, even though I had a Top Secret Crypto security clearance. The directive to me was an informal order from my commander, and in the military you are supposed to follow orders. As disturbing as the directive was, from that point on there was no further discussion or mention of this situation between my commander and I, or with anyone else. That was the requirement. In retrospect, this is the major thing that has served to initially reveal a policy of deception in a grand cover-up by the Air Force. For me, knowing that the Air Force has continued since that time with the deception, by instructing many airmen involved with such incidents that "It Never Happened," these very same words, there is absolutely no doubt that this has been a primary means to keep everything out of the public eye.

The most troubling aspect to all this is that our nuclear-tipped ICBM weapons were severely compromised. They were drastically interfered with by some outside force, and the Air Force was powerless to do anything about it except to hush people up. Those hardened and secured missile sites, built with the latest technology, were no match for some other kind of intelligence using infinitely greater and far more superior technology. This is the hard and startling fact, which the Air Force and other secretive organizations of the government face, and will continue to face. It is unlikely they would want this revealed.

If Truth were revealed by high authority, it would not be a question of whether heads would roll. It would be a question of whether the public could withstand an announcement that flying objects are absolutely real and controlled by superior intelligence, or that U.S. atomic weapons are easily vulnerable to being toyed with by forces beyond government control. The government would not want to face public pressure and questioning about the wisdom of having atomic weapons exposed, as they obviously are, to a superior force. The government would not want to answer whether we can continue to have a secure nation without a strong nuclear deterrent against those who might threaten harm. This is something that secretive government agencies would rather not deal with, and it is definitely something that the government does not want the public to be aware of.

We launch control officers, who were involved with this particular incident, were left with a confounding situation having to do with required perfection in performance. It was very perplexing and troubling to us that a comprehensive review was never conducted of our handling of the incident, or how we might have managed it differently. We were never instructed on what to do if such an incident should ever happen again. In all other irregular situations or technicalities, we expected follow-up questioning, feedback, and then a briefing, or at least some sort of retraining. In this case, we were all left in limbo, and left on our own to conjure in our minds how other similar situations might unfold and be handled. Other missile crews were also left in the dark, with no knowledge of our incident, or the fact that they might potentially be involved in a future incident. This serious lack of follow through by the Air Force was more than bewildering, and it served to keep this situation on our minds. Unknown to us at the time, and long afterward, these incidents repeatedly occurred at Minot AFB, and at all other ICBM bases. And they continue to periodically occur around ICBM missile facilities to this day!

Unique Incident?

For me, the passage of time has diluted memory from that particular day. Perhaps a better way to put it is that the experience has melded with many other life experiences since that time. Nevertheless, particular memories of the incident that made a hard impression continue to stand out, and remain quite vivid.

The haunting periodic flashbacks of my experience can be likened to a long unsolved mystery, or a cold case crying out for further investigation. I continue to wonder why I and others were told to keep silent. Was it really a unique situation that I must honor with silence for the rest of my life? Was I never to get answers or an explanation? Perhaps there was something more to it, and I fell into something quite unbelievable and quite disturbing if publicly revealed. Were there serious national security implications? What did it all mean? According to the way the Air Force handled the situation, I knew that circumstances involved with the incident were very unusual, which left many unanswered questions, but I didn't have anything substantial to provide a real understanding of what was behind it all.

In the mid-1990s, after many years of pondering, it occurred to me that I needed to find a confirmation of my experience. In that regard, I wondered about the chance of other former Air Force launch officers coming forth and revealing their unique but similar experience. What was the possibility of others associated with my incident coming forth and speaking out? Could I locate my former crew commander after thirty years? With the "new" Internet now being used as a worldwide chain of information and communication, I wondered if I could possibly find a clue.

When I finally got around to taking a cursory glance on the Internet, I found a huge number of stories about flying object encounters of one kind or another. But they all seemed so fantastic, obtuse, and unbelievable, which came from people with no credentials. It was an obvious waste of time, so I then set aside the notion of searching the Internet further.

But then, after the passing of a few more years, in 2001, I suddenly found something on the Internet that was so profound, and so totally startling, that I could hardly believe what I was looking at. Amazingly, it gave me an immediate confirmation of my experience. It revealed something that was impossible for me to deny. It was one of those "Oh My God" moments!

ༀ

I call it a life-changing experience because of the profound way my Air Force incident affected me. But it was not necessarily that my active life, or living my life, changed per se. It was the fact that I received a new perspective on "life" itself. More than that, however, I came to realize that I had fallen into a trap where there appeared to be no escape. I knew the Truth, and I knew others knew, but I could not talk about it and neither could they. I knew that the Air Force, and probably other government agencies, were covering up a magnificent secret, which was something they dared not expose to the public. But what exactly was it all about? I didn't have the answers, and they probably didn't either, although they probably knew much more than I.

Today, there are many former Air Force people who continue to hold the secret of an unearthly experience. They include the eleven other people involved with my incident. For many of them, maintaining the secret can be a frustrating endeavor, especially when knowing that the secret is something important to all of humankind. Unfortunately, many airmen from my time in the Air Force are now deceased, including my former commander. It can only be hoped that others will come forward to confirm their experience, but they need to do it before it becomes too late for them also.

CHAPTER FIVE

Revelation and Confirmation

The Captain Robert Salas Incident

I shouted out, "Oh my God!"; "Wow!" I was totally stunned and overwhelmed by my discovery. My outburst was followed by immediate feelings of tremendous relief and joyous exhilaration, all because of pent-up frustration in harboring a haunting, monumental secret for so many years. It finally gave way to exuberant vindication.

This happened in 2001, when I again attempted to find information on the Internet about other Air Force officers who may have been involved with my flying object experience. It didn't take long to come upon a particular incident that took my breath away. It was an incident reported by retired Air Force Captain Robert Salas, and he described an incident nearly identical to mine. It was so coincidentally similar I could hardly contain my emotion upon reading his account, which I immediately read several more times. It was an account that left me absolutely stunned, because I had little confidence I would actually find such information, although I had great hope. Now, Truth was staring me in the face, and I was suddenly overcome with a sense of reassurance. It was confirmation of what I experienced so many years ago, and I was suddenly enveloped by a miracle. It was a miracle that released the secret I harbored for so long. Someone else exposed the secret for me.

A great burden was lifted from my mind. The secret was in the open, and I now felt freedom to unshackle myself from it–or to at least inform my new wife of four years. I would no longer be alone with my memory, and I could hardly wait to tell her. But then it occurred to me that she might not have the same reaction, or any reaction at all. I had kept the secret from her too, and I previously led her to believe there were no secrets between us. I would have to dispel that idea, but I would also have to share a secret that would be difficult to digest.

The revelation by Salas was particularly significant, due to the fact his incident occurred at a LCF associated with Malmstrom AFB in Montana, and it occurred within a few months of my incident at Minot's November Flight. We were at different air bases in adjacent states, at about the same time period, and we were uniquely impacted in the same manner. When searching the Internet, I was looking for someone connected with my particular incident, but I never expected to discover a completely separate, but similar incident elsewhere. It provided even more confirmation that my incident was not just a single happening, and it indicated a pattern of intrusion by something from out of this world. It dispelled any possibility my experience was a figment of imagination, or a fantasy gone wild after so many years.

I always knew I was involved with something incredible and unbelievably mysterious, but shallow thoughts and ideas suggesting implausible explanations for my incident were beginning to tease my mind after so many years. I knew my experience was real, but the absolute lack of confirmation was starting to play games with me. Now, there would be no doubt, because Robert Salas' incident provided the solid and valid confirmation I was looking for.

Salas, with James Klotz, published a book in 2005 titled *Faded Giant*, which detailed Salas' experience. Salas has also participated in many interviews, and has appeared three times at the National Press Club in Washington, D.C. Some of what he related at the National Press Club, on 27 September 2010, is the following:

> *In 1967, I was a first lieutenant stationed at Malmstrom Air Force base, Montana. I was a Missile Launch Officer...and in March 24th, 1967 I was on duty at what we called Oscar Flight. It is an underground capsule, a hardened site about 60 feet underground. We had security guards topside. The main guard is called a Flight Security Controller...Sometime in the evening hours on March 24th I received a call from one of my topside guards, a Flight Security Controller, stating that they had been observing strange lights in the sky, making odd maneuvers and wanted to report it...I kind of dismissed the call. He called back about 5 minutes later. This time he was screaming into the phone saying that they're looking at an object, a red glowing object hovering just above our front gate. This object was about 30 feet in diameter. He could not make out too much of the details of the object, only that it was pulsating...He had all the guards out there; he*

was very frightened, wanted me to give him direction. I think I said something like make sure nothing comes inside the perimeter fence. He immediately hung the phone up...I went to wake my commander... who was taking a rest break, started to tell him about the phone call, and just as I told him, our missiles began going into what's called a no-go condition or unlaunchable; essentially they were disabled when the object was still hovering over our site, the Launch Control Facility...We also had some security lights, meaning security incursions at some of the launch facilities, so I called the guard back upstairs and directed that a security team be sent out. At this point the guard told me the object had left at high speed, again silent, no noise...The main indication we got from our equipment was guidance and control system failure...We were relieved the next morning...reported to our squadron commander ...I asked him specifically if it could have been an Air Force exercise, and he assured me it was not an Air Force exercise. There was also a member of the Air Force Office of Special Investigations in the room. He ordered us to not ever talk about this...

Releasing My Secret

When looking back now, it did take quite a few years for that astounding revelation on the Internet to appear in front of me, but Truth finally revealed itself– thanks to Salas. It then became a matter of getting comfortable with overwhelming certainty of the revelation, and the strong urge to share it. It was still tempered, however, by the oppressive order I received long ago to keep my lips zipped. But the euphoria in realizing that Truth had sprung loose was overpowering, and I needed to do something about it.

First, I quickly decided to speak to my wife, Diana. With self-contained excitement, and great anticipation, I showed her the website describing the incident of Robert Salas. In a way, I was letting Salas reveal the secret for me. I told her, "This also happened to me," and I went on to tell her that the Air Force told me to keep silent, and instructed me, "As far as you are concerned, It Never Happened." To be honest, her reaction was, "Oh really?"–it was a matter-of-fact kind of acceptance, with no follow-up questioning. Previously, I thought my revelation would not affect her in the same way, but I was thinking she might have some sort of interest, or more of a response. I knew, however, that reality of what I was telling her would be difficult to comprehend or sink in. And with her response, I fully realized what I would be up against in talking with friends and others in relating my extraordinary experience.

I also knew I needed to make contact with Salas, to touch base with him and let him know he was not alone. I felt there was now a real friend out there to reach out to and relate with. I searched the Internet in an attempt to find contact information, but I failed in that endeavor. Several years later, in 2010, Salas discovered me and sent an email. In my response, and with a feeling of close kinship regarding a mutual experience, I told him I was glad to have found a "real friend" in this situation. But afterward, I thought my comment probably seemed rather odd, since we were both strangers to each other. After that, we exchanged emails a couple times, but we never engaged in any substantial dialog until 29 April 2013 in Washington, D.C. It was as if we both realized that the UFO question was evident to each of us, and that the phenomenon was probably pervasive. It was only a matter of waiting for disclosure by high authority. I sincerely thanked him for his contribution in coming forth and revealing Truth, because it allowed me to reconcile a haunting memory.

How Could It Be Otherwise?

The truly interesting part of the Salas incident, and my similar incident at Minot, is the number of things that corresponded. The same is true to a certain extent with other incidents I've come across. First of all, topside personnel, without exception, were very frightened during their close-up observation of the object. Encountering the object was outside their realm of an everyday life experience, and it intruded into a sense of reality. The object made odd maneuvers with strange flashing/pulsating/glowing lights. It hovered silently next to the main front gate of the LCF, which was when the launch control console below ground displayed unexpected indications showing outlying missiles malfunctioning with guidance and control system failure. When it was over, the Air Force OSI told everyone involved to never speak of the incident again.

Located underground in a hardened capsule, we launch officers never observed the objects personally, or directly, since we were required to remain at our duty stations. Estimated size of the objects could only be based on descriptions given by airmen located in the LCF building topside. In my incident, the size of the object was based on descriptions given to me, but I cannot confirm that it was the same size or same kind of object in Salas' incident, although there does seem to be similarity.

Through the years, before Salas' revelation, I often wondered about chances of such an incident happening again. The incident in my case was so significant, and so overtly threatening, I could think of no reason why it might not reoccur, and it was easy to think it might. There was no way, however, to find out, since the Air Force was squelching the Truth and keeping everyone silent. I could only hope that someone involved in my particular incident would speak out, and I'm still hoping for that. Like everyone else, however, I was cooperating in that silence without saying a word to anyone. Because of the amazing and stunning account by Salas, which uniquely replicated my experience, my long wait for a confirming answer was over. For me, it became evident, Truth became known, and I knew it so very well. How could it be otherwise?

The Captain Val Smith Incident

Since my discovery of Salas, I have found many airmen from my days at Minot who experienced incidents with flying objects. I've also discovered that incidents have occurred at every ICBM missile base in the country, including other bases with nuclear connections. And the incidents continue to this very day, as witnessed by many airmen at various facilities. In discovering this, I've been amazed at the proficiency of the Air Force in corralling and controlling the secret, not just within the general Air Force population, including individual units at my wing at Minot, but with high level government officials and the general public as a whole.

With one exception, I previously had no knowledge, or even a serious suspicion that others with me in the three squadrons at Minot, other than those involved with my incident, harbored any such secret. The exception was a clue I received in local news at the time, but the identity of the Launch Control Center, and crew names, were not available. Even though I was in possession of weekly and monthly crew schedules, allowing me to identify crew members assigned to various facilities, I was not permitted to talk about flying objects because of my order of silence. My lips were zipped, and there was nothing I could do to ferret out information without putting myself at risk.

The incident I received a clue about occurred on 25 August 1966, which was probably within a month of my incident. It was "shouted-out" in a 6 December 1966 issue of *The Minot Daily News* (see next page), where a bold headline stated: **"Minot Launch Control Center 'Saucer' Cited As One Indication Of Outer Space Visitors."**

Evidently, editors of *The Minot Daily News* received prior information of an article to be released on 17 December 1966 by the *Saturday Evening Post* magazine, which was in regard to a flying object incident at a Minot AFB missile site. If names of airmen had been disclosed and made available on the incident, both publications would have attracted considerable attention to the UFO question, both in the town of Minot and on base. People would have been talking and asking questions.

Most interesting to me, however, was my discovery a few years ago of a missileer, Captain Val Smith, who was listed in a document found in BLUE BOOK files. It identified him as the officer involved in the incident described in *Post* magazine.

THE MINOT DAILY NEWS

Established 1884 ★   Minot, North Dakota, 58701, Tuesday, December 6, 1966   56 Pages   Price 10 Cents

Minot Launch Control Center 'Saucer' Cited As One Indication Of Outer Space Visitors

More than three months after it reportedly occurred, an Unidentified Flying Object sighting in the Minot area has been blown up as the lead-off topic in an article entitled "Are Flying Saucers Real?" which appears in the latest issue of Saturday Evening Post magazine.

This sighting, says the Post article, occurred at a Minot Air Force Base Minuteman missile launch control center. Date of the incident is given as Aug. 25, 1966.

Base information officers confirmed such a report was made. It was never released in Minot, but was sent to Wright-Patterson Field in Ohio. From there, it presumably was channeled to the Secretary of the Air Force's office—and he, again presumably, released it to the magazine.

The article is by Dr. J. Allen Hynek, identified as chairman of Northwestern University's astronomy department and an Air Force consultant on "flying saucers" from 1948 until this year.

The Minot incident is detailed in this manner:

A launch control center officer, who was in an underground capsule, discovered static was interfering with radio transmission. While he was attempting to clear the problem, personnel on the surface reported seeing a UFO with a red light high in the sky. At the same time, a radar crew picked up the object at a height estimated at 100,000 feet.

Static stopped when the object climbed, the report maintains. After climbing, it began to swoop and dive, then apparently landed some distance away.

"Missile-site control sent a strike team . . . to check. When the team was about 10 miles from the landing site, static disrupted radio contact with them. Five to eight minutes later, the glow diminished and the UFO took off. Another UFO was visually sighted and confirmed by radar. The one that was first sighted passed beneath the second. Radar also confirmed this. The first made for altitude toward the north and the second seemed to disappear with the glow of red."

It also is Hynek's assertion that a similar incident occurred near the Minot base a few days earlier. The article does not quote any names of persons involved, but states:

"A police officer—a reliable man—saw in broad daylight what he called 'an object on its edge floating down the side of a hull, wobbling from side to side about 10 feet from the ground. When it reached the valley floor, it climbed to about 100 feet, still tipped on its edge and moved across the valley to a small reservoir.

"The object, which was about 30 feet in diameter, next appeared to flatten out and a small dome became visible on top. It hovered over the water for about a minute, then moved to a small field, where it appeared to be landing. It did not touch the ground, however, but hovered at a height of about 10 feet some 250 feet away from the witness, who was standing by his parked patrol car. The object then tilted up and disappeared rapidly into the clouds. A fantastic story, yet I interviewed the witness in this case and am personally satisfied that he is above reproach.

Hynek also details what he terms "one of the most puzzling cases that I have studied," this occurring in 1953 in the Bismarck area.

It started, he claims, when persons northwest of Rapid City, sighted objects, which also appearing on radar screens at Ellsworth AFB. An F84 jet dispatched aloft reportedly radioed the object moved twice as

See MINOT SAUCERS CITED—Page 2

Front Page Headline of *The Minot Daily News*
December 6, 1966

Val Smith was in my squadron, and I knew him, but because of our order of secrecy, we never knew the other was involved in an incident.

Curiously, and despite this particular outburst of media information, it did not create a splash of attention among those I associated with on base. Everyone who experienced an incident was forever under strict orders to keep quiet, and not discuss the subject. Those who had no such experience did not know what to think, especially if they were not involved in flying object scuttlebutt, which was discouraged. In my own mind, the article describing the incident was of great interest, and I was anxious to hear more information, while also hoping for more media attention, which never came. For a while, I contemplated that increased media attention might result in release of information about my incident, which gave hope that I would be released from Air Force imposed silence, but it never happened. I have to admit, however, that thoughts occurred to me about approaching the media with my story, but those were whimsical thoughts fraught with serious consequences.

The *Saturday Evening Post* article, which *The Minot Daily News* received advance information on, was written by Dr. Allen Hynek after he investigated Smith's incident. This was an incident that Hynek normally would not have responded to because the 4602nd Air Intelligence Service Squadron (AISS) was considered "primary responder" at the time, along with the OSI. Hynek, however, did get notice of the incident, and also another significant one in the area at the time, and he considered it worthwhile to personally investigate. Smith's incident, in particular, along with several others, eventually led Hynek to take critical notice and change his attitude toward the UFO question.

The *Saturday Evening Post* article by Hynek stated, in part, the following:

> *On August 25, 1966, an Air Force officer in charge of a missile crew in North Dakota suddenly found that his radio transmissions was* [sic] *being interrupted by static. At the time, he was sheltered in a concrete capsule 60 feet below the ground. While he was trying to clear up the problem, other Air Force personnel on the surface reported seeing a UFO–an unidentified flying object high in the sky. It had a bright red light, and it appeared to be alternately climbing and descending. Simultaneously, radar crew on the ground picked up the UFO at 100,000 feet. When the UFO climbed, the static stopped. The UFO began to swoop and dive. It then appeared to land ten to fifteen miles south of the area. Missile-site control sent a strike team (well-armed Air Force guards) to check. When the team was about ten miles from the landing site, static disrupted radio contact with them. Five to eight*

minutes later the glow diminished, and the UFO took off. Another UFO was visually sighted and confirmed by radar. The one that was first sighted passed beneath the second. Radar also confirmed this. The first made for altitude toward the north, and the second seemed to disappear with the glow of red. This incident, which was not picked up by the press, is typical of the puzzling cases that I have studied during the 18 years that I have served as the Air Force's scientific consultant on the problem of UFO's.

This unique account by Hynek of Smith's incident is significant in a number of respects, other than the fact that it is one of many that gave him particular reason to ponder true reality of the UFO question. The key word he uses for these kinds of incidents is "puzzling," and this became his moniker for incidents where data collected did not correlate with naturally known phenomena. The fact that this incident took place within a highly sensitive area of Air Force responsibility provided particular reason for Hynek to take very special notice, and give critical attention to it. Of particular note to him was the fact that many ground observers were involved with this incident, along with radar observations from "South Base."

The radar station at South Base, located on a hilltop sixteen miles south of the town of Minot, was operated by the 786th Aircraft Control and Warning Squadron, and it functioned as a Ground Control Intercept (GCI) and warning station of the Air Defense Command (ADC) network. Advanced radar installed there enabled determination of height and direction of targets. Airmen operating this facility communicated with both the North American Air Defense (NORAD) Command, and also with radar technicians at Minot Air Force Base. Both bases in this incident were able to obtain fixes on the objects, and obtain corroborating data from airmen on the ground. The antics of two flying objects swooping around at altitudes from 100 thousand feet to 40 thousand feet, and then diving quickly to near ground level within seconds, coincided with recorded radar observations and visual sightings by ground observers. This gave great cause for serious consideration by Hynek.

Captain Smith reported that his HF radio transmissions were interrupted by static when the object was overhead, and close to the LCF, but when the object climbed, the static stopped. His VHF radio also experienced strong interference when attempting to contact his security strike teams, which were headed toward the object after it appeared to have landed. Although his radio equipment experienced difficulty during the incident, it worked perfectly before and after.

Smith was in contact with his FSC upstairs, as well the Wing Command Post and Base Operations by telephone. When radio transmissions were not being interrupted, he was involved with relaying information received from the field to Base Operations, who were communicating with radar technicians.

Unknown to me at the time, the Base Director of Operations at Minot AFB, Major Chester Shaw, submitted a secret report of the incident to Project BLUE BOOK on 30 August 1966, which was in accordance with "Air Force Regulation 200-2." This regulation, titled "Unidentified Flying Object Reporting," established responsibility and a procedure for reporting information on flying object sightings, and also included restrictive guidelines and stiff penalties for releasing pertinent information to the public. In his report, he described two round objects, which were observed for three and a half hours in clear night skies. They were as big as a B-52 on radar, and they shined brightly with red, greenish-yellow, and white colors. They were seen visually from the ground, with observations verified by radar. It was this document (see next page) that I previously referred to, which listed Missile Combat Crew Commander Captain Smith as the officer on duty at Mike Flight LCC. And it was this report that Hynek evidently obtained, which prompted him to visit Minot AFB.

Smith's normal crewmate was DMCCC First Lieutenant Daniel Grossman, who has since indicated to me that he was in New York City getting married at the time of the incident. Getting married was a highlight of his life, but he also said that he was sorry to have missed the excitement of the flying object incident. He also told me the following:

> *The Aug 66 "visit" was witnessed by the topside crew. Even though it was nighttime they could tell that the object was huge, silent, with some lighting, but no actions were taken by it. The main base dispatched some F-106s to chase the UFO to no avail. It took off quickly.*

The normal duty assignment for Smith and Grossman was at Lima Flight, but on this particular occasion, Smith was assigned to Mike Flight with a different deputy crew commander.

According to the usual practice of secrecy, I was never able to talk with anyone involved with this incident at the time, and I was not aware of Dr. Hynek being on base, or the significance of his coming our way in his investigation. I did not know who Hynek was, and I was unaware of his particular role, except for what I read in *The Minot Daily News* article, which was published about three months after the incident.

Although I do not recall hearing anything of Smith's incident within the wing, because I was staying clear of such things, I do remember questions such as, "Did you see the headlines yesterday? Do you know what it's all about?" And

DEPARTMENT OF THE AIR FORCE
HEADQUARTERS 862ND COMBAT SUPPORT GROUP (SAC)
MINOT AIR FORCE BASE, NORTH DAKOTA 58701

REPLY TO
ATTN OF: BDO

30 AUG 1966

SUBJECT: UFO Report

TO: AFSC (FTD)
Wright-Patterson AFB, Ohio 45433

In accordance with AFR 200-2 as changed, the following information
is submitted:

    a. Description of the Object(s)

        (1) Shape: Round.

        (2) Size: At largest point, size of a B52 aircraft.

        (3) Color: Red, Greenish yellow, White.

---

    k. SHAW, CHESTER A.; JR., Major, USAF, Base Director of Operations.
Comments: Capt Smith (Missile Combat Crew Commander) on duty at Missile
Site (MIKE Flt) sixty (60) feet underground indicated that radio transmission was being interrupted by static, this static was accompanied by
the UFO coming close to the Missile Site (MIKE Flt). When UFO climbed,
static stopped. The UFO appeared to be S.E. of MIKE 6, range undetermined.
At 0512Z, UFO climbed for altitude after hovering for 15 minutes. South
Radar base gave altitude as 100,000 feet, N.W. of Minot AFB, NDak. At
this time a strike team reported UFO decending, checked with Radar Site
they also verified this. The UFO then began to swoop and dive. It then
appeared to land 10 to 15 miles South of MIKE 6. "MIKE 6" Missile Site
Control sent a strike team to check. When the team was about 10 miles
from the landing sight, static disrupted radio contact with them. Five
(5) to eight (8) minutes later, the glow diminished and the UFO took off.
Another UFO was visually sighted and confirmed by radar. The one that
was first sighted passed beneath the second. Radar also confirmed this.
The first, made for altitude toward the North and the second seemed to
dissappear with the glow of red. A3C SEDOVIC at the South Radar base
confirmed this also. At 0619Z, two and one half (2½) hours after first
sighting, an F-106 interceptor was sent up. No contact or sighting was
established. The Control Tower asked the Aircraft Commander of a KC-135
which was flying in the local area to check the area. He reported nothing.
The Radar Site picked up an echo on radar which on checking was the KC-135.
No other sightings. At 0645Z discontinued search for UFO.

    l. None.

CHESTER A. SHAW, JR., Major, USAF
Base Director of Operations

First and last page of Minot AFB report citing Capt. Smith's incident.

I remember banter among some missileers regarding a restricted project called Project BLUE BOOK, but I never took an opportunity to learn what it was all about. It is interesting, however, that Project BLUE BOOK was mentioned in the *Saturday Evening Post* article, which exposed the project to the public. The project was set up by the Air Force to keep tabs on flying object activity, and to function as an open door for flying object reports. Investigative "expertise" was available in Project BLUE BOOK, but not for anything deemed top secret. The project operated primarily as a "front" for debunking sighting reports while collecting sighting information from the public and general military. The public, however, was continuously given the impression that the Air Force was absolutely uninterested in flying objects, and BLUE BOOK was mostly kept confidential.

Hynek evidently communicated with Major Shaw, and he possibly interviewed Smith, as suggested to me by Grossman. Unfortunately, Hynek's article in the *Post* did not mention Smith's name, and the incident vanished from my mind with time, and no thought was given to it again.

Val Smith passed away in January 2011, and it was shortly afterward that I discovered his name in BLUE BOOK documents. After conducting research in the attempt to locate Smith, I then discovered I missed an opportunity to contact him. It would have been quite interesting to compare notes, and perhaps determine who his actual crew mate was at the time, provided he might still remember.

In January 1967, Hynek was still involved with this particular case as evidenced by a secret letter he received from First Lieutenant Roger Meyer, a missile security operations officer of the 862nd Combat Defense Squadron at Minot AFB. Meyer personally investigated this incident and responded to further questions from Hynek regarding the case.

I have since discovered at least seventeen names in BLUE BOOK files on this incident. Visual ground observers included personnel topside at Mike Flight LCF, maintenance personnel at Mike Flight's LF-4, and at least nine people at LF-6. Also involved were airmen at Base Operations, at least four radar technicians at South Base, and two air surveillance technicians at 28th Air Division Headquarters who directed launch of an F-106 interceptor from Minot AFB piloted by Captain Burg. As far as I know, there has been no research conducted that I am aware of to locate any of the individuals involved with this incident, some of whom must still be alive. In regard to this, *The Minot Daily News* article concluded with the following statement:

> *An attempt today to get names of base men involved in the sightings met with the response: "Sorry, any information will have to be released by the Secretary of the Air Force."*

Of specific interest concerning this incident is the color of the object stated in the Minot article, which is the same as mentioned in the Robert Salas incident, where it is described as red and glowing. *The Minot Daily News* article reported:

A launch control officer, who was in an underground capsule, discovered static was interfering with radio transmission. While he was attempting to clear the problem, personnel on the surface reported seeing a UFO with a red light high in the sky. At the same time, a radar crew picked up the object at a height estimated at 100,000 feet.

The color of the object in my incident was not mentioned to me, to my recollection, but the red or orange color is a very familiar description mentioned in a great many accounts. Dr. Hynek also stated:

...since 1964 there has been a sharp rally in the number of puzzling sightings. The more impressive cases seem to fit into a pattern. The UFO's had a bright red glow...When the objects at last began to disappear, they vanished in a matter of seconds.

Robert Low Visits Minot With Hynek

In November 1966, when Hynek visited Minot Air Force Base to investigate Smith's incident, he was given less than a warm welcome, partly because of the ultra-sensitive nature of the top secret incident he was about to investigate. But it was also because he brought with him Mr. Robert Low from the University of Colorado. Low was instrumental in implementing a special contract, on 6 October, between the Air Force and the University for study of flying objects. The Air Force, however, had special covert intentions in regard to that study, and it was necessary to keep Low away from reality of flying objects. But it is noted that Low had a background in the intelligence field, and there is speculation he was connected with the CIA

The Air Force did not want Hynek to determine Truth of Smith's incident, because it was considered top secret, and it didn't matter that Hynek was chief scientific investigator of flying object incidents for the Air Force. As far as the Air Force was concerned, Hynek was not authorized to be in-the-know on this incident, which was no different than us missileers who were ordered to keep silent. And the Air Force also didn't want Robert Low involved for the same reason. Hynek, on the other hand, saw a beautiful opportunity to get Low involved with this puzzling incident, which could lead to potential discovery of flying object reality.

Previously, after Major Shaw's report to Project BLUE BOOK on Smith's incident, a secret memo dated 12 October 1966 from Colonel Jerome Jones, the Deputy for Technology and Subsystems of the Electronics Directorate in the Foreign Technology Division (FTD - same as T-2, TID, and ATIC), was addressed to Major Quintanilla, head of Project BLUE BOOK. It stated:

> *1. This office has no explanation for this incident. It seems possible that ball lightening...might well cause static sufficient to interrupt communications—but not for 3 ½ hours.*
> *2. It is suggested that this is an ideal case with which to inaugurate the $300,000 contract with the University of Colorado.*

In regard to item number one, it must be noted that Captain Smith's incident involved visual sightings by very credible people, and Colonel Jones was not willing to use the "ball lightning" excuse for what occurred. The excuse was extremely far-fetched when considering reported circumstances of the incident.

In regard to item number two, the "contract" to study flying objects was previously approved by the University of Colorado, and it resulted in the University's creation of the Condon Committee. As mentioned, Robert Low was involved with this, and the committee ultimately produced the *Condon Report* in 1969, which recommended closing Air Force investigation of flying objects by Project BLUE BOOK. It put all future flying object investigative efforts undercover and underground. One can debate the "exact" meaning of "inaugurating the $300,000 contract," but it probably meant that Smith's incident would be a good place for the Committee to start its study efforts. And Jones was probably not aware of Air Force reluctance to get the committee involved with this particular incident.

In regard to the Condon Committee and Robert Low, the appearance of Low at Minot is quite significant for a number of reasons. Previously, on 9 August, which was prior to Smith's incident, Low wrote a memo discussing his thoughts regarding the University taking on the study project for the Air Force, and he attempted to placate a few of his university staff who were against it. He stated:

> *I can't imagine a paper coming out of the study that would be publishable in a prestigious science journal. I can quite easily imagine, however, that psychologists, sociologists, and psychiatrists might well generate scholarly publications as a result of their investigations of saucer observers...the trick would be, I think, to describe the project so that, to the public, it would appear a totally objective study but, to the scientific community, would present the image of a group of nonbelievers trying their best to be objective but having an almost zero*

expectation of finding a saucer. One way to do this would be to stress investigation, not of the physical phenomena...if we set up the thing right and take pains to get the proper people involved and have success in presenting the image we want to present to the scientific community, we could carry the job off to our benefit.

It's quite obvious, in light of this memo, that Robert Low harbored preconceived ideas about how the flying object study would be conducted, and the result from it. He was preparing university staff for the fact that identification of flying objects would not be allowed. This was because some staff were embarrassed about unworthiness of the study. They did not think it appropriate for a university to study such fantasy, which had no basis in reality. As a university administrator, part of Low's job was to pursue grants for the University in order to finance its staff, and it was important that staff recognize that proving "reality" of flying objects was unlikely, and that other results from the study would provide definite benefits.

With Low and Hynek visiting Minot AFB after the University contract was authorized, there was reason for both to be shunned by the air base, probably with direction from Air Force Headquarters. There was reluctance to allow them, and especially Low, to converse with credible Minuteman missile launch officers and other airmen. But one can also speculate, for good reason, that Low was motivated to badger Hynek and impede his investigation for "benefit" of both the Air Force and University of Colorado. Hynek may not have been prepared for Low's desire to exhibit strong negative attitude toward flying objects, but Hynek would soon become aware of that attitude.

Air Force Put on Notice by Hynek

At the time Hynek wrote his *Saturday Evening Post* article, he was not pleased with Air Force handling of the UFO question, and possibly with his treatment at Minot AFB. This makes it tempting to believe that Hynek gave *The Minot Daily News* a forewarning of the *Saturday Evening Post* article, but the *Post* possibly informed the *Daily News* as a matter of courtesy. Nonetheless, Hynek pulled no punches with his article, and he let loose on the Air Force, which included some of the following:

...The public, I am certain, wants to believe–what can be believed– about the "flying saucer" stories that seem to be growing more sensational all the time. With all loyalty to the Air Force, and with a deep appreciation of its problems, I now feel it my duty to discuss the UFO mystery fully and frankly. I speak as a scientist with unique

experience. To the best of my knowledge, I am the only scientist who has spent nearly 20 years monitoring the UFO situation in this and other countries and who has also read many thousands of reports and personally interviewed many sighters of UFO's...Gradually, I began to accumulate cases that I really couldn't explain, cases reported by reliable, sincere people whom I often interviewed in person. I found that the persons making these reports were often not acquainted with UFO's before their experience, which baffled and thoroughly frightened them...privately, I was becoming more and more concerned over the fact that people with good reputations, who had no possible hope of gain from reporting a UFO, continued to describe "out-of-this-world" incidents...

Additionally, Hynek mentioned in his article that he was being publicly accused of selling out as a scientist because he "didn't admit that UFO's existed." And he stated he was becoming known as "the Air Force's stooge, its tame astronomer, a man more concerned with preserving his consultant's fee than with disclosing the truth to the public."

He went on to say he wrote a letter to the Air Force stating:

...I feel it is my responsibility to point out that enough puzzling sightings have been reported by intelligent and technically competent people to warrant closer attention than Project Blue Book can possibly encompass at the present time...

Smith's incident, and other various incidents occurring at Minot around this particular time, including the one I was involved with, was probably reverberating with high officials at Air Force Headquarters, and also with the Control Group. Hynek was investigating at Minot AFB not long after my incident, although he never found out about it. Subsequently, he was also banned from the base and not allowed access to airmen. Very likely, he was not happy with that, which gave him further reason to question his investigative role and reality of the UFO question.

The Don Flickinger Incident

Within that *Saturday Evening Post* article, Hynek also described another incident that occurred on 19 August 1966, which was a few days prior to Captain Smith's incident, and it involved a police officer he interviewed. Although not named in his article, this officer was U.S. Border Patrolman Don Flickinger, and this particular incident happened near Donnybrook, North Dakota, within the

Mike Flight area of control, and near the Inter-squadron Connection for missile communications.

In the article, Hynek mentioned that Flickinger saw, in broad daylight, a flying object moving toward him down into a valley. Flickinger estimated it to be about thirty feet in diameter, and he noted it had a small dome on top. This incident was also recorded in Project BLUE BOOK files, courtesy of Hynek, where it mentioned that the object hovered about ten feet above the ground and within 250 feet of Flickinger while he was standing next to his parked patrol car. The object then tilted up and suddenly zoomed into the clouds. Hynek indicated he was personally satisfied that this officer was above reproach.

In discovering this incident in BLUE BOOK files, I found that Flickinger's name was mentioned. And this set me to wondering whether I might be able to locate him after almost fifty years. After considerable investigation, and much to my surprise, I traced Flickinger to a town in Montana, and arranged to meet with him on 12 June 2016. My wife and I met with him at a restaurant in town, and we found him to be a very engaging gentleman. With vivid memories of his incident, he was more than anxious to relate his encounter to us. It was surreal to meet this man after so many years, especially after reading his anonymous account by Hynek in the old *Saturday Evening Post* article.

The following is what Flickinger related about his incident: At the time, he was an officer in the U.S. Border Patrol, and was on his way with his partner and two prisoners in his patrol car to deliver the prisoners to Canadian officials at the Canadian border, at Portal, North Dakota. They were traveling up Highway 52, heading northwest, and they were about a mile or two past the small town of Donnybrook. It was then that one of the prisoners noticed, and commented, on a silvery object silhouetted three-quarters of a mile away at the top of a sloping hill to the right.

Highway 52 and the town of Donnybrook are situated in a shallow valley within a mile-wide drainage containing the tiny Des Lacs River, which is bounded by grassy sloping hills on each side of the drainage. Flickinger's partner replied to the prisoner that the silver object was probably a small grain storage bin or silo, which is a common sight in the area. After a moment, however, both prisoners replied that they never saw a silo "fly." At about the same time, they came upon ten or twelve cars and a semi-truck parked beside the road, with people standing and looking toward the top of the slope bordering the valley. The silvery disc-shaped object was moving down the slope, close to the ground, and Flickinger quickly parked his car. They watched as the object slowly headed in their direction.

The concerned and nervous prisoners in the caged backseat then wanted to be let out, so they were handcuffed together, and they all stood outside the car watching. Movement of the craft toward them appeared slightly erratic, and it approached soundlessly. Also, when it passed over a small body of water, the water surface became slightly ruffled. The disk was about thirty feet in diameter, and slightly convex on both top and bottom surfaces. There was a small dome on top, and a dark circular area underneath. When it came quite close, it lifted to within 100 feet above them and the others who were watching. It glided back and forth above people and cars, and then started spinning, tilted on its side, and shot instantly skyward through the clouds.

Flickinger said he called Minot AFB and talked to someone about the incident, and also provided details, but he heard nothing further from the base until Hynek called and said he was coming out to visit. This was probably at the end of October or early November. Hynek undoubtedly decided to make his trip to Minot because of the significant information he received on both the Smith and Flickinger incidents.

Together, Hynek and Robert Low went with Flickinger to the site of Flickinger's sighting, and they walked to the top of the hill where the object was initially sighted. Hynek took pictures of the surrounding area, and he noted ground effects where the object rested.

Flickinger indicated he did not have a good initial impression of Low because of his constant negative attitude, and increasingly caustic remarks. Flickinger then became so distressed with Low's comments and innuendo that Hynek needed to admonish Low. Flickinger's parting impression of Low was that he was an unbelievably "despicable character," and that he "never met such a disgusting person before or after."

Later, Hynek made a second visit to Minot and stayed with Flickinger. He was there to look into another sighting encounter that Flickinger found out about that involved four hunters who witnessed a flying object, which is another very interesting encounter. On this second visit, Hynek was not allowed access to Minot AFB, but he was able to interview the hunters.

Flickinger indicated he kept in close contact with Hynek and helped him many years later in the investigation of strange and suspicious cattle mutilations.

ঔৣ

My discovery of retired Captain Robert Salas, and the incident he was involved with, was a fantastic circumstance. It helped me alleviate a haunting memory, and it allowed me to release my secret. It was a confirmation of Truth, which I sought for many years. The only thing that could top this would be for the missile crew I replaced at November Flight, or even some of the airmen who were topside during the incident, to come out of the shadows and release their secret. I am aware, however, that two of the site managers who served at November Flight are deceased, and there are probably others who are no longer living. As previously stated, my commander at the time is also deceased. But many others who served at Minot AFB during this period should be able to provide additional confirming information, and I encourage them to step forward.

Other than Salas, my discovery of previously secret documents in Project BLUE BOOK files that originated from Minot AFB, and connected with articles in the *Saturday Evening Post* and *The Minot Daily News,* provided important confirming information that Captain Val Smith, a fellow missileer from my squadron, was involved in a flying object incident. This definitely gave me pause for contemplation, and provided an understanding and insight to the fact that the Air Force was heavily involved in protecting and maintaining a magnificent secret. This provided additional prompting and incentive for me to learn all I could in order to confirm a massive Air Force cover-up.

The thing I didn't realize when initially beginning investigation of Air Force involvement with flying objects, is how terribly contorted and convoluted the situation would become with Air Force attempts to control and conceal their secret. But those Air Force actions would also prove to be tightly interwoven, leaving no doubt of extreme Air Force duplicity. Confirmation of my experience, which was provided by Salas and Smith, is all I have needed to press forward with my investigation and validation of Truth. And the appearance of Robert Low in this chapter, in association with Hynek, is just a "taste" of what eventually contributes to confirmation of Air Force duplicity for everyone else.

In recent years, other missileers who were with me at Minot AFB have started to speak out, and others have confided to me that they were involved with flying object incidents. I now realize that many others are still on the fence. Some hold back from going public for what they think is good reason. When you have been dependent on an Air Force pension, it is quite prudent to hesitate for more than a moment. When you have been told in no uncertain terms to keep quiet about an incident, or you sign a document agreeing to a compliance of silence, it is quite prudent to hesitate for yet another moment. For those who are speaking out, it's only because courage has overbalanced trepidation. The only real benefit to them is knowing they have made an attempt to bring Truth to the world. It is interesting, however, that in taking the risk, none have ever encountered the slightest reprisal or provocation from the government in relating their experience and revealing Truth. In the next chapter, we will hear from some of them directly.

CHAPTER SIX

More Confirmation

The Lieutenant Dave Schuur Incident

First Lieutenant Dave Schuur harbored his experience in silence for many years. It was not until he heard of others speaking out that he decided to make his experience known. I was with Schuur at Minot AFB and knew him, having brushed shoulders with him from time to time. He was in the 740th Missile Squadron, and like me, he was sworn to secrecy about his incident. We never knew at the time that we both shared something in common.

In February 2010, I received a surprise phone call from Schuur inviting me to a 2010 Minot missileer reunion. This was a surprise because I had not been in contact with anyone since those days at Minot, but I indicated to Schuur that I could not attend the reunion. I was not aware that reunions were taking place, and this was to be the twelfth reunion. He and his wife, Judy, were involved with organizing periodic reunions, and maintaining a list of former Minuteman missileers from Minot. He found my name in a squadron personnel listing from the 1960s, and he then searched for current contact information on me. In a subsequent exchange of emails, I replied to him and stated:

> *...one other thing, which I hesitate to bring up, do you (or any others) ever recall anything involving UFOs at the LCFs?"*

This was my first communication about the subject with anyone from my missileer days. When bringing it up, it was done with much hesitation, since the order of silence was still imbedded within me. Totally unknown to me, and much to my surprise, I just happened to pick the right person to ask my question, and Schuur replied with the following:

> *I was particularly interested in your topic of UFOs. I had a situation at Echo capsule which I provided to Robert Hastings. As to my "experience," I was essentially told that if I was interested in an Air Force career, it never happened. In a nut shell, we had all the missile indicator lights on the commander's console, including "Launch," come on sequentially for several missiles. During this time, top side security reported sighting a very bright light over several LFs. We activated the "Inhibit" switch several times. After a few seconds, the indicators on one missile returned to normal, then another missile would light up and return to normal. This occurred for several of our missiles. Unfortunately, I took my warning serious and never documented anything, so only have memory to go on. This was in the 1966-1967 time frame.*

I was astounded by his response. I was currently aware of the Salas incident, but I never previously attempted to find and reach out to anyone who was with me at Minot, primarily because I lost all contact with everyone I knew from those days. This unexpected communication from Schuur was another "Wow Moment" for me. Again, I received confirmation of my experience, but this time it was from a fellow crew member in a sister squadron at Minot. I wrote back to Schuur expressing my profound appreciation for his reply, and then I described my incident to him.

His mention of Robert Hastings is significant because Hastings is a well-known and respected UFO researcher, who specializes in investigation of flying object incidents, and connecting them to nuclear weapons. He is the one who wrote the Foreword to this book, and he has become a very trusted friend. Hastings soon contacted me, in June 2010, and I provided him with a narrative of my incident, which he recorded.

In September 2011, Dave Schuur and I met each other at the next Minot missileer reunion in Bountiful, Utah. In meeting again after forty-five years, it was consoling to find someone from the past I could relate to regarding an experience from so long ago. We were able to compare notes from our memories regarding our incidents, and he mentioned that he told his wife, Judy, of his incident immediately upon arriving home from his tour of duty at the launch center. She also confirmed this to me when I talked with her. For me, this was a very worthwhile reunion experience, and it would later pay off with additional missileer contacts.

The Lieutenant Larry Manross Incident

In Schuur's email to me, he went on to state that, "I have had a couple other former Minot crew members contact me confidentially that they also had similar experiences." He then included an email response he received from another crew member, whom he didn't name, but who was in my squadron. The email stated:

> *I was in the 742nd, but as you know many times we worked in other capsules. I found your story very interesting because the capsule I was in had a very similar experience. At the time, I was the junior officer so I had the early sleeping shift and slept through it, but when awakened by my commander he told me the incredible story of the missiles lighting up etc. He said base radar was tracking something and the security team upstairs had gone into a defensive position with guns drawn and that is when the command console lights all came on. To be honest, I was a bit skeptical when I woke up, but the security team was pretty jacked up the next morning when we went topside after being relieved. They described their defensive actions.*

I later learned that this particular incident, forwarded to me by Schuur, was experienced by First Lieutenant Larry Manross. I then managed to locate and communicate with Manross who recognized my name, and also the picture I sent him. His return email stated:

> *Like you, it is pretty hard to remember details, but the fact is, there was an incident. In fact, I can remember the chatter going on at the time, and there were apparently multiple incidents. I heard a lot of it probably because I found it so interesting and kept asking questions. In fact in our squadron, at certain launch control centers during the summer (if there had been "activity"), the security team actually claimed they would sit on top of the roof watching for UFOs. There were also stories (unsubstantiated) about a security team that was camping out at the missile site because of an electronic malfunction, which witnessed an object hovering over the blast door* [the overhead multi-ton door over the missile]. *"*

Unlike me, Manross obviously had his ear close to the ground at the time, and he was tuning into confidential talk taking place among personnel he associated with in the missile wing. He could not remember for sure who his commander was during the incident, but it was not his regular commander, or an officer he ever crewed with again. It may have been someone from another squadron. He said that he remembered being told by a senior officer that, unless authorized, he should never talk about his incident, but Manross pressed on anyway by quietly asking questions and learning some of what was happening. For him, the incident he was involved with was so profound and so startling that he needed to find out more of what was going on, which he did. In my particular case, it was my commander who passed word to me that I should keep silent, and I was very careful to make sure he continued to maintain trust in me, primarily for the sake of my Air Force career.

Manross was taking a scheduled rest break when his crew commander became overwhelmed with serious and unexpected indications on the launch control console. The console lit up like a "Christmas tree," which required necessary and frantic actions while communicating with security, base headquarters and others. Things happened so fast that there was no time to rouse Manross from his rest period, but shortly after, Manross was awakened and given a review of what happened. He had some skepticism when initially becoming aware of the situation, but he then sensed that his commander and topside security were engaged in some very serious and heavy action. It became very evident to Manross that his commander and personnel topside were in a state of distress at the time of the incident, when the flying object "buzzed" the LCF.

Manross could not remember the particular flight he was at during the incident. Like me, and with the passing of so much time, memory blended in with what was normally just routine duty. After so many years, many details simply blend with all the many "duty tours" we pulled underground. Exceptions are the particular and special events that standout in memory from what was non-routine. Alert tours were pretty much routine, but where there were significant events or happenings, the mind would replay them in vivid flashbacks. They were the memories that did not escape from the mind.

Larry Manross also mentioned to me that he received an interesting piece of information in later years from a family friend, who was an Air Force lieutenant colonel. The Colonel said that a close acquaintance of his, who was head of Project BLUE BOOK (probably Major Quintanilla), informed him that BLUE BOOK was literally camped out at Minot during the time we were experiencing the incidents. This piece of information fits in well with what is known of Hynek taking an investigative interest at Minot. It also explains how the OSI seemed to be on top of things, while keeping Hynek at bay and managing to zip lips. It is quite evident that Minot was a hotbed of activity involving unearthly flying objects, and that the head of Project BLUE BOOK was probably well aware of what was taking place. In consideration of this, it is disturbing to now realize that we missileers were relegated to the status of silent pawns in an Air Force cover-up!

The Lieutenant Paul Johnson Incident

Missile crews would often be flown to and from Launch Control Facilities via helicopter, especially those facilities furthest from base. But the weather was often a controlling factor that sometimes prevented choppers from being used. They were noisy, vibrating contraptions, and our skillful expert pilots would sometimes maneuver them as if they were still flying in Vietnam. Otherwise, our transport was the standard blue station wagon. My preference was to use the station wagon, probably because of the high speed helicopter crash landing I once experienced on the airbase runway.

After completing an alert tour and returning to base one day, I encountered my friend, Lieutenant Paul Johnson, who just returned from Kilo Flight. He proceeded to tell me of his thrilling experience from the day before when his chopper lost power on the way to the LCF, and it became necessary to auto-rotate into a forced landing. Before contacting ground, however, the engine came alive and the unplanned landing was avoided. Johnson was quite shaken by his experience, and he wanted to tell me more, but he indicated he was not allowed to discuss it further. I surmised at the time he was simply referring to a technical situation, or a pilot issue.

In our later years, Johnson and I lived within thirty miles of each other, but we maintained little contact during the last fifteen years, except for exchanging Christmas cards. After deciding to speak out on the UFO question, I made it a top priority to visit with him and his wife, Cindy. My wife and I then dropped in on them when we happened to be in their area. We enjoyed reminiscing about the "good old days" at Minot, but then I brought up the subject of my experience at November Flight. When I finished, Johnson then reminded me of his scary auto-rotation experience in the chopper. He said there was more to it that he never told me about, because he was ordered to keep silent. He said that when the chopper's engine shut down, there were a couple of flying objects "escorting" the chopper. One large object was off to the side, and one was directly in front when the chopper's engine lost power. The object in front immediately landed on the ground ahead of them, while the chopper auto-rotated downward. Before meeting with the ground, however, the pilot was able to regain power, and they were able to continue on their way.

Presumably, the two flying objects managed to shut down the engine of the chopper, and then allowed it to start up again. This is reminiscent of many stories I have since heard where vehicles of security strike teams became mysteriously inoperative when in close proximity to an LF, where an object was seen hovering over the missile site, or positioned nearby. And this is reminiscent of situations where radio communications became disabled by interference, or missiles were taken out by possible EMF effects by an object.

In talking further with Johnson, he told me of another incident he experienced that closely matches an incident recorded in archives of the National Investigations Committee on Aerial Phenomenon (NICAP). This incident occurred on 5 March 1967, just days before the incidents on 16 and 24 March associated with the missile field at Malmstrom AFB described by Robert Salas. Here again, radar at South Base tracked an object descending over a Minot Minuteman LF. Security strike teams were dispatched to the area, and they reported a disk-shaped object ringed with "bright flashing lights" moving slowly and hovering about 500 feet above the ground. At about the same time, a radar technician at South Base received a call from a missile maintenance crew reporting they were observing flying objects making odd maneuvers in the area. The objects quickly dropped down from very high altitude, to about five thousand feet, and then slowly descended to about 500 feet. The objects then rose back up and repeated. One then dropped down as if to land, and South Base obtained a radar fix on it. Minot AFB also obtained a radar fix and identified it as a solid return from a real object. The object then maneuvered out over the LCF for a moment, and then it suddenly climbed straight up at great speed, and disappeared. This happened as a F-106 aircraft was scrambled from the airbase to intercept.

When I visited with Paul Johnson and his wife, he told me of this incident, which particularly caught my attention with mention of the object having "bright flashing lights." This was the description given to me in my incident. In addition, Johnson mentioned that his FSC above ground requested permission to fire on the object, which was briefly hovering over the LCF. Johnson replied to the FSC that, "It probably would not be a good idea!" This statement was briefly mentioned in Chapter 2, and was included in Johnson's obituary.

The Airman Wilbur Gunther Incident

Upon further research and investigation on the Internet, I came upon Airman Wilbur (Jim) Gunther, a former security strike team member and FSC, whose normal duty was at Minot's November Flight. I found Gunther on an Internet military blog, and he mentioned a flying object incident he was aware of while stationed at Minot AFB, but he didn't go into detail. He also posted an old photo taken of the main entrance to Minot AFB, which amazingly pictured me and my car receiving a salute from a guard (pictured in Chapter 4). I could hardly believe it when I saw the picture, because I remember when it was taken. The guard asked if I would back my car up and remain still for a moment while his buddy took the picture. When I saw the picture, I knew I must make contact with Gunther. After more investigation, I managed to discover his current phone number and email address, and I gave him a call. He indicated he was at Minot AFB during part of the time I was there, but he was not aware of my particular flying object experience. Usually, security airmen would work three days at an LCF, and then relocate to base for another three or more days before returning. Because of that, Gunther would not necessarily hear or know of my incident if he had not been there at the time.

In a subsequent email received from Gunther, he said he was "reluctant to divulge information on secret material," but he did relate an experience that obviously made a significant impression on him. He said that it, "Forever etched an imprint on my mind." He stated:

> *I can still remember sitting in the AP* [Air Police] *truck with my strike team partner (name was McKosy), looking at the radio, and hearing one of my buddies, having "lost his cool" screaming over the radio that there was an orange UFO hovering over the site in alarm. I don't recall which Oscar site this was. Later I heard that they had to come out and redo the Target and Alignment* [T&A] *the next day.*

This incident occurred before Gunther became an FSC at November Flight, and while he was part of a two-man strike team investigating security violations. He and his partner were listening to radio communications from an Oscar Flight strike team, who were not far away and observing a hovering object directly over an Oscar LF. Gunther's air police friend was clearly in distress while observing the object. Evidently, the object must have taken the hardened and secured underground missile "off alert," probably with a "guidance and control system" malfunction. The mention of "Target and Alignment" was a maintenance procedure necessary to bring a missile back to "alert status," which was accomplished the following day.

In another email, Gunther again referred to the same incident, and stated the following:

> *The only UFO incident that occurred when I was there (that I was aware of) was at Oscar Flight when an OZ/IZ* [Outer Zone/Inner Zone] *alarm went off at one of the sites, and the young strike team member using the radio that night "lost his cool" when he got there and I heard him screaming that there was a UFO directly over the site. Though I can't remember this kid's name, he was a friend of mine and I later spoke to him about the incident. He told me it was an orange glowing saucer shaped object that left shortly after he arrived. I am not sure but I believe this incident occurred in late 1965 or early 1966.*

Gunther's mention of an "orange glowing saucer-shaped object" has a familiar ring to it. As mentioned previously, this was a very common description that I have heard over and over again.

There are many recorded incidents where security or missile maintenance teams viewed a glowing object up close while it hovered over a LF. Some security personnel became so traumatized after their initial experience that they later invented problems with their vehicle to delay arriving at a LF, especially if they knew an object was reported in the area. In some cases, strike teams simply refused to approach a LF, and remained parked at a distance. Others became so terrified and frightened that they refused to enter the missile field again, and they applied for other duty.

Robert Salas related that one of the frightened airmen involved in his incident later phoned him at home with a plea to talk about the experience, but Salas refused because of the oath of silence he made. To this day, Salas has serious misgivings, and he regrets not talking with the airman who came face-to-face and stood within a matter of feet of the terrifying situation. Salas never heard from or saw the airman again.

The "Anonymous" Incident

One the most significant flying object incidents occurring at Minot AFB when I was there involves a launch officer who prefers to remain anonymous. I have total respect for his desire to keep his identity secret, as well as details of his incident. It is, however, truly frustrating not to have a detailed contribution from him about what occurred, which would add to what was taking place at Minot when we were there. He did, however, make himself known to me, which provides a strong indication of his desire to have Truth revealed, and to become part of our growing cadre of Minot missileers who are stepping out from the shadows. During our short conversation, he stated the following:

> It was a very, very serious incident. When I returned to base after my incident, I was directed to sign a document stating I would remain silent!

At the beginning of this chapter, in the review of Dave Schuur's incident, I mentioned that Schuur was arranging periodic Minot missileer reunions. After attending the reunion in Bountiful, Utah, in 2011, and one in Houston, Texas, in 2013, I volunteered to host a reunion in Seattle, Washington, which took place in September 2014. During that reunion, I reserved an evening to give a special presentation to thirty missileers, along with their wives and friends. With complete backing from my reunion committee of four other Minot missileers, I described and reviewed the unearthly experience I had at November Flight in 1966.

My presentation started by asking if anyone remembered anything about flying objects being discussed at Minot AFB. Seven people raised their hands. I then asked if any missileers were involved in a flying object incident while "pulling alert duty" at a LCF. Three people raised their hands. Two of them later stated to me that their security strike teams reported the sighting of a flying object hovering over a LF after they were directed to investigate a security violation. This must have been similar to Jim Gunther's incident. The third individual ("Anonymous"), indicated he was involved in an incident that resulted in missiles going off-alert, but he did not provide further details.

When a Boeing Company videographer showed up the next day at one of our functions, I selected three of our missileers for a video interview. They were people I knew who might provide a good perspective on what it was like in the very early days of Minot Minuteman missile duty–Boeing was the major contractor for Minot Minuteman missile sites in the early 1960s. "Anonymous" was one of the missileers I chose for the interview. But just prior to his interview, he asked me if he should reveal his flying object experience. I suggested it might not be appropriate to relate it at this particular venue, since it would not fit in with the kind of interview Boeing was specifically looking for.

Because I was the host for the reunion and spending a great amount of time conducting and managing activities, I was never able to question "Anonymous" further before he departed the reunion. I am sure that my presentation took him by surprise, and that he was seriously considering his options in speaking out. I then sent him an email, and I also mailed a letter to him asking if he could provide me with details of his incident, but I received no response. He evidently decided the risk was too great for him to talk further about his incident.

I again met up with "Anonymous" at the next reunion in Charlottsville, Virginia, in September 2016. When finding an appropriate moment to talk privately with him, "Anonymous" made it clear to me that he could not tell me anything further about his incident. It was an awkward moment, but he did say that when he returned to base from his alert tour after the incident, he was ordered to sign a document committing him to forever remain silent. And he also emphasized that his incident was extremely serious, and that he was terribly disturbed by the total lack of follow through by the Air Force. He received no advice from the Air Force on what should be done if such an incident should ever happen again. This was unsettling to him, just as it was for me in my incident. He latter stated that he worked at the Pentagon for many years after his duty at Minot, and this must have added to his further trepidation in speaking out about the secret he harbored.

After initially learning about my experience from the presentation I gave, my missileer friend was definitely thinking about revealing Truth of his incident, but he then had second thoughts after giving it additional consideration. Perhaps I should have advised him, in the interview he gave to Boeing, to reveal details of his incident. I do, however, have great empathy for him, and I understand the very uncomfortable conflict he endures.

It's been very troubling not to recall the officer crew members my commander and I relieved at November Flight. The inability to recall, however, was possibly due to the fact we relieved a crew that was not from our squadron (the 742nd). With some exceptions, most of us were not personally familiar with crews outside our own squadron, and "Anonymous" was from the 740th, but he also stated he "thinks" his incident occurred at a 740th LCF. It is quite likely, however, that he was part of the crew that my commander and I relieved at November Flight, and I am very much inclined to believe he was, but his help is needed to verify it. When a "crew scheduling document" from 1966 is finally located, however, it should resolve the matter.

෨෧

There is much more that can be said in regard to incidents occurring at Minot AFB, and about the extensive Air Force cover-up that took place while I was there. In addition, there are many other significant incidents that have occurred at Minot since I left. The preceding, however, is all the confirmation I have needed in order to realize that my incident was not unique, and that there were many additional missileers, and others on base, who were wrapped up in a colossal secret. Many have taken their secret to the grave, while some continue to hold on to their secret to this very day. With the passing of more than fifty years, however, there can be no ethical, moral, or critical reason for keeping the secret from humankind any longer.

I have definitely received confirmation of my experience, and I definitely know that many of us missileers at Minot experienced something from out of this world. With all the secrecy that was involved, there can be no doubt the Air Force was keeping the UFO question under wraps. It was something we were all instructed to forget as if "It Never Happened."

Because others with me at Minot took the liberty to speak out, by relating their experience, Truth of the UFO question was exposed, and definitely confirmed for me!

CHAPTER SEVEN

Truth vs Deception

Alarm Bells?

Truth was confirmed to a number of Air Force missileers who experienced the "unknowns." They were instructed, or ordered, by the Air Force to keep silent about flying objects that were actively tampering with, compromising, and disabling "protected" nuclear-tipped ICBM missiles.

If the public were to find out, "alarm bells" would be expected to reverberate with questions on secrecy, moral responsibility, and national security! In a democratic nation, it would be expected that a concerned public would demand an explanation from elected public officials. It's a question of "truth versus deception." Meanwhile, flying objects continue to probe, harass, and evade the military with impunity, and a select few within government are totally engaged and vitally concerned with maintaining the secret.

From my vantage point, I would have thought the Air Force or some government agency would become seriously alarmed by the unusual "situation" that I and others experienced. Our highly designed fail-safe ICBM missile facilities were the ultimate in American engineering and technology. Buried deep underground, the LCCs and associated LF silos were designed to absorb nearby nuclear blasts and still deliver retaliatory strikes with powerful ICBMs that could hit their targets on another continent with very great accuracy.

The massively hardened LF silos containing Minuteman missiles were designed to be impenetrable to any intruder, and it required a long drawn-out and complex process for entry and access. LFs were built for survivability, and designed so missiles could still be launched if a parent Launch Control Center was out of commission. But the human factor was also considered. The Minuteman system design did not allow for a single individual to perform a launch without cooperation from a fellow crewmate, and another Launch Control Center.

In addition, special attention was given to launch officers entrusted with maintaining and controlling the weapon system. Officers were carefully screened, scrutinized, and provided with months of intensive training before given access to highly sensitive and critical duties. A Personal Reliability Program (PRP) put officers under constant watch for signs of instability. The Air Force depended on these stringent and necessary safeguards in order to maintain high reliability in both systems and personnel.

With all the high technology and high quality built into the Minuteman system, one wonders about those alarm bells at Air Force Headquarters when missiles at Minot's November Flight became disabled due to a flying object tampering with and disabling them. Alarms were certainly going off on the launch control console at November Flight, as observed by frantic officer crew members. There is no doubt about that.

The high technology and quality management built into Minuteman systems stands in sharp contrast to how Air Force leadership reacted to the exceptionally grave circumstance. Having been one of the launch control officers involved in the situation, where nuclear-tipped ICBMs were easily taken down and put out of commission, it seemed clear at the time that there would be extreme concern on the part of the Air Force to immediately resolve the situation, or inform launch officers of progress toward that end. One might even envision a great deal of scrambling at very high levels, and talk verging on panic among high government leadership and Air Force officials. In fact, I assumed this was going on. It was unthinkable that the Air Force would not seriously put forth a concerted effort, with upmost priority, to resolve the situation and eliminate a potential repeat of the problem. But it did happen again, and again, and again in a continuation of incidents. No doubt it's still happening today!

Incidents that I and others experienced were seemingly ignored by the Air Force, and we were simply left with the instruction "It Never Happened." The Air Force continued to play this game with us, by providing no information or training on how to handle similar situations should they occur again. There was no further talk, and we were left to determine on our own how we might handle a potential future situation.

Secrecy and Silence

It was a mystery to us, but we knew that some exotic object, controlled by some exotic intelligence, had the ability to knock out our secured and hardened ICBMs. We also knew the Air Force knew this, otherwise we and the entire situation would have been put under a microscope. We knew what happened, and the Air Force obviously knew what happened, probably from a previous history of observation and inability to control such incidents.

Because a cloak of secrecy was placed around it all, with everyone silenced, there can be no doubt the Air Force must have known about other flying object incursions. Certain people in high places were definitely in-the-know and intensely concerned, and this resulted in a serious effort to conceal and cover-up the situation. They, however, did not have all the answers, especially with preventing and controlling unwanted intrusions disabling our nuclear-tipped missiles, and they did not want undue attention brought forth. All talk needed to be immediately subdued.

One can easily imagine that my incident at November Flight would go insanely "viral" up through Air Force ranks and filter into the public domain. If that happened, those in the media would have a field day and go berserk. In knowing that the November Flight incident was not unique, and that other incidents were occurring at Minot, and at other locations, it is hard to figure how the Air Force could keep the damper closed. How could they prevent word of such incidents leaking out and spreading far and wide? It is especially hard to imagine because of the number of people involved, not just those directly impacted such as myself, but also the many other personnel who were charged with passing critical information up the chain of command, or bringing missiles back to alert status, or conducting investigations. I cannot help but wonder what would have happened if the local Minot TV station had learned that the strange lights, reported and observed by residents in the town of Mohall, had actually interfered with the nearby Minuteman Launch Control Center, and put its connected missiles out of commission.

It was easy for the Air Force to maintain secrecy about incidents. It simply issued immediate and strict orders to those involved to remain silent. Everyone connected with our missile wing had a security clearance, and everyone knew it was imperative to keep one's mouth shut, not just with classified material, but also with verbal orders to keep silent. Failure to do so would spell the end of one's career, or worse. In some cases, we were simply told by a superior officer to zip lips. In other cases, it was necessary to sign a document swearing an oath of silence. In a majority of cases, the order for silence came down from the local Air Force Office of Special Investigations. It was usually delivered by someone dressed in civilian clothes with a badge under his coat, or in his wallet, or by a phone call.

So it was with me. I was told upon arriving topside from our underground LCC that I was to forget everything. I was to never say another word about the incident. Being instructed that "It Never Happened" was dismaying, but I had my orders, and I was proud of my integrity in being honorable and trustworthy. I never questioned the order, except to ask where it came from (the OSI). It would have been imprudent for me to challenge it further, since I valued my career and status. I never questioned what kind of security classification the incident had, and I was never told what the security classification was. I assumed of course that it was super sensitive, but I was never told whether it was secret, top secret, or exactly what it was all about, or why my commander and I were to pretend it never happened. That is why I continued to hold the secret for almost forty years, and long after I separated from the Air Force. That is why so many others did the same, and continue to do so. After a while, as the Air Force must have hoped, the incidents would become old lost memories. They did become old, but not lost.

Moral Responsibility

To be realistic, it's the aspect of not being "truthful" that has been bothersome in later years. There is always the conundrum of keeping true to a secret, but what if that secret involves a lie? You then ask yourself whether you have really been truthful, or have enabled perpetration of a lie toward society and the whole human community. How long can you assist in hiding a lie that would be of monumental interest to the scientific establishment, and to all of humankind? It then becomes a question of balancing moral responsibility toward a greater cause.

I have always held my government in highest regard, and with utmost respect. I fly the flag, my patriotism is beyond question, and I continue to participate with the national Air Force Association, which has a goal of promoting and advocating for strong national defense and a strong United States Air Force. I thoroughly enjoyed Air Force life and the camaraderie that came with it. I worked with very respectable, conscientious, and competent people who possessed high moral values, and who worked hard to serve the Air Force in the best way they could. They all possessed college degrees, and were working to further their education. They were regular, normal Americans who were serious in their work ethic, and dedicated to home, family, friends, and their nation. Granted, this may not necessarily be a general characterization of many in the Air Force, but it certainly was in my experience, and the environment I was part of.

Because of the great life I had in the Air Force, I've never hesitated in telling young people that joining the military, especially the U.S. Air Force, can be a great career decision, or a short-term benefit either way. Contrary to the sometimes negative characterization of the armed forces being focused on waging wars, and being an aggressive militaristic machine bent on death and destruction, it was a very positive professional experience for me, which was all about defending freedom and maintaining peace in the world. That is what we did, that is what everyone wanted to do, and that is what we were prepared to do. When I and fellow missileers went "on-alert" in our Minuteman Launch Control Centers, with our finger on the nuclear trigger, we showed our teeth of deterrence, and we had no doubt that any would-be aggressor posing a threat to our existence would regret it. But our intent was to let world peace and freedom reign, and that is what we definitely preferred.

When I was told to never utter another word about my incident, I maintained trustworthiness in that regard. I didn't speak another word. I didn't speak to my commander again about it. I didn't talk to my wife about it. I talked to no one, because doing so was off limits. That is the way it had to be, even though flash-back memories of my incident would continue to haunt me for many years afterward. That's the way it was for many former Air Force officers who are now speaking to the Truth, and that is the way it remains for many who are still reluctant to do so.

The aspect of truthfulness, however, has come to my attention in the last few years when I learned that some Air Force officers, and other very reputable people, were speaking up about incidents they harbored quietly for many years. The incident experienced by Robert Salas became a tremendous grand awakening for me after some thirty-five years, and it made me realize that Truth was overdue. A long protected lie needed to be brought into the open.

Revealing Truth is risky business, especially for reputable, credible, reliable, and responsible people. They risk their cherished status in life with nothing to gain, and with knowledge that skeptics, debunkers, and detractors will come their way with disparaging remarks. Many more credible people, however, will continue to awaken and speak out, with knowledge that the magnificent secret they harbor and protect must be exposed. It's a moral obligation that takes hold, punctuated by knowing the difference between Truth and Deception.

So, what is the justification for continued deception by high Air Force officials? Well, they will not talk about it. Ironically, however, my "integrity" is now reinforced and justified in proclaiming Truth. My integrity is the primary reason the Air Force originally put faith and trust in me, but maintaining that integrity now requires that my silence have a voice. I have realized that I was "privileged" to know the Truth, while others remained ignorant and oblivious to something important to all of humankind. The disconcerting part of this is that the Air Force knows a whole lot more about the UFO question than I do. But I do have first-hand experience with the fact that the Air Force does not want the American public and the rest of the world to know! Where is the justification for that?

I'm now aware that there are some in the Air Force who desire to maintain their silence, including some who were with me in the three missile squadrons at Minot AFB. They have yet to stand up and be counted, and I know who several of them are, but I understand why they haven't come forth. For them, maintaining one's integrity, and protecting a hard-earned station in life, overbalances a moral obligation to speak to Truth. I respect this, but I also hope there might be some reconsideration by those people, or an effort to leave behind some sort of confirmation, declaration, or affidavit of their experience – just like my old friend Paul Johnson. If there are any airmen out there who would like my confidential counsel, I would be more than willing to listen, and perhaps offer advice.

National Security?

Although the Air Force has been quiet on the subject over the last few decades, the UFO question is considered a very serious matter at high levels, and a working plan has been in place for many years for internal reporting of sightings and encounters. Quick notification to higher headquarters is provided through a short chain of command, while much scrambling takes place to silence witnessing personnel. The Air Force simply instructs people to keep lips zipped, and it informs a curious public who report or inquire about flying objects that:

> *The Air Force has no interest in the matter since it has been previously and adequately proven that flying objects do not exist or pose a threat to national security.*

Interestingly, and strangely, the awkward ambiguity posed by the above statement might lead one to ponder it a bit further.

∽✍

Some of us missileers were involved in incidents and encounters with exotic flying objects that were not of this world, and controlled by some sort of unknown intelligence. We have vivid memories of our particular experience, and understand the significance of what happened. Many of the incidents with Unearthly Flying Objects were observed concurrently, and very close-up (within feet) by personnel on the ground, while at the same time also observed from the air (by F-106 aircraft), and recorded on radar. Documents available from Air Force investigative projects, such as BLUE BOOK, and documents released via FOIA, have verified this.

Very significantly, some of the flying object encounters and incidents tampered with and compromised our nuclear-tipped Minuteman ICBMs, which made them unlaunchable. And that really does bring up the subject of national security. This vulnerability of our nuclear ICBM forces should concern everyone!

Serious alarm bells still ring within Air Force Headquarters and other government agencies in regard to the UFO question. But do those alarm bells resonate within the American public, and the world? Certain agencies in the U.S.

government certainly hope not, otherwise the public would have been informed long ago about Truth of the UFO question. The fact is, deception continues to reign within the Air Force and other secretive government agencies.

PART TWO

Era of Public Awakening

❦

In the past, Truth was out there, but who knew? Some would say it began as an apparition within the fanciful and imaginative minds of science fiction writers. But no one really knew or had the slightest idea, although it could be said that flying object sightings and encounters have been occurring for eons.

Sightings of mysterious flying objects began to get serious when the public began to take notice. Reports circulated of people sighting strange, unfamiliar objects behaving in an unearthly manner. Some were explainable and others not. Many were explainable as being hoaxes and pranks perpetrated by people with devious motives, or by people wanting to see their names published. Many sightings were simply misidentifications of birds, airplanes, balloons, or natural phenomena, but a certain few of them seemed to be truly mysterious, with an apparent origin from out of this world.

An initial review and understanding of the UFO question now begins with a look at the modern era of public awakening, and a review of early classic flying object sightings and encounters, along with an introduction to the incident at Roswell, New Mexico. It is this that provides a backdrop to statements made by CIA's first Director of Intelligence, Roscoe Hillenkoetter, and his vain attempt to "suggest" a massive cover-up by the U.S. Air Force.

Most everyone has heard of the Roswell incident, but very few people are familiar with details surrounding the episode, all of which the Air Force strongly denies. It's generally looked upon as a fabrication originated by "believers" in extraterrestrials, and centered in the town of Roswell, New Mexico. This is where an alien flying object supposedly crashed many years ago. Although there are many books on the subject, most people are not interested in reading about such fantasy, or have others think they might be interested. It is only the tabloids that have kept Roswell alive, along with a few researchers, and the town itself. Today the town capitalizes on the "fantasy" during the first week of July every year. There are several days of celebration, with a parade, an alien costume contest, carnival, and vendors hawking memorabilia and tasty food. It has become a very popular and entertaining festival.

Most people do not realize, however, that the U.S. Air Force is responsible for covering up a truly historic and momentous occasion that took place there in the early days of July 1947. And there is good reason to celebrate Truth of what happened at Roswell, despite what the Air Force denies. When one closely looks at the details, it becomes evident that a great many people were involved within government (Washington, D.C.), the military (Army Air Force), and the town itself. The coordination of managing and pulling off the cover-up is truly amazing and revealing, and those details are about to be reviewed.

Of course, the perpetual skeptics and debunkers tend to have a heyday with their myopic look at the "total picture" of Roswell, but they cannot take away from the countless credible witnesses who have spoken out on the magnificent event that took place.

In the end, the Army Air Force knew they pulled-off a great "escape from Truth," which has lasted for about seventy years. But Truth will be restored with new confirming evidence, with public education about the cover-up, and with official "disclosure."

Roswell will one day be considered a great historical happening in the affairs of humankind, and the Air Force and U.S. government will one day owe a huge apology to all those who were involved and forced to keep their lips zipped.

CHAPTER EIGHT

Public Awareness

1946 - 1947

Foo Fighters

Throughout recorded history, many stories of flying object sightings are mentioned. Researchers point to ancient writings and other evidence that leads one to surmise that flying objects, and those piloting them, have been around for a very long time. In less ancient times, during World War II, particular notice was made of "Foo Fighter" balls of light phenomena, which were observed by air combatants on both sides of the war. The objects became a concern because of their apparent intelligent maneuverability, and close monitoring of combatant aircraft, but concern about them dissipated with end of the war.

Swedish Ghost Rockets

In August 1946, the famous Ghost Rockets of Sweden caught public attention, and they also created concern among military personnel. Preferring to be a neutral country during the war, Sweden wanted to remain that way when its Soviet neighbor became more antagonistic or unfriendly toward Western nations, and the Swedes were careful in masking contacts with the United States and other nations. But they were reporting mysterious objects darting through the sky, which were similar to meteors, but different.

First noticed in May 1946, sightings of the objects quickly increased in number, and speculation took hold that the hundreds of observed Swedish sightings were Soviet-made devices, possibly A-9 rockets with wings. It was also speculated that they were launched from the Soviet captured area of Peenemunde in Germany and other places around the Baltic, but no "confirming" evidence was ever produced to substantiate this.

The Swedish military carefully considered that the objects were guided rockets, which sometimes maneuvered in wide turns with a burning light, and with a flat trajectory that left a trail. But little trace of them was found; most appeared to completely disintegrate before hitting the ground. A few were observed to fall into lakes, with no evidence found. A few rare impact sites were also discovered that revealed small pieces of slag, but with little or no metallic material.

It is quite possible that Sweden was reluctant to suggest that the Soviets were playing a game of intimidation and harassment with the Ghost Rockets, and they held back critical information on their findings. The other side to this is that many people in Sweden and the United States were convinced that the objects were not of this earth, primarily because of their curious nature and a lack of specific tangible evidence that would identify them as man made.

Era of Awareness

In 1947, an unprecedented wave of sightings began. In early June, sightings increased significantly in the United States, with most occurring in the Midwest and western states. They reached a peak in the latter half of the month in the Pacific Northwest, and this might be why Army Air Forces Chief of Staff General Carl Spaatz was in the region at the time. Reportedly, he was on vacation. In early July, sightings increased exponentially until 8 July, and that is when sightings were reported from all over the country. After that, sightings declined substantially, although sighting reports continued to a certain degree.

Thus began the modern era of public awareness involving the UFO question. It was attracting public attention to a great extent. A majority of sightings were of the usual misidentifications of airplanes, birds, balloons, natural phenomena, or perpetrated hoaxes. A great many puzzling incidents, however, were reported by reliable, reputable, and credible people, which generated much government concern. This concern was compounded, dare I say it, when extraterrestrial evidence was recovered, which resulted in a government cover-up of massive proportion. It was then that an unplanned government within the government was organized to manage the huge secret. Those words, a "government within the government," are a bit shocking to contemplate, as previously mentioned in Chapter 2.

Yes, evidence was recovered! It happened just when extensive organizational changes were about to occur within government as a result of the National Security Act of 25 July 1947. Changes in intelligence operations, and other government agencies, were reflected in separation of the Army Air Forces (AAF) into separate components, which gave birth to the U.S. Air Force on 18 September 1947.

The Kenneth Arnold Incident

One of the first incidents catching serious public and government attention involved private pilot, Kenneth Arnold, who reported his incident on 24 June 1947. He was thirty-two years old at the time, and a very accomplished pilot. He took his first flying lesson at age sixteen when attending high school at Minot, North Dakota. He was also one of those having credentials of a reliable, reputable, and responsible individual. He established a very successful business, founded the Idaho Search and Rescue Pilots Association, and became a Deputy Sheriff in an aviation posse. He also became a Deputy U.S. Federal Marshal flying prisoners to detention centers. In 1962, at the age of forty-seven, and recognized as a respected citizen and public servant with impeccable credentials, he ran for Lieutenant Governor of Idaho.

At the time of his incident, he was returning home to Boise, Idaho, in his small search/rescue plane after a business trip to Chehalis, Washington. His return would include scheduled stopovers in Yakima, Washington, and Pendleton, Oregon. Upon leaving Chehalis, he decided to divert his flight plan slightly northward to look for evidence of a missing C-46 military transport carrying thirty-two officers and enlisted men, which went down seven months previous in the Cascade Mountains on the southwest side of Mount Rainier. After searching for about an hour, and positioned about twenty miles southwest of Rainier near Mineral, Washington, he decided to turn east and head directly for Yakima.

It was then that Arnold spotted a string of nine objects traveling southward in echelon formation at supersonic speed, and passing in front of and behind various peaks. He calculated their speed at about 1,700 miles per hour by timing their passage through those landmarks. To give this speed some perspective, Chuck Yeager was credited as the first man to break the sound barrier in his rocket-powered plane on 14 October 1947, and he attained a speed of 662 miles per hour in level flight at forty thousand feet.

Arnold opened the side window of his plane to get a better view, removed his sunglasses, and followed the progress of the objects, while observing their shape and behavior. When first seen, the objects were traveling southward and silhouetted between him and Mount Rainier in the northeast, and then in front of him between snow-covered peaks toward Mount Adams. Flashes of light reflected from the objects in the mid-afternoon sun, while they exhibited erratic motions like saucers skipping or bobbing across water at a consistent relative elevation.

When Arnold landed in Yakima, he could hardly wait to tell his good friend waiting at the airport. Then upon landing in Pendleton, he found that his story preceded him. When the press got hold of his report, the objects were referred to as "flying saucers," or "flying disks," in the manner described by Arnold, thus coining those terms. Despite the fact that several witnesses on the ground were able to confirm his sighting, and that two AAF intelligence officers confirmed Arnold's character and integrity, the AAF officially listed his sighting as a "mirage."

Although Arnold's sighting was given much publicity at the time, there were many other sightings documented from all over the world. Some were sufficiently disturbing and important enough to catch attention of military investigators. The press, however, paid little attention and were attracted instead to sightings involving misidentifications of aircraft, balloons, and natural aerial or astronomical phenomena. This encompassed more than 80 percent of all reports. A great many of those reports were featured on the front page of newspapers, with quirky articles. This reporting verged on idiocy and fantasy, and it was apparently intended to attract reader attention. It's quite likely, however, that the press didn't take most of the reports seriously, due to lack of substance and detail.

The Roswell Incident

On the heels of Arnold's sighting, the famous Roswell incident in New Mexico occurred in early July 1947. The *Roswell Daily Record* newspaper, on 8 July, carried bold headlines stating *"RAAF Captures Flying Saucer On Ranch in Roswell Region."* The acronym RAAF stood for Roswell Army Air Field, and this announcement was carried by several major newspapers across the nation (see following two pages). It originated from RAAF public information officer Lieutenant Walter Haut in a press release authorized by Colonel William Blanchard, who was commander of the 509th Bomb Group at Roswell Field. It was released after Major Jesse Marcel, a base intelligence officer for the 509th, spent the entire previous day retrieving scraps of debris from a reported crashed flying object at a sheep ranch about seventy-five miles northwest of Roswell.

Of particular note is the fact that the 509th Bomb Group at Roswell was the unit responsible for dropping the atomic bombs on Japan that ended WWII, and this unit was attached to the most ultra-secure and only base in the United States capable of delivering nuclear weapons to an enemy target. The base was also situated not far from other highly secret, sensitive, and secure locations involved with rockets and nuclear technology. They included White Sands Missile Range, Kirtland Field (now Kirtland AFB), Sandia National Laboratory, Los Alamos National Laboratory, Alamogordo Field (now Holloman AFB), Trinity Atomic Test Site, and others.

A few hours after the press release, Eighth AAF Commander General Roger Ramey at Fort Worth Field, Texas, stated in a follow-up press release that the recovered object was actually the remnants of a fallen weather balloon. Of course, there was no way Roswell personnel were so incredibly naïve, or dumb,

Front page Headline of Roswell Daily Record, July 8, 1947.

American League Beats Nationals, 2 to 1, in Tight All-Star Game
— DETAILS AND WIREPHOTO, PAGE 14

DISK LANDS ON RANCH IN N. M.; IS HELD BY ARMY

NIGHT FINAL EXTRA

The Seattle Daily Times

MAin 0300   SEATTLE, WASHINGTON, TUESDAY, JULY 8, 1947.   24 PAGES   PRICE FIVE CENTS

FAIR; 75° TO 80°
Seattle and vicinity: Fair and continued warm today and tomorrow. Highest temperature both days, 75 to 80 degrees. Low tonight, 50 to 55. Gentle northerly winds, increasing to moderate during afternoons. (Complete weather report, Page 4.)

Published Daily and Sunday and Entered as Second Class Matter at Seattle, Washington. Vol. LXX, No. 159.

Front page Headline of The Seattle Daily Times, July 8, 1947.

to misidentify and "capture" the remains of a weather balloon and radar reflector consisting of a few pieces of tinfoil and sticks. Weather balloons were frequently launched from Roswell, and it would be unthinkable for such capable, professional, and technically involved people to label such familiar wreckage as "saucer debris," and then provide the "saucer" statement to the press without being positively assured of their actions. The fact that higher headquarters, in the person of General Roger Ramey, took on the job of refuting the announcement, rather than RAAF retracting it, gives considerable cause to believe that a major cover-up was initiated. An RAAF retraction would have carried much more weight if there had been a simple statement from Roswell Field, but so much concern and fuss was generated within high levels of government and the AAF, that a high level denial was deemed imperative.

Years later, Major Marcel, who was involved in recovery of the material with ranch manager Mac Brazel (who made the original discovery), stated with much publicity that the military had covered up the recovery of an alien spacecraft. Marcel was the first significant player in this incident to stray from a strict constraint of secrecy. He, however, was far from the last person to do so. A great many other former AAF and civilian personnel followed his lead in providing Truth to the American people. Despite this, however, the general public today remains quite ignorant and uninformed regarding this event.

Several retired generals have spoken out, and other witnesses have provided deathbed statements regarding the cover-up, which gives credence to Major Marcel's account. And this mystery has continued to remain on the fringe of public awareness, even though all evidence was sequestered away and all information about it quickly squelched by the AAF. Although hundreds of government employees and civilians were involved, or aware of the incident, all were told in no uncertain terms (even with some threatened) to keep lips zipped. Included were those at Roswell Field, local people in town, and many within the general military. In the end, it is quite conceivable that only the president and a few top level officials remained in charge to manage the situation with a small contingent of people. In contemplating that a hasty deception was involved, it can be imagined that much scrambling was taking place at the Pentagon among high officials. It was necessary to keep the situation quiet, while taking control of the magnificent secret and the object that appeared from the heavens. Much detail on this is described in the following two chapters.

U.S. Government and World Change

At the time of Roswell, there was great concern about a series of political events taking place around the world. Mao Tse-tung's communist revolution was under way in China, which was a great worry. There was concern over the expanding control of Eastern Europe by the Soviet Union, including strengthening of their partition in Germany, which would eventually lead to a blockade of Berlin. There was a massive intelligence effort to determine the extent of Soviet progress in production of an atomic weapon, and much effort was taking place to determine how the United States might defend against an atomic or other surprise attack from Soviet forces.

There were also many changes taking place within U.S. government, with much debate on how national security should be managed, especially with the Soviets becoming a serious threat. On 8 May 1947, President Harry Truman appointed Admiral Roscoe Hillenkoetter to be the third successive Director of Central Intelligence (DCI) in heading the Central Intelligence Group (CIG). Then, with ratification of the National Security Act on 25 July 1947, Hillenkoetter assumed control of the new Central Intelligence Agency (CIA) on 18 September. This was essentially a renaming of CIG, and it happened on the same day the AAF was separated into individual services of the Air Force and Army. Stuart Symington became Secretary of the Air Force on that day, and James Forrestal was installed as the nation's first Secretary of Defense on the day prior.

With so much going on in the world, and in U.S. government, it was not a time to announce recovery of an Unearthly Flying Object, which could potentially result in public panic about a landing of aliens from outer space. It is easy to imagine that government officials would want to keep a very tight lid on Roswell, and they likely felt they had no choice in the matter. Assuming that Roswell and other recoveries were considered extraterrestrial, there was no way the military would want to announce their inability to control the flying objects, and it also meant that any recovered alien technology would require great security and protection.

Forgotten Hero - Roscoe Hillenkoetter

CIG Director Roscoe Hillenkoetter, the primary intelligence advisor and close confident to President Truman during Roswell, and then the nation's first CIA director, stated the following many years later:

> *The Air Force has constantly misled the American public about UFOs
> …I urge Congressional action to reduce the danger from secrecy.*

Unfortunately, this very significant statement from Hillenkoetter was ignored, with no acknowledgement or consideration by anyone. In regard to his mention of "the danger from secrecy," he knew that the secret of Roswell and the UFO question would eventually come out in the open, and he knew it must be revealed quickly to avoid public mistrust of government. This particular event was much too colossal for the government to sequester from the public and world. He also recognized that the manner in which the secret was protected might develop into a situation where control by responsible government could be lost. This is also what President Dwight Eisenhower alluded to when he spoke of danger involving secrecy within the "military industrial complex."

At another time Hillenkoetter stated:

> *Unknown objects are operating under intelligent control. It is imperative that we learn where UFOs come from and what their purpose is.*

A *New York Times* article dated 28 February 1960 mentioned that Hillenkoetter sent a letter to Congress stating the following:

> *It is time for the truth to be brought out in open Congressional hearings... behind the scenes, high-ranking Air Force officers are soberly concerned about the UFO's...But, through official secrecy and ridicule, many citizens are led to believe the unknown flying objects are nonsense...to hide the facts, the Air Force has silenced its personnel.*

That 1960 article, stating that "the Air Force has silenced its personnel," was true at Roswell, true in my Air Force experience in 1966, and still true today. Hillenkoetter made the statement while he was a member of the board of directors (1957 to 1962) of the National Investigations Committee on Aerial Phenomena (NICAP), which was a civilian nonprofit research group that contributed much toward flying object investigation over the years. Before joining NICAP, Hillenkoetter completed his service with the CIA in October 1950, and then returned to the Navy until his retirement in 1957.

Because Hillenkoetter was connected previously with the CIA, one might contemplate that he deftly infiltrated NICAP, as did other CIA operatives later on. But Hillenkoetter was very serious about Truth being disclosed on the UFO question. He tried to garner interest of Congress and the American public without incriminating himself, which was the situation until he became quiet on the entire issue when he suddenly separated from NICAP. Speculation then rose about the possibility of him being threatened and silenced by the CIA. At that particular time, he was participating in a plan to bring the UFO question before Congress. No doubt he was mindful of what happened to his friend and first Secretary of Defense, James Forrestal, which will be discussed in Chapter 11.

Hillenkoetter was not the only high-ranking official to speak out on the UFO question. There have been many others who have done so, but the press and media have gingerly treaded around similar credible proclamations for decades, and continue to ignore them without serious consideration. The unique thing about Hillenkoetter is that he was one of the higher officials involved with Roswell, and he ended up campaigning for Truth. Could it be that it was not until he was about to announce choice details on Roswell that he was threatened and forever silenced? If so, it's time to learn more about Roswell.

ক্ষ

Throughout history, even ancient history, there is ample evidence and documentation that entices speculation regarding flying object sightings in the past. In the last century, especially around 1947, public awareness increased substantially, particularly with the sighting by Kenneth Arnold and the news it generated. But the Roswell event took people by surprise, and also the government. Bold headlines suddenly appeared, with information originating from the Army Air Force, and those headlines stated that a "flying saucer" was "captured" and in government "possession." This was not frivolous or quirky front-page news. It was information that was initially allotted careful attention and consideration by the Army Air Force prior to its release. But then it was hastily retracted by higher headquarters. Was there an unfortunate or dumb mistake involved with initial release of the information? Or was this the beginning of a colossal and hasty cover-up by the United States government?

CHAPTER NINE

Discovering Roswell

July 1947

Saucer Captured!

It would be easy to consider a report of an extraterrestrial object crashing into a remote area of New Mexico in July 1947 as some kind of science fiction, or a hoax pulled off by some prankster for the thrill of it. However, it was strange indeed that the Army Air Force would announce that one was actually "recovered," or "captured!"

In reviewing this incident at Roswell, it must be remembered that it was originally announced in sensational headlines in major newspapers, at least in the central and western parts of the country. During this particular time, it was quite common for newspapers to include fanciful accounts of flying object sightings on front pages, but this incident created bold headlines in major newspapers (see Chapter 8), and it caused quite a stir. The naïve media and public, however, quickly accepted an unprecedented retraction from the AAF only a few hours later. It was a repudiation that stands out as a hastily arranged deception. The AAF not only retracted their first announcement, but they denied it with a statement produced from higher headquarters claiming that only a small tattered weather balloon was recovered. Evidently, those at higher headquarters were more adept at recognizing a common weather balloon and its remnants than highly-trained and technically capable people at Roswell Army Air Field (RAAF).

As with other fanciful reports and sightings, the incident at Roswell quickly became an inconsequential event, and it was quickly forgotten by the general public. But it was not forgot by those at Roswell who became involved with the incident, which took place over a period of many days.

The sheer number of individuals involved provides credibility to the incident. It would be impossible for so many people from various walks of life to fabricate and orchestrate the myriad individual pieces of a scenario that fit so neatly together. The magnitude of it all is awesome. With retrieval of documented activities of higher officials, and confirmation, statements, and testimony of so many other individuals previously silenced (estimated at more than five hundred), the Truth has been revealed. The general public, however, really doesn't know much about Roswell because mainline media has avoided the subject. Only tabloids have managed to keep the name of Roswell alive.

The remarkable thing about the affair is the great and tremendous impact it had on so many people, and how totally revealing and significant it proved to be in later years. When taken altogether, the many stirring accounts from many witnesses form an incredible and vivid picture of an unearthly crash and recovery.

Information covered here, and in the next chapter, provides only highlights from the vast amount of information that has come to light, although one should be aware that errors may be buried in some details. When taken and viewed in total, however, the great mountain of available material provides a very revealing picture of the incident at Roswell, as well as an overwhelmingly clear picture of government and AAF duplicity in a grand cover-up.

Names of people mentioned are very important to note. Many are key players that provide a further understanding as to what transpired. Connections between people, and their interactions together, reveal not only internal workings of their roles at Roswell, but also a sense that many "in-the-know" continued long after in a supporting role. Only a very few government officials, however, may have remained as a "controlling group" to maintain the cover-up. A vast majority continued with regular jobs, and taking orders while sworn to silence, or being threatened with consequences if they didn't keep silent.

Crash Notification

At the time of the Roswell crash, Lieutenant General Hoyt Vandenberg was Deputy Chief of Staff of AAF, and the one primarily in charge. There is also speculation that his boss, AAF Chief of Staff General Carl Spaatz, was involved with Roswell to some extent, and definitely in-the-know. Spaatz was looking forward to retirement within a year, and was on the West Coast, supposedly on vacation, where other major flying object sightings were reported at the time.

It was probably sometime on Sunday, 6 July, at the end of a Fourth of July weekend, that Vandenberg was notified of the Roswell incident. It is not known how he was notified, but Major General Clements McMullen, who was the designated Chief of the Strategic Air Command (SAC), at Andrews Field (now Joint Base Andrews) near Washington, D.C., was notified of the incident on that day. This was two days prior to the "recovered disk" press announcement. And one might surmise that McMullen was the one who notified Vandenberg.

Previously, McMullen was commander of Eighth Air Force, which was headquartered at Fort Worth Field (renamed Carswell AFB, but now named Fort Worth Joint Reserve Base) in Texas. He was promoted in January and given total responsibility for SAC by his new boss, General George Kenney. As Chief of SAC, part of McMullen's responsibility included Eighth Air Force and its three bomber groups, which were now managed by McMullen's replacement, General Roger Ramey. One of the bomber groups was the 509th at Roswell Army Air Field (renamed Walker AFB in January 1948, but closed in June 1967). As previously mentioned, this was the only U.S. strategic unit with atomic weapons capability, and it was the unit that dropped the atomic bombs on Japan in WWII under command of General Nathan Twining's Twentieth Air Force.

On 6 July, when McMullen was notified about Roswell, General Ramey and the Commander of Fort Worth Field, Colonel Hewitt Wheless, were away all day at an air show in Denton, Texas. Substituting for Ramey at Fort Worth was Colonel Thomas Dubose, Ramey's deputy chief. With Ramey temporally out of contact, it's quite likely that McMullen received word directly from Roswell about the crashed disk, especially due to the extreme nature of the incident.

After his notification, McMullen communicated and discussed with Colonel Dubose the recovery of crash debris at a particular remote area northwest of Roswell. Dubose was directed to help coordinate delivery of a sealed satchel of debris from the crash site to Fort Worth Field, where it would then be redirected to McMullen at Andrews Field. Dubose was instructed that this subject was more than top secret, and that absolutely no one else should be informed who didn't have a need to know.

The satchel was delivered via "colonel carrier," which was how most super-secret material was handled. General McMullen's intention was to later transfer the material to Wright Field (now Wright-Patterson AFB) in Ohio, and to Major General Benjamin Chidlaw, who was Deputy Commander for Operations at Air Material Command (AMC), which was under command of Lieutenant General Nathan Twining. Chidlaw was designated to eventually receive the material because Twining had suddenly departed for the New Mexico area.

Vandenberg was in Wichita Falls, Texas, on Friday, 4 July, attending an Independence Day event at Sheppard Field (now Sheppard AFB), and he possibly left there on 6 July. On 7 July, several Texas newspapers stated the following after Vandenberg stopped briefly at Dallas, Texas:

Lieutenant General Hoyt S. Vandenberg, deputy commander of the US Army Air Forces, stopping off briefly at Hensley Field in Dallas, said the AAF was receiving thousands of queries about the flying disks.

It is remarkable that Vandenberg made this particular statement to the press about "flying disks," and it's easy to speculate he was rushing back to handle the situation he was now aware of. It's also known, however, that many sightings were reported across the entire United States at this time. Many sightings were centered in the Texas and New Mexico areas, as well as the Pacific Northwest, and there is little doubt that the public was hyped-up about flying object sightings. Many people across the country had been alerted previously to Kenneth Arnold's sighting near Mt. Rainier in Washington State, and they were turning their attention skyward with many fanciful reports.

Vandenberg may have arrived in Washington, D.C., from Texas late in the evening of 6 July, or in early morning hours of 7 July. At 2:30 P.M., on Monday, 7 July, Vandenberg cancelled a dentist appointment and personally went to the airport to pick up Stuart Symington, the AAF assistant secretary of war for Air. The two of them returned to Symington's office at the Pentagon. Vandenberg then met briefly with Major General Emmett O'Donnell, who was Director of Information for AAF, and Vandenberg followed that with a brief meeting with Major General Curtis LeMay, who was Chief of Staff for AAF Research and Development. Following these meetings, Vandenberg met again with Symington. These unscheduled meetings, squeezed between his scheduled ones, suggest that Vandenberg was involved with something fairly important. In that regard, these meetings were in concert with his other activities and meetings that would take place over the next few days.

Joint Research and Development Board

The following morning, on Tuesday, 8 July, Vandenberg cancelled a previously scheduled meeting to attend a hastily arranged meeting at 8:30 A.M. with the Joint Research and Development Board (JRDB). This significant meeting is pivotal to an understanding of what transpired at Roswell, and with the UFO question in general. It is therefore important to take a quick look at this organization.

The JRDB was under auspices of the Joint Chiefs of Staff, with Dr. Vannevar Bush as its chairman. It was created on 6 June 1946 from a combination of three organizations: the National Defense Research Committee (NDRC), the wartime disbanded Office of Scientific Research and Development (OSRD), and the Joint Committee on New Weapons and Equipment (JWN). Each of these was organized and directed by Vannevar Bush, and each played a significant role in weapons research for national defense. Bush was also chairman of the Military Policy Committee, which oversaw the ultra-secret Manhattan Project in development of the atomic bomb.

JRDB board members were General Jacob Devers of the Army Ground Forces, General Carl Spaatz of the Army Air Forces, Assistant Secretary of the Navy W. John Kenney, and Admiral DeWitt Ramsey. This board presided over a very large organization filled with the brightest and most technically educated and talented people in the country. It was served by a six-person Executive Council headed by Dr. Lloyd Berkner, who controlled a Management Division, a Planning Division headed by David Langmuir, and also a Programs Division headed by Ralph Clark.

The JRDB primarily operated through committees, from ten to sixteen, which included panels concerned with specific technical matters. Committees were staffed by several hundred scientists and engineers, including Dr. Karl Compton, who was a physicist and president of MIT; James B. Conant, who was a chemist and president of Harvard; J. Robert Oppenheimer, who directed development of the atomic bomb; Isidor Rabi, who was known for his work in nuclear magnetic resonance (receiving a Nobel Prize for that); and Julius A. Stratton of MIT.

As leader of the JRDB, Dr. Bush was also involved as president of the Carnegie Institution, and was looked upon as a significant scientific administrator. There was no other man so involved with the nation's highest secrets, except for a select few who worked directly with him. Two of those mentioned above, and deserving special attention, are Dr. Lloyd Berkner and Ralph Clark. They would play extensive and continuing roles associated with the UFO question, and with the ongoing cover-up. Most other people mentioned, however, will also be recognized as playing certain roles beyond Roswell.

In attendance at the JRDB meeting on 8 July, along with a few board members, was General Curtis LeMay and his deputy, Major General Laurence Craigie. It is important to note that Craigie attended this meeting because he had just returned from a very quick trip to Roswell, after being sent there by General LeMay the previous day, probably after LeMay met with General Vandenberg. Craigie spent several hours at a second remote Roswell crash site while learning firsthand all he could. What he learned and personally observed there must have been exceedingly startling, and unbelievably significant to behold. His visit was to a different site than that previously mentioned, but it was believed that this site was related to debris from the first site where material was collected and sent to General McMullen. Attendance by Craigie at this hastily scheduled JRDB meeting is very significant, and the meeting could one day be recognized as a very historic one in regard to the UFO question, but only when Truth is ultimately disclosed and revealed by official authority.

Crash Site Discovered

According to reports of various researchers, it appears that a low-flying object may have rebounded from the ground, or somehow experienced a low-level explosion in a severe electrical thunderstorm during the night of Wednesday, 2 July. This was about seventy-five miles northwest of Roswell as the crow flies, and about thirty miles southeast of the very small settlement of Corona. There, on the morning of Thursday, 3 July, ranch manager Mac Brazel of the Foster Ranch was on horseback searching ranch property for sheep scattered by the storm of the previous night. His curiosity and astonishment was piqued when he came upon an extensive and elongated debris field displaying material of an unusual nature. It was very different from a downed weather balloon he had previously seen, and it was quite disturbing to come upon this scattered mess. He picked up samples to carry with him, and he stored larger pieces in a cattle shed.

In the course of the next few days, Brazel showed the material to friends and distant neighbors, who noted it looked quite strange indeed, and they suggested he contact authorities about his discovery. Many were now aware of the mysterious flying objects being reported, and his friends thought the material might possibly have something to do with the objects. Brazel's particular interest, however, was finding who was responsible for the extensive debris scattered in his field, and having it removed.

On Sunday, 6 July, Brazel drove to the town of Roswell, which was several hours away over very rough country roads. He stopped to show friends in town the debris he carried with him, and then he went to the office of Chaves County Sheriff George Wilcox with debris in hand. When Wilcox examined the material, he thought it worthwhile to contact Roswell Field to see if they could identify it. Major Jesse Marcel, the base's intelligence officer, then showed up at the Sheriff's office and examined the material. He was sufficiently impressed by the strange debris to notify his commanding officer, Colonel William Blanchard.

I am guessing here that some of Blanchard's conversation with Marcel may have been in regard to the exact location of the debris site, what the debris field looked like, and whether main wreckage of a crashed aircraft was seen or located. It may have been known or assumed, however, that no aircraft was in the area, or reported missing or damaged. He then instructed Marcel to go back to the ranch with Brazel and take someone from the counter-intelligence unit with him to survey the debris area, and to collect as much material as possible for identification. This would be standard procedure.

The person who went with Marcel was Captain Sheridan Cavitt of the Counter-Intelligence Corps (CIC); he was dressed in civilian clothes, which was customary for his unit. He was not part of the 509th Bomb Group, but part of a separate command on base where he and his unit would become part of the Air Force Office of Special Investigations about a year later. In separate vehicles, Brazel, Marcel, and Cavitt arrived quite late in the day at a small old outbuilding on the ranch, surrounded by barren, grassy, and desert scrubland. There, they bunked up for the night with plans to set out the next morning to inspect the debris area.

Debris material left at the sheriff's office by Brazel was soon delivered to Colonel Blanchard. When Blanchard saw the material, he determined for himself that it was strange indeed. General McMullen was soon contacted at SAC Headquarters, and McMullen then contacted Dubose at Fort Worth Field to help coordinate transfer of the material from Blanchard. When the hand-carried, sealed satchel of debris arrived at Fort Worth from Roswell, it was turned over to Fort Worth's Acting Base Commander, Colonel Alan Clark, who then continued on to Andrews Field in a B-26, and delivered it to McMullen.

On Monday, 7 July, after spending the previous night at the ranch, Marcel and Cavitt scoured the debris area and picked up as much material as they could stuff into their two vehicles. Brazel was soon joined by Walt Whitmore of Roswell's radio station KGFL, who hoped he could obtain an interview with Brazel in town, and he offered him the opportunity to stay at his home overnight. Evidently, the debris Brazel showed friends in town had come to Whitmore's attention.

As Marcel and Cavitt continued to scour the area, they were unable to locate any substantial body of wreckage in the very large and elongated debris field, but they did recover many smaller scraps by hand. Their effort took a majority of the day, and they returned back to town in darkness during early morning hours of 8 July. Cavitt then continued on to the base, but Marcel's intention was to briefly stop off at his home in town for a rest.

Upon arriving home, Marcel woke his wife and young son, Jesse Marcel, Jr., and showed them the curious fragments of debris he had recovered. He knew this was something special, and he wanted his family to have a look at it. After a short rest, he then continued back to Roswell Field, which was about five miles south of town. He would arrive just in time to attend a regularly scheduled early morning staff meeting. But this would turn out to be more than just an ordinary meeting.

Second Crash Site Discovered

Early on 7 July, when Marcel and Cavitt were about to collect debris at the ranch, the unexpected once again fell into the lap of Colonel Blanchard. He was informed that massive wreckage from an unfamiliar flying object was located some miles away from the site where Marcel was at, but closer to Roswell. Amazingly, there were bodies found scattered within and away from the wreckage. There were three to five of them, with one surviving. But they were not of this Earth!

There can be no doubt that Blanchard immediately relayed this information up the line, and both McMullen and Vandenberg were informed. As noted earlier, Vandenberg had recently arrived in Washington, D.C., and met with AAF Secretary Symington, General LeMay, and others. He probably also met with McMullen, who showed him crash debris he received overnight. With an additional report from Blanchard about the second site, LeMay dispatched his deputy, General Craigie, to Roswell to obtain a firsthand report and quickly return. Craigie's pilot was Lieutenant Ben Games.

The day before, on 6 July, after Blanchard saw debris from Brazel and notified higher headquarters, he ordered Provost Marshal Major Edwin Easley's military security out to guard and block all roads and passages to desolate areas north and northwest of Roswell leading off from U.S. Highway 285, and also roads and trails branching south from State Route 42. He must have also ordered search aircraft to look for possible wreckage. On 7 July, after Blanchard received a report of wreckage and bodies located at the second site, those primary highways, and also Pine Lodge Road (the southern boundary), were then completely blocked by security units, and a security corridor was established. Now, it was a full-scale recovery effort, with more ground search and security teams ordered to the sites.

After General Craigie arrived at Roswell Field from Washington, D.C., both he and Blanchard must have headed out into the field to control and monitor the situation. It was a wild and unforgettable experience to behold, but Craigie was there just long enough to see what he needed, and then he headed back to Washington, D.C.

The frenzy of activity continued through the night, and for the next several days. It was a very intense and massive operation, with many people involved. Security enforcement was extraordinary in its application, because of the amount of debris needed to be protected and recovered at the two sites, including the partially intact but wrecked object, and the valuable unearthly bodies found at the second site.

Historic Meeting

On Tuesday, 8 July, General Ramey and Colonel Dubose flew from Fort Worth to Roswell Field to attend the early morning general staff meeting. Most importantly, this meeting was held in conjunction with the previously mentioned JRBD meeting in Washington, D.C. Major Marcel and Lieutenant Cavitt provided a briefing on what had been found at the first site, and Blanchard provided a review of the second site, with debris material passed around for inspection at both meeting locations. Lieutenant Walter Haut, the public information officer for Roswell Field, was also present and was informed he would need to be prepared to issue a press release.

In retrospect, if not for the secretive cover-up, this connected meeting between Washington, D.C. and Roswell could be considered a truly momentous occasion in annals of world history, and counted as the foremost human awakening event of all time. But it still could!

The material passed around at the two meetings confirmed a colossal discovery that required top priority handling. The debris material was quite unlike anything created with human technology. And the strange bodies made it imperative that crucial protection be implemented like nothing before envisioned. Discussion must have centered on how to ensure complete security in removal of debris, and how to secure the unearthly bodies, including the one still alive. What would be done with it all? Apparently, there was a consensus that much of the debris and wreckage would be sent to T-2 Intelligence at Wright Field for analysis. But it was also necessary to inform higher political authority of the situation.

Aside from security, how would secrecy be handled? Undoubtedly, aside from the astounding nature of what had come down from the heavens, discussion at the concurrent meetings must have centered on whether to make the situation public or not, and how the press and the many people involved would be handled. Many people at Roswell Field, and some in town, now knew of the event and were told in no uncertain terms to keep quiet, but many were not yet silenced, and word was still spreading.

The preferred plan was to cover it up. From a government standpoint this made sense, because so much more information was needed about these Unearthly Flying Objects, and it wasn't known how the public might respond to the fact that a superior intelligence, with superior technology, had made an appearance on earth, and that bodies were recovered. Given that so many people in town and on base were now aware of what was going on, including some within the local press, it would make sense to divert attention from the more important crash site containing the crashed object, and just acknowledge the first site on the Foster Ranch, which many people were already aware of. Such a plan would simply acknowledge a situation where strange debris was recovered from a more remote and less accessible site. On the other hand, many people were involved with recovery of the crashed object and bodies at the second site, including civilians providing special assistance.

Consideration needed to be given as to whether secrecy could be maintained without putting the government in a position of false denial, and with many knowing otherwise. It became a conundrum of being truthful and fully informing the public, or admitting partially to some activity occurring, or putting up a smokescreen on it all while instituting a total clampdown on everyone. The ability to make a decisive and collaborative quick decision on this was hampered by relatively poor communication between Washington, D.C. and Roswell. Communications technology was rather primitive compared to that available today. In any case, the highest authority in the land needed to be notified and consulted, while also increasing and maintaining the security of the crash sites. Also, a press release would be necessary to balance escaping rumors, and Blanchard, with Haut's assistance, would handle that.

In the meantime, Walt Whitmore, of radio station KGFL delivered Brazel to Roswell Field after hearing that the AAF was looking for him. Shortly after that, Whitmore received a call from the Federal Communications Commission (FCC) in Washington, D.C., and was severely warned not to air his recorded interview with Brazel. This was followed by a call from New Mexico Senator Dennis Chavez in Washington, D.C. warning the same thing. Later, Whitmore's recorded interview with Brazel was confiscated. For the next week, Brazel was incommunicado and officially detained as a "guest" of Roswell Field, while he was being subjected to severe intimidation. He was strongly encouraged to remain quiet about what he "thought" he found in regard to the debris, and he was instructed to inform others that what he previously told them was erroneous. Others in town, such as Sheriff Wilcox and Whitmore, were similarly pressured.

Press Release

After the morning staff meeting, Ramey and Dubose departed Roswell for Fort Worth Field. Shortly after that, Blanchard provided wording to Lieutenant Haut on the now famous press release, which would be delivered to two newspapers and two radio stations in town. The Associated Press would later quote Lieutenant Haut with the following:

> *The many rumors regarding the flying disk became a reality yesterday when the intelligence office of the 509th Bomb Group of the Eight Air Force, Roswell Army Air Field, was fortunate enough to gain possession of a disc through the co-operation of one of the local ranchers and the Sheriff's Office of Chaves County.*
>
> *The flying object landed on a ranch near Roswell sometime last week. Not having phone facilities, the rancher stored the disc until such time as he was able to contact the Sheriff's office, who in turn notified Major Jesse A. Marcel, of the 509th Bomb Group Intelligence office.*
>
> *Action was immediately taken and the disc was picked up at the*

rancher's home. It was inspected at the Roswell Army Air Field and subsequently loaned by Major Marcel to higher headquarters.

Much can be questioned about the content and intent of this press release. Some speculate that Blanchard erred in formulating its content, but the release must have been written as intended, and worded according to circumstances at the time. The essence of the message was that a "disk" was recovered at a ranch, and that Marcel "subsequently loaned" it to higher headquarters. With so many people involved with recovery, news of the flying object crash was spreading, and it was imperative that this press announcement be submitted quickly. But major concern still centered on obtaining approval from highest authority, probably the president, in regard to secrecy and a possible cover-up. This required time to accomplish.

Close examination of the press release reveals that editorial privilege was taken with facts melded together from both recovery sites. Certainly, the use of the word "disk" would not apply to scraps of material found at the first site if a wrecked disk was not also found at the second site. It is significant that the word "disk" was used instead of "small widely scattered pieces of debris," and it is significant that bodies were not mentioned, even though bodies might have been expected from a wrecked flying disk. When it stated that the disk was "subsequently loaned by Major Marcel to higher headquarters," the only material sent to higher headquarters at the time of the press release was material originally brought in by Brazel two days prior, which was delivered by courier to General McMullen. It's possible, however, that a contingency plan was in the works for another special delivery of debris. The press release was truthful to a certain extent, but confusion factors were built-in, which was partially intended as a "gloss-over," as well as calculated misrepresentation to buy more decision time for responsible authority, and a possible cover story.

It is clear that Blanchard suffered no consequence from his news release, which he was given authority to provide. Under normal circumstance, the release of such a "fantasy" would jeopardize Blanchard's assignment in commanding the only nuclear bombing group in existence. But he continued to be very handsomely promoted over time, just like many others involved with Roswell. In fact, he ultimately obtained the military's highest rank of four-star general, which was an accomplishment that very few could dream of receiving. He was well-rewarded, and there can be little doubt that he was given an incentive to guarantee his continued support, protection, and maintenance of the UFO question.

Around noon or shortly before, Haut delivered the news release to two local radio stations (KGFL and KSWS), and two newspapers (the *Roswell Daily Record* and the *Roswell Morning Dispatch*). After receipt by local media outlets, it was expected that the news release would be dispatched to major wire services such as

United Press (UP) and Associated Press (AP) for worldwide distribution.

Cargo Loaded for Wright Field and T-2

Shortly after 1:00 P.M., a flight plan was secured for a B-29 named "Dave's Dream" to depart Roswell at 2:00 P.M. for Wright Field. The intent was to transport debris that Marcel and Cavitt gathered the previous day, and possibly wreckage from the second site also. This was not a pre-planned flight, and it was quite extraordinary, because it would not be manned by the plane's normal crew. Five of the plane's crew were assigned to scour the two crash sites for debris, including a few hundred other personnel from base. Only a few from the normal crew, including Corporal Milton Sprouse, were left on base in case the plane was needed for flight preparation. The plane was then taken over, but manned by high-ranking officers and noncommissioned officers from Roswell Field. The pilot was Base Deputy Commander Lieutenant Colonel Payne Jennings, and the copilot was Base Executive Officer Lieutenant Colonel Robert Barrowclough. Also on the crew were two other officers, and four ranking sergeants from base.

Prior to loading the plane, it was surrounded by security people. Cargo and armed guards were then loaded into the plane's bomb bay. This was not an ordinary flight to say the least, but it was a special historic one where unearthly items required strict handling and extraordinary high security protection.

The news release, originated by Blanchard and delivered by Haut, was already received by major wire services for distribution before the B-29 would take off for Wright Field. When the Associated Press (AP) received news on Roswell, they began releasing periodic short bulletins on their network, and this would ultimately provide a significant timeline for events to follow.

৯০৫

The Roswell incident, as noted in the foregoing, was a very big deal to everyone involved. It reached highest levels of government and caused much tumult, commotion, and concern, which was not worthy of an alleged recovery of a small tattered weather balloon. It also involved huge numbers of people, including government officials, military personnel, and civilians. Major players in this scenario knew they were dealing with a highly critical situation, which included very unusual items from another world. And it needed to be very carefully managed within a relatively short period of time. How they would handle it from this point on, however, would determine success in covering up the most magnificent secret of all time. Would they be able to manage this without slipping up in the process?

CHAPTER TEN

Roswell Uncovered

July 1947

Complicated Critical Communication

In the previous chapter, the general scenario of the Roswell incident was introduced and reviewed, which showed a huge amount of attention paid to the incident by high-level officials of the military and government. Also examined were details about people who were "hands-on" at Roswell. It was a very significant happening, where effective communication was necessary between officials at AAF Headquarters at the Pentagon, Eighth Air Force Headquarters at Fort Worth Field in Texas, and Roswell Field in New Mexico.

Communication was complicated because officials in those areas needed to be on the same page to properly coordinate matters. It didn't help that electronic communications were somewhat primitive compared to those of today. It was a time of party lines, manual switchboards, and circuits that could easily become overloaded with traffic. Broadband digital technology did not exist. Any major event could wreak havoc with communications availability, and with priority lines crucial to running the country. This would soon become readily apparent to those at the Pentagon, Fort Worth, and Roswell.

Wire services and media organizations were hindered as well. Only slow-speed teletype existed, which also used simple telephone lines. Getting the news out was a protracted process involving piecemeal release of various snippets of information. That is the way information about Roswell was transmitted across the country.

In regard to this communication process at the time of the Roswell incident, Associated Press (AP) bulletins released on its wire service will now be reviewed.

These bulletins provide great clues on what was taking place between the three areas involved. In fact, many researchers have latched onto these bulletins to analyze and study them in the attempt to prove or disprove the Truth of Roswell. The same will be done in the following, but directed to a more concise analysis of the timeline, which reveals a chaotic and obvious cover-up of Roswell and the UFO question by General Vandenberg and General Ramey.

All the AP bulletins are given in Eastern Standard Time (EST), but equivalent Fort Worth and Roswell times are also provided for reference. During this period in the United States, a few localities were observing Daylight Savings Time (DST), such as Washington, D.C. But both Fort Worth and Roswell were observing Standard Time. In this analysis, three time zones are reflected, but DST is not involved.

These AP bulletins are significant in revealing the scenario of AAF cover-up, and what transpired in pulling it off. It must be noted, however, that the following can be confusing to follow without paying careful attention, but this is reflective of how Air Force Headquarters also became confused in a devious plan to deceive the public, and to cover-up the incident at Roswell.

4:26 P.M. EST AP Bulletin

The first AP bulletin, at 4:26 P.M. EST (3:26 P.M. Fort Worth time, 2:26 P.M. Roswell time), stated the following:

> *ROSWELL, N.M., July 8. – The army air forces here today announced a flying disc had been found on a ranch near Roswell and is in Army possession.*

It was then that the world began to learn of the found "disc." This stark announcement, however, seemed unreal, and the message lacked substance. Most media news organizations receiving it were not yet prepared to react, because of the need for more information. The AP, while continuing with periodic bulletins on its wire service, was preparing to investigate the news further before consolidating a more complete report of its own.

The startling announcement was followed minutes later with similar news from United Press (UP), and then phone lines started ringing. Inquiries regarding the announcement inundated AAF Headquarters in Washington, D.C., Eighth Air Force Headquarters at Fort Worth Field, and also Roswell Field, but there seemed to be no central person prepared to answer or handle the calls. No spokesman took responsibility for the situation, and it is likely that Colonel Blanchard at Roswell Field was reluctant to do so. His superiors would have preference over what could and should be said, and he didn't want to contradict what they might be offering on the situation. Like everyone else, he probably wanted the phones to just stop ringing.

The press announcement likely created a huge crisis in Washington, D.C. because of the commotion it generated, and it probably put those in high authority on edge. Roscoe Hillenkoetter, Director of the Central Intelligence Group, was probably already involved in an attempt to learn all he could, as well as the FBI, and their agents were probably preparing to converge on Roswell. AAF Deputy Chief General Vandenberg was undoubtedly concerned by the commotion, and President Truman may have been informed and assured that the situation would be quieted in short order. Communication by telephone, however, was now becoming very difficult.

It was considerably past the scheduled time for the loaded B-29 to depart Roswell for Wright Field (at 4:00 P.M. EST, 3:00 P.M. Fort Worth time, 2:00 P.M. Roswell time), and serious questioning was probably taking place on how to quiet the situation caused by the press announcement. The loaded plane was ready to leave, but inquiries from AP, UP, and many other media sources were flooding in. There must have been talk about what further response was needed to satisfy the press and calm the situation, but the inherent communication barrier was affecting the decision process, and enhancing turmoil and concern. True significance and the potential consequence of Lieutenant Walter Haut's press release was beginning to sink in at high levels, and one can speculate that debate was still taking place on how secrecy of the Roswell incident should be handled. One can also speculate that some officials were partial to revealing Truth of the matter, but revelation of Truth would be a presidential decision.

At about 5:00 P.M. EST (4:00 P.M. Fort Worth time, 3:00 P.M. Roswell time), Major Jesse Marcel was hastily summoned by Colonel Blanchard to help escort the recovered debris to Wright Field. It's possible that Marcel was on "standby," and that his departure would not take place unless it was determined it would be necessary for him to "loan" material "to higher headquarters," which would not be Wright Field. He was to hand-carry a small box containing special debris collected, possibly the same material he showed his wife and son in early morning hours.

5:11 *P.M. EST AP Bulletin*

At 5:11 P.M. EST (4:11 P.M. Fort Worth time, 3:11 P.M. Roswell time), another AP bulletin was issued, which suggested the AP was in the midst of their own investigation, and making calls to find out what was going on. It simply stated:

The War Department in Washington had nothing to say immediately about the reported find.

Information From Ramey - 4:30 P.M. CST

At about 4:30 P.M. Central Standard Time (CST) in Fort Worth (5:30 P.M. EST, 3:30 P.M. Roswell time), in the midst of heavy phone traffic, a reporter from the *San Francisco Examiner* managed to talk with General Ramey at Fort Worth, which was about an hour after the first AP bulletin. Ramey told the reporter he believed the "object was probably the remnant of a weather balloon and radar reflector."

This was the first indication that AAF Headquarters and Ramey had chosen a plan of action to deflect attention from the crashed "disc," and it was the first real indication of how the Roswell incident would be covered up. A critical decision was finally made regarding secrecy of the Roswell incident, and there would be no more talk regarding a recovered disc. Truth was not going to win out. But it was not clear whether the destination of the B-29 was changed, and it's possible it was still destined for Wright Field to unload its precious cargo. But it does appear that the B-29 was possibly in the air at this time, or Ramey would not have suggested what the object "probably" was.

5:53 P.M. EST AP Bulletin

When the AP next made contact, an AAF Headquarters spokesman provided additional details received from General Ramey at Fort Worth. The AP relayed this information at 5:53 P.M. EST (4:53 P.M. Fort Worth time, 3:53 P.M. Roswell time) with the following statement:

> *In Washington, D.C., Brig. Gen, Roger Ramey said today a battered object described as a flying disk, found near Roswell, New Mexico, is being shipped by air to the A.A.F. research center at Wright Field, Ohio.*

It is clear from this that AAF Headquarters at the Pentagon was aware of Ramey's and Blanchard's intent to transfer the "flying disk" from Roswell to Wright Field, but this information was somehow delayed by the AP, or perhaps the AAF spokesman was not aware of current information that Ramey expressed in his 5:30 P.M. EST phone conversation with a San Francisco reporter.

One can speculate whether the statement "is being shipped" equates to a statement of intent, but it may mean the B-29 was actually in the air, probably before 5:30 P.M. EST (4:30 P.M. Fort Worth time, 3:30 P.M. Roswell time). Either way, this AP bulletin demonstrates the thinking of Ramey, and especially those at AAF Headquarters in Washington, D.C. This is quite important, because it reflects

Ramey's original intent.

The American Broadcasting Company (ABC) News Radio reported they contacted Wright Field, and "they were expecting a shipment soon" from Roswell. Also, an ABC "Headline Edition" radio report stated the following:

> *"...has been inspected at Roswell, New Mexico and sent to Wright Field, Ohio for further inspections."*

The AP then received information from UP, which confirmed the B-29 was heading to Fort Worth Field instead of Wright Field. It stated:

> *The United Press quoted officers at the base as saying that the disk was flown in a B-29 Superfortress to higher headquarters.*

Destination plans for the B-29 were definitely changed at the last moment. Could it be that plans were changed just before the B-29 took off from Roswell, or maybe just after? Perhaps when Marcel first boarded, he was prepared to proceed to Wright Field in order that critical cargo might first be off-loaded, and then he would continue on to Fort Worth where he would later meet with Ramey. Perhaps after takeoff, or at the last moment, it became more urgent that Marcel and the B-29 proceed directly to higher headquarters at Fort Worth.

6:02 P.M. EST AP Bulletin

Nine minutes after the 5:53 P.M. EST bulletin, AP put out another bulletin at 6:02 P.M. EST (5:02 P.M. Fort Worth time, 4:02 P.M. Roswell time) stating the following:

> *Ramey, commander of the Eighth Air Force with headquarters at Fort Worth, received the object from the Roswell Army Air Base. In talking by telephone to A.A.F. headquarters at Washington, Ramey described the object as of "flimsy construction, almost like a box kite." It was so badly battered that Ramey was unable to say whether it had a disk form. He did not indicate the size of the object.*

This particular AP bulletin is very significant in regard to Truth about Roswell, and it will be referenced further.

Because the plane was supposedly already at Fort Worth, per the 6:02 P.M. EST message, one wonders if the 5:53 P.M. message was considerably delayed, or what? It probably was, but if the plane left about 5:30 P.M. EST (3:30 P.M. Roswell time), it could not possibly have arrived at Fort Worth according to the 6:02 P.M. EST message (4:02 P.M. Roswell time). There is considerable strangeness about this bulletin, and things just don't add up.

Taken together, the 5:53 P.M. and 6:02 P.M. EST bulletins from AP suggests that plans to transport recovered Roswell debris to Wright Field were definitely changed at the last minute, and it confirms that an abrupt turnaround was made in how the Roswell incident would be handled. It is seen that a plan was in the works to relegate the "battered object described as a flying disk" being sent to Wright Field (as described in the 5:53 P.M. EST bulletin) to an object "of flimsy construction, almost like a box kite" that supposedly arrived at Fort Worth (as described in the 6:02 P.M. bulletin).

Ramey and others were previously aware of the nature of the unearthly object. Its wreckage and debris would still be in the process of recovery for many days. The debris material, wreckage, and bodies were previously inspected and confirmed by Ramey at a meeting with other officials earlier in the day at Roswell, and this was well before the B-29 ever left Roswell for what was to be a flight to Wright Field. Wright Field was the original intended destination of the B-29, because it was the home of Technical Intelligence (T-2) within Air Material Command (AMC) headed by General Nathan Twining, and this is where it was appropriate to transport such highly sensitive debris for storage and study. Wright Field was expecting to receive the material, and this is where Ramey and Blanchard originally intended to send it.

6:11 P.M. EST AP Bulletin

At the heart of the Roswell cover-up is a subsequent bulletin issued just minutes after the 6:02 P.M. EST AP bulletin, and it contradicted and nullified all previous AAF Headquarters statements. As published by *The Seattle Daily Times*, the AP issued the following at 6:11 P.M. EST:

> *Later A.A.F. headquarters officials denied they had any information at all on the matter. They said they were trying to call Roswell by telephone to find out what happened.*

Perhaps not everyone at AAF Headquarters in Washington, D.C. had current information, or full knowledge of what was going on between Fort Worth and Roswell, especially with phone lines tied up. Nonetheless, it is troubling that AAF Public Information Officers (PIOs), the only people allowed to speak with the public or press, provided conflicting statements, and professed ignorance after the fact. It suggests that mistakes were made, and it then became necessary to stonewall.

It must be noted that all information for AP bulletins came directly from AAF Headquarters in Washington, D.C., and not personally from Ramey at Fort Worth, although he was likely "quoted" by AAF Headquarters on his status reports, or on an agreed "story line."

Of particular significance, is the timing of this last AP bulletin, which coincided exactly with the time Vandenberg felt compelled to immediately go to the press room of AAF's Public Information Office and handle a critical communications mistake. It was a time of panic and chaos in the press room!

One could speculate that AAF Headquarters previously prepared or scripted their statement in regard to the 6:02 P.M. EST AP bulletin, and that it was intended to be a "forthcoming press release," which was often a standard procedure. However, they released it prematurely to the AP. It is quite likely that information given to the AP for this bulletin was actually intended to be provided an hour later, at around 7:00 P.M. EST (6:00 P.M. Fort Worth time, 5:00 P.M. Roswell time). The context and timing of the last statements by AAF Headquarters (at 6:11 P.M. EST) provides credence to this, and it strongly suggests that a colossal error was made at AAF Headquarters due to confusion and chaos. AAF Headquarters was in the midst of a massive crisis, and Vandenberg was at the center of it in AAF's press room.

It was previously noted that Blanchard summoned Marcel at about 3:00 P.M. Roswell time. This means the B-29 probably departed at about 3:15 P.M., or just prior to 3:30 P.M. Roswell time (4:30 P.M. Fort Worth time, 5:30 P.M. EST). According to the 6:02 P.M. EST AP bulletin, the B-29 arrived prior to 5:02 P.M. Fort Worth time (4:02 P.M. Roswell time), which implies that the flight time from Roswell was only about a half hour, maybe slightly more.

Considering that the average cruising speed of a B-29 was 220 mph, with a maximum speed of 357 mph (at 30,000 feet), it was impossible to fly a distance of about 415 miles in the time implied by the 6:02 P.M. EST AP bulletin. At a greater than average speed of 270 mph, it would have taken the B-29 about an hour and a half to reach Fort Worth, but B-29s were normally kept at a slower cruising speed so that their temperamental magnesium engines would not risk fire. Also, the plane remained at a low altitude, and at a slower speed to protect the guards and valuable cargo, since the cargo bay was not capable of being pressurized.

When looking at a realistic flight time for the B-29, and a projected arrival time of about 6:00 P.M. at Fort Worth (7:00 P.M. EST), the AP bulletin was issued approximately an hour before the plane could possibly have landed with its precious cargo. Therein lies the mistake of the 6:02 P.M. EST AP bulletin, which provided fictitious information stating what Ramey "received" ("almost like a box kite") upon the supposed "landing" of the B-29 at Fort Worth. And this explains the "ignorant" response from AAF Headquarters in the 6:11 P.M. EST AP bulletin, and the necessity for Vandenberg to rush to AAF's press room to stop the release of further information.

After the 6:02 P.M. EST AP bulletin was released, evidence suggests that AAF Headquarters did not want to admit that the B-29 carrying Marcel was not yet at Fort Worth. There was no way to retract their previous statement without tipping their hand and facing questions they did not want to answer. The only way AAF Headquarters could respond to further inquiries was to profess ignorance, and stonewall for at least another hour. The massive number of phone calls, inquiries, and messages pouring in and inundating Haut's, Blanchard's, and Ramey's offices added to the confusion and chaos. AAF Headquarters likely knew those phones were tied up at Fort Worth and Roswell, and probably hoped they would remain that way.

The AP's 6:02 P.M. EST bulletin stated what Ramey "received" before the plane actually carrying it landed at Fort Worth. It then proceeded to describe it in terms of Ramey's supposed "inspection" upon "arrival." Obviously, there was a plan conceived prior to the B-29 landing at Fort Worth, and before Ramey was able to conduct a subsequent press conference in his office. Could this be considered incriminating evidence of duplicity by AAF Headquarters and provide reason to condemn all aspects of subsequent AAF statements and comments on Roswell, and comments by the Air Force years later?

Information From Fort Worth Field - 6:30 P.M. EST

Fort Worth's intelligence officer, Major Edwin Kirton talked with the *Dallas Morning News* at about 6:30 P.M. EST (5:30 P.M. Fort Worth time). In a published article the following day, the newspaper quoted Kirton with the following:

> *There is nothing to it. It is a Rawin high altitude sounding device...it will not be necessary to forward the object to Wright Field, as originally planned.*

This information from Major Kirton was provided while the B-29 was still in the air heading for Fort Worth. It would still be another half hour before the B-29 and Marcel would arrive. The above statement confirms a change in the flight plan, and the decision to discredit the recovery of a "disk."

Ramey's Ruse

Instead of arriving at Fort Worth just prior to 6:02 P.M. EST time, Marcel arrived around 7:00 P.M. EST time (6:00 P.M. Fort Worth time), which was just enough time for him to proceed to Ramey's office and participate in Ramey's ruse. He did not have to wait for an hour or more before joining Ramey in his office, where Ramey would give the performance of his lifetime regarding a tattered weather balloon (a Rawin sounding device). Ramey was prepared, and he previously announced that a plane would soon be arriving with debris carried by Marcel. He summoned Warrant Officer Irving Newton from the base weather office, and also notified local press that there would be an imminent announcement in regard to the flying disk.

One can speculate on what was going through Marcel's mind at this particular moment. He once thought he was going to Wright Field, but now he found himself at Fort Worth Field. He probably was not prepared for what would actually happen in Ramey's office, especially after handing over his secure box of special debris to Ramey. One might further speculate that Ramey did not want to give Marcel a chance to object to his ruse, and Marcel did not want to risk standing up to the general in front of the press, and those present, while poised with weather balloon remnants.

When the Fort Worth *Star-Telegram* reporter James Johnson arrived at Ramey's office, and then left with his story and photographs, the rest became history. Marcel was pictured with weather balloon remnants spread out over brown paper wrappings on the floor, and so were Dubose and Ramey.

An interesting aside to this is that one of the pictures taken of General Ramey with weather balloon remnants shows him casually holding a telex message, inadvertently, and partially turned toward the camera. Using modern computer technology, the message has been analyzed and deciphered as a message to Vandenberg with very telling information regarding shipment of "victims" and a "disk."

Marcel was well-acquainted with weather balloons, and technically very capable, and he must have been chagrined to come all the way to Fort Worth just so General Ramey could paint him as a doofus. Marcel must have known the difference between what he collected at the Foster Ranch, what he hand carried in his box, and what was displayed on the floor before him. He probably wondered why the deception could not have been carried out just as easily at Roswell, but he probably realized that this could not have been done because of ongoing recovery and the commotion taking place there. It's certain that AAF Headquarters and General Ramey didn't want the media and press to be anywhere near Roswell, due to ongoing recovery activity. The secret needed to be quickly contained and protected.

7:29 *P.M.* EST AP Bulletin

About three hours after AP's initial bulletin regarding a recovered disk, another AP bulletin appeared at 7:29 P.M. EST (6:29 P.M. Fort Worth time), which put finality to the Roswell situation with the following:

> *Roswell's celebrated 'flying disk' was rudely stripped of its glamor by a Fort Worth army airfield weather officer who late today identified the object as a weather balloon.*

The B-29 then returned to Roswell Field without Marcel, and without proceeding to Wright Field. Prior to leaving, however, a large crate was off-loaded from the plane onto a flatbed truck, and other packages were loaded onto a B-25 for transfer to Wright Field. The small box that Marcel took to Ramey's office disappeared when he was distracted by Ramey, and it was quickly and purposely substituted with weather balloon remnants just prior to Ramey's dramatic display of them. Marcel's small box of debris was probably taken to the B-25, or perhaps it was kept by Ramey as a souvenir. On the trip back to Roswell Field, everyone on the plane was instructed that the crashed disk, and associated debris, was now a damaged weather balloon, and they were instructed that there would be no more talk about a crashed disk. With the ruse in place, there was no way this particular plane would continue to Wright Field, because any reason for doing so was eliminated.

The next day, on 9 July, another B-29 was flown to Fort Worth from Roswell carrying an even larger crate under very tight security. This plane, which was named "Straight Flush," was piloted by Captain Frederick Ewing, and it was another hastily arranged flight. The cargo was escorted and monitored by a contingent of security guards, and it's speculated it contained unearthly bodies. Marcel, after finally getting a night's rest, then boarded this empty plane for the flight back to Roswell. According to Haut, Marcel was visibly upset upon his return, and strongly complained about the staged event that he experienced at Fort Worth Field.

Roswell Put to Bed

On the day prior, when Marcel's B-29 departed Roswell for Fort Worth, Blanchard left his office and took Haut with him to get away from ringing phones. They went out to a hangar on base to see additional debris returned from the second crash site. Blanchard then monitored loading of a C-54 with the material, which was then flown to Wright Field by Captain Oliver "Pappy" Henderson. After that, Blanchard told Haut to hide out and take the rest of the day off. Blanchard then became unavailable, supposedly on a 21 day leave, although one might speculate he was setting up a base camp and managing protective recovery of debris at the two sites. Colonel Jennings was left in charge at Roswell Field upon Jennings' return from Fort Worth.

Reportedly, the majority of crash debris and wreckage was transferred to Wright Field, but some of it was transferred from Roswell and Fort Worth to other ultra-high security locations in New Mexico, Texas, and the Washington, D.C. area for storage and analysis. The deceased unearthly beings required special handling, and immediate transport for critical study. They were supposedly shipped to several specific locations. The one still alive was provided extra special attention, and reportedly secured temporarily at one of the most secret and isolated facilities in New Mexico.

To complicate matters further, the crash at Roswell was not the only one that occurred at this particular point in time. Another happened at a location some 200 miles west of Roswell, at a locality called the Plains of San Augustin. This incident also required immediate attention, and very complex recovery logistics. Several bases in the New Mexico area became involved with it.

The chaotic plan, and hasty deception to put the Roswell incident "to bed," may have been Ramey's idea, but it was probably done at the behest of those above him, and orchestrated haphazardly at the highest levels of government. Because of a deluge of phone calls resulting from news of the "recovered saucer," high officials were forced to make a hasty decision regarding secrecy and a cover-up, and they must have been quite nervous in contemplating the potential ramifications if their secret was discovered.

This, however, would not be the last heard from Ramey. Two years later, in late July 1952, Ramey would be involved in another major effort to mislead the public on the UFO question. It seems that he was the one specific individual the Air Force could rely on, and trust, for this particular duty. It leads one to speculate on what thoughts he may have had in his later years, especially when reflecting back on the fact that he was responsible for misleading the public about a magnificent secret. Did he have a twinge of regret?

Truth of Roswell

The Roswell incident is either a fantastic, highly-fabricated, intricate, and complicated story, or it presents a picture of Truth. The thing that must be recognized, aside from analysis of AP bulletins, is that all of what has been presented in this and the previous chapter was derived from the voices of a great many people, whose recorded actions and statements translate to a clear picture of AAF duplicity. Many of the people involved, after holding their tongues for some thirty years or more, decided to come forward and speak to the Truth without benefit to themselves, except to clear their conscience. They did not contrive together to construct the story of Roswell, but they all contributed individually, which eventually developed into a recognizable picture of what took place at Roswell, Fort Worth, and AAF

Headquarters.

The editor of the *Roswell Morning Dispatch*, Art McQuiddy, who had been a good friend of Colonel Blanchard, tried several times to get the "real story" from him. Eventually, in a relaxed moment a few months later, Blanchard said to him:

> *I will tell you this and nothing more. The stuff I saw, I've never seen anyplace else in my life.*

About thirty years after Roswell, Marcel admitted in a television interview that there was a cover-up. He also left a signed, witnessed, and notarized affidavit regarding what he knew, both from what he personally observed, and what he learned from others. As a retired colonel and one of those most intimately involved, he was very vocal in his later years about the incident.

After Marcel passed away on 24 June 1986, his son, Jesse Marcel Jr. continued to decry the deception, which culminated in his deposition before six former members of Congress on 1 May 2013. Current members of Congress were, and still are, reluctant to participate in such a forum for obvious political reasons. Marcel Jr. passed away just over three and a half months later, on 24 August 2013. He performed his last active duty service in Iraq in 2004 as an Army flight surgeon holding the rank of colonel.

Another person who confirmed the existence of a cover-up in the Roswell incident is retired Brigadier General Thomas Dubose. Dubose was photographed in Ramey's office with the weather balloon remnants. In his later years, he disclosed that he received a phone call from General McMullen, after the initial AP bulletin. McMullen requested that General Ramey eliminate the saucer story, and "get the press off our back." Dubose acknowledged that the staged event in Ramey's office was a cover-up, and he backed this up with a sworn affidavit. In a home video made just before he died in 1992, General Dubose stated:

> *It was a cover story, the balloon part of it; it is the story that was to be given to the press and that is it...*

Dubose was one of the highest ranking individuals to ever comment on the incident, although not the only general, and he is one of a great many military and civilian people to come forward with the Truth. Many other individuals, including verified key players, have bravely come forward with disclosure about the incident, and many have affirmed their statements with sworn affidavits. Among them is Walter Haut, who left an affidavit regarding his participation, including what he saw in the flight hangar that he and Blanchard visited. Many others have provided frightening and emotional stories of being threatened, even with their lives, to keep silent.

There were also those who were too frightened, or not motivated enough to speak out, such as Captain Cavitt, who was with Marcel in the recovery of Roswell debris. In later years, when researchers began to investigate Roswell, Cavitt denied he was ever at Roswell, although his military records prove otherwise. His connection with the Army Air Force CIC, and then with the Air Force OSI, likely constrained his participation in revealing Truth. Today there are many who can relate with Cavitt. They are so strongly tied to an Air Force lie that they cannot escape. For some, it must be a terrifying position to be in. They know the Truth, but if they contemplate revealing it, there could be trouble in store for them. One can only thank them for service to their country, but also hope they may one day find courage to express Truth. Their country will thank them a whole lot more!

ক্ষ৵

The toll taken at Roswell includes many innocent people who were coerced, threatened, harassed, ridiculed, and purposely made fools of by the government. The government, and AAF, chose deception rather than an honorable and truthful approach to the UFO question, and that choice continues to be maintained by the U.S. Air Force to this day.

The Air Force continues to maintain the secret of Roswell, and there is no end to their battle in suppressing Truth, which is repeatedly demonstrated and confirmed in following chapters of this book. Even though denials are made, Truth often prevails, and reveals itself in ways never imagined by those attempting to suppress it. Years later, many people previously involved at Roswell decided to reveal Truth. There are also former Air Force officers and missileers who were once silenced, but who are now also deciding to reveal the Truth.

AP bulletins regarding the Roswell flying disk provide all the information necessary to reveal the Truth of Roswell. Bulletins released by the AP show the mishandling of statements at AAF Headquarters, which demonstrates confusion and indecision by high officials who were formulating a plan to deceive the public. This revelation of AAF duplicity in the cover-up of Roswell validates the confessions, testimonies, and affidavits of all those who spoke out to reveal Truth. Could it be possible that they will one day be vindicated as genuine American heroes?

PART THREE

Active Government Interest

Army Air Force announcement of the Roswell incident initially created bold headlines, but then it was quickly dismissed by higher AAF authority. It was accomplished amidst frantic and haphazard action at AAF Headquarters, while a naïve and trusting public accepted AAF's deception without question–there was no possibility the government could contrive a lie against its own citizens. More than thirty years later, however, astute researchers pieced together a very clear picture regarding the reality of Roswell. It was backed up with witness affidavits and a great number of eyewitness accounts by the people of Roswell, both civilian and military. A naïve public, however, remained ignorant.

After success with the lie, then came the challenge of managing and controlling the monumental secret. This needed to be accomplished while creditable observers continued to report sightings of flying objects, and while AAF investigators (with no clue about Roswell) studied those reports for answers. This was happening while the government itself was heavily involved with organizational change.

As AAF investigators actively studied sighting reports, they began "knocking on the door of Truth." Eventually, they determined that flying objects were quite "real!" Indirectly, this confirmed the reality of Roswell, although it was inconclusive that the objects were from another world. A stamp of approval was provided by General Twining, when he pronounced in a secret letter that flying objects were "something real and not visionary or fictitious." And this provided additional emphasis for determining the origin of flying objects.

It was a very intense time of active government interest, which the public knew nothing about. But the Control Group was now involved with a balancing act. Any and all AAF investigative activity would need to be carefully monitored in order to protect the magnificent secret. It was a matter of protecting and preserving the secret of Roswell, and Truth of the UFO question.

CHAPTER ELEVEN

Controlling the Secret

July 1947

A Clear Picture

At the time of Roswell, in July 1947, the general public knew only what the press told them. First, that a "flying disk" was retrieved from a ranch outside of Roswell, and then an announcement a short time later that it was only a simple weather balloon. A trusting public took it all in stride, although many wondered how there could have been such gross misidentification. Many at Roswell Field and in the nearby town of Roswell knew otherwise, but all were instructed and threatened to keep lips zipped.

When I think back on this incident, I recall it as a child living in Seattle, Washington. I was about a month shy of seven years old, but I distinctly remember my parents commenting with awe and wonder about what they heard on the radio and read in the local evening paper. No one doubted the report, and there was great anticipation to learn more. It was fantastic news with much excitement in the air, but it was all deflated by the weather balloon report. Even at my young age, I remember curiosity and anticipation for more news about the "saucer." On hearing additional news and comment the next day, I wondered how the military could have made such a blunder, but my thoughts were probably a reflection of what my parents and others were expressing. I remember a twinge of disappointment with the last press announcement, but it was also difficult to imagine recovery of an unearthly object. It didn't fit with reality of everyday life. The exciting news quickly vanished from the mind, and the public quickly lost interest and forgot, but those managing the cover-up were relieved.

More than thirty-five years later, despite a continuing cover-up by the Air Force, facts on the Roswell incident began to leak out in the 1980s. It happened when suspicious researchers began to interview local Roswell townspeople and former military. Like pieces of a jigsaw puzzle, facts were uncovered and the amazing larger picture became clear. Of course, after so many years, scattered omissions and inconsistencies would be expected, along with certain details lost with time. A great mountain of evidence, however, was generated from many witness accounts, including recorded personal interviews and affidavits, all which provide a clear, recognizable, and undeniable picture. Only where scattered holes exist in some areas do critics and skeptics bore in to expose and examine, with subjective rhetoric, on what they believe are misconceptions and inaccuracies. They ignore the big picture, and focus instead on a few missing pieces, or confusing facts. The greater picture, however, cannot be altered or rendered unrecognizable.

Years later, Air Force Headquarters participated in an effort to discredit Roswell with two sizable publications. The first was issued in 1994, and titled *The Roswell Report: Fact vs. Fiction in the New Mexico Desert*. The second was issued in 1997, and titled *The Roswell Report: Case Closed*. Both reports are very detailed, and filled with extraneous and bogus information to camouflage the real story of Roswell. Interviews with living witnesses were totally avoided, while false out of context information was provided with extensive dissertation by the authors, who were Air Force intelligence agents. The false information, however, was adequately and properly pointed out by researchers. Air Force claims were taken apart, which left the published reports as outstanding examples of continued Air Force duplicity. The publications were provided through the approval and courtesy of Air Force Secretary Sheila E. Widell (6 August 1993 - 31 October 1997).

Enforced Silence

Major players in the Roswell incident, both civilian and high-ranking military, opened up years later to expose the Truth. The incident becomes significant because of the great number of witnesses who were involved in recovery of the crashed object. Their independent accounts make it impossible to avoid the fact that an astounding event did take place. Testimony of each generally corroborates that of others, and when it is all pieced together, it reveals how the cover-up by the AAF was constructed and maintained.

The magnitude of the Roswell incident was massive, with many people involved and coerced to remain silent, and this causes one to ponder how silence could have been preserved for so many years. However, when one considers the enforced silence in other massive projects, such as the Manhattan Project in development of the atomic bomb, or other "black projects" involving tens of thousands of people, it becomes understandable.

Unfortunately, because of AAF's successful ability to stifle the local public and military personnel at Roswell, the secret remained undetected for many years until researchers began to take a more careful look. A few brave souls took a chance to speak out to investigators to relieve their consciences, which resulted in a flood of others who felt less threatened and willing to do the same. They won their freedom from haunting memories, with no other benefits gained. Most simply wanted the Air Force and government to come clean and admit to unjust treatment and guilt forced upon them by harboring a government lie.

Today, there are still many who want the government and Air Force to come clean with the secret, including many who continue to have their lips zipped by the Air Force. The Air Force, however, continues to maintain an unrelenting cover-up. But it is not just the secret of Roswell. It is the secret of the UFO question, and also a great number of other very important incidents.

Management of Reality

Prior to and during Roswell, a great wave of sightings occurred all over the world. Many hundreds were sighted in the United States by very credible witnesses, including military personnel. This was the beginning of serious government consideration of the UFO question, primarily because the general military became very concerned about reality of the objects. As a result, efforts were initiated to organize, investigate, and learn more about what was happening, but it was with an eye toward avoiding public concern.

Pentagon officials were initially very concerned by the fact that flying objects appeared to be both real and unidentifiable as to domestic origin. Then, in the Roswell incident, reality was suddenly revealed in a tangible way, and it required the AAF to sequester evidence and seal lips of people involved. The cover-up was created to shield the public from learning of the government's inability to deal with the objects, which exhibited vastly superior technology within a framework of high intelligence. It was necessary to hide away the recovered valuable evidence and associated high technology. But it was still necessary to determine what the objects were all about. It was the harassing nature of the objects, and their intimidation of the military in various incidents that brought this about.

During Roswell, the Commander in Chief of the AAF was General Carl Spaatz. He previously led a great bombing campaign against Germany in WWII, and he directed General Jimmy Doolittle of the Eighth Air Force and General Nathan Twining of the Fifteenth Air Force. General Dwight Eisenhower was Army Chief of Staff, which became his position after President Truman chose General George Marshall to be Secretary of State in early 1947.

General Hoyt Vandenberg took primary control over the UFO question during Roswell, and was assuming many duties from Spaatz who was scheduled to retire within a year. Vandenberg was previously in charge of the Central Intelligence Group (CIG) before leaving that position in May, and he was replaced in CIG by Admiral Roscoe Hillenkoetter. Presumably, Vandenberg and Hillenkoetter were in close contact at this time in order that Hillenkoetter might be brought up to speed on intelligence issues, including the Roswell incident.

With General Spaatz seemingly in the background, Vandenberg took charge of managing the Roswell secret, but there are many questions regarding General Spaatz and the role he may have played in regard to Roswell. Questions remain about his supposed vacation in the Pacific Northwest during reported sightings, and indications he was also in and around the area of New Mexico shortly after the Roswell incident. Information regarding Spaatz is sparse, but it seems probable that he was more involved with the UFO question at this particular time than appears on the surface.

Decision Time on Roswell

On Tuesday, 8 July, just prior to the momentous and hastily arranged meeting of the JRDB involving Roswell, Vandenberg met briefly with General LeMay, and also Dr. Edward Bowles, who was a distinguished MIT physicist and consultant to Secretary of War Howard Peterson. Both LeMay and Bowles were involved with Project RAND, which was a think tank they helped form in 1945. Vandenberg's purpose in meeting with them was to solicit all the assistance he could muster, and to also inform those who might assist with decision-making on Roswell.

After the JRDB Meeting, Vandenberg met with Secretary Stuart Symington. Quite likely this was to brief him on the JRDB meeting, and to discuss the security situation on the Roswell disc recovery. He also wanted to solicit advice regarding secrecy, and on instituting a possible cover-up. It would be necessary to pass decision-making up the chain of command and to the president where the "buck" would stop. Then, around 4:00 P.M. EST (2:00 P.M. Roswell time), phone calls were coming in as a result of Haut's press release regarding the "captured saucer." Now, Vandenberg was likely in contact with Symington again, with General McMullen at SAC Headquarters, with General Ramey at Fort Worth, and probably with many others, including President Truman. It was decision time, and time was at a premium to make a final decision regarding secrecy and a potential cover-up.

A B-29 at Roswell Field was scheduled to take off for Wright Field with recovered debris and wreckage, but frantic discussion ensued on how to curtail what they now saw as a major tactical blunder with Haut's press release. If higher officials surmised that putting crash debris under high security was enough, they now understood that the press and public would need to be "properly" informed. At 5:00 P.M. EST, Major Marcel was summoned to board the plane at Roswell Field, which was previously scheduled to depart at 4:00 P.M. EST. Evidently there was great commotion and distress involved in making a decision on the situation, and plans were changed at the last moment. Marcel was then on his way to Fort Worth instead of the intended destination of Wright Field. Then, sometime around 6 P.M. EST or shortly after, Vandenberg rushed to the AAF Press Room with realization that serious miscommunication had taken place in the press office. It was chaos.

Managing the Roswell Secret

In regard to Vandenberg, there can be no doubt that the series of events occurring over the last few days gave him cause for losing sleep. The extraterrestrial reality must have had an effect on him, and also on many others. Along with other pressing military concerns, Vandenberg now had additional tasks of monitoring recovery of crash site debris, and insuring its proper handling and processing. There was also a tremendous secret to control. It was not just a matter of keeping a major secret from the public, but it was also a matter of managing information flow, and limiting the need-to-know to just a few select people in government. But he was not the only one with this responsibility. There were others in high places, above and alongside him, who would also be involved, including President Truman.

On Wednesday, 9 July, Vandenberg met with Symington and General James "Jimmy" Doolittle. It is noted that Doolittle was a primary confidant of President Truman, and was previously involved with investigating "Foo Fighter" balls of light phenomenon (observed by WWII air combatants) and also "UFO Ghost Rocket" reports from Sweden in August 1946. At the time, there was a general consensus within military and intelligence agencies that Ghost Rockets were definitely of Russian/German origin. Reportedly, Doolittle was convinced that the Swedish incidents were of extraterrestrial origin, but this topic has been considerably muddied. Years later, CIA agent Frederick Durant stated that Ghost Rockets were not of Russian or German origin, although one can reasonably assume he was involved with CIA disinformation.

After meeting with Symington and Doolittle, Vandenberg met with Army Chief of Staff General Eisenhower and General Lauris Norstad. Norstad was Director of Plans and Operations, and charged with supervision of psychological warfare activities for the War Department. Subsequently, Vandenberg called and talked with President Truman, which was just after Truman met with Senator Carl Hatch of New Mexico. After that, Vandenberg met again privately with Symington, then with the Joint Chiefs of Staff, and then with Symington once again.

As a result of Ramey's deception in the final press release, Vandenberg and others were encouraged by the cooperation of a docile press and a trusting public, which had played into their hands. Their panic of the previous day had subsided, but there was still a need to tie up loose ends, complete the recovery process, and decide on how to manage scattered evidence from several crash sites. This would take time to resolve.

On Thursday, 10 July, Vandenberg met with General LeMay, Major General Leslie Groves, and General Robert Montague. Both Groves and Montague were from Sandia Base, which was located near Kirtland Field at the southeast edge of Albuquerque, New Mexico. General Groves was the head of AAF Special Weapons Project at Sandia Base, and he worked with Robert Oppenheimer of Los Alamos to facilitate the update and production of atomic weapons. General Montague was Commander of Sandia Atomic National Laboratories at Sandia Base, and it's speculated that this ultra-secret facility received a very sensitive recovery from Roswell. After this meeting, Vandenberg and Doolittle met with President Truman. Presumably, Truman wanted to hear in person from Vandenberg regarding current status of recovery at the crash sites, disposition of extraterrestrial entities, and security plans moving forward.

In-the-Know

At this time, one can consider who may have been privileged to know about Roswell. There is no doubt that many people in the town of Roswell, and those stationed at the air field, knew of and participated in the incident, which was borne out by later admissions and affidavits. Names and details can be easily obtained, although they and their stories will not be covered here. Those already mentioned are rancher Mac Brazel, Sheriff George Wilcox, Walt Whitmore, Major Jesse Marcel, Colonel William Blanchard, Captain Sheridan Cavitt, Lieutenant Walter Haut, Colonel Payne Jennings, and Colonel Robert Barrowclough. All were connected to the town or air field at Roswell.

People connected to Fort Worth Field included Colonel Alan Clark, Colonel Thomas Dubose, and General Roger Ramey.

People in Washington, D.C. included General Laurence Craigie, General Curtis LeMay, General Clements McMullen, General George Kenney, General Jacob Devers, Admiral DeWitt Ramsey, General Emmett O'Donnell, General James Doolittle, General Leslie Groves, General Robert Montague, Dr. Edward Bowles, General Dwight Eisenhower, General Lauris Norstad, Colonel Robert Landry, Navy Assistant Secretary John Kenney, General Carl Spaatz, General Hoyt Vandenberg, Secretary Stuart Symington, and President Harry Truman.

There are also others previously mentioned who were likely in-the-know about Roswell, either because of close connections, or circumstances of association.

They include General Benjamin Chidlaw, General Nathan Twining, Lieutenant Ben Games, Dr. Lloyd Berkner, Ralph Clark, Dr. Howard Robertson, Dr. Karl Compton, Senator Dennis Chavez, Secretary of the Navy James Forrestal, Assistant Secretary of War Howard Peterson, Undersecretary of War Robert Patterson, and Director of Intelligence Roscoe Hillenkoetter. These names are not necessarily complete, because a great many others must have also been involved to one extent or another.

Defense Secretary James Forrestal

One of those mentioned above, James Forrestal, was secretary of the Navy in the closing year of World War II, and he was tasked with supervising demobilization efforts that followed. Initially, he earned a good reputation in overseeing that process, but this ultimately deteriorated into major conflicts with angry citizens, and with congressmen protecting vested interests in their states. After appointment to Secretary of Defense by Truman, Forrestal received additional displeasure because of lack of needed respect, or support, from the military service secretaries whom he had no authority over. Air Force Secretary Symington seemed to have most of Truman's ear. On top of this, Forrestal was very concerned about a brewing Soviet menace taking place, and he was heavily conflicted with what he saw as a serious loss of funding to keep the nation's defenses intact. He was constantly fighting a budget war in an attempt to maintain an adequate defense posture, while also dealing with inter-service rivalry. With such a huge amount of responsibility in many areas, he worked exceedingly hard, and he spent many long hours working each day on his problems.

It was a towering burden for Forrestal in dealing with problems of the world, the politics of office, safeguarding the UFO question, and carrying defense of the nation on his back. After about a year and a half of this, in March 1949, Forrestal was forced out by Truman. Upon reelection to a new term the previous November, Truman wanted his friend Louis Johnson to take over Forrestal's position. After many years of public service, Forrestal was suddenly left jobless at the end of March, but the mountain of problems he felt responsible for remained with him. Then, within a couple days, Forrestal quickly descended into a serious mental breakdown, and he was admitted to Bethesda Naval Hospital in early April.

Believing that Forrestal was on the road to recovery, his brother called the hospital on 21 May 1949 to inform its staff that he would be taking Forrestal out of the hospital the next day. He wanted him ready for discharge. His brother, was very concerned with Forrestal's safety, and also suspected that Forrestal was being held against his will. His objective was to take Forrestal to a countryside location to continue with recovery. Also at the time, Forrestal's wife and son were in France looking for an out-of-the-way place to reside. It would be a secluded place where Forrestal could quietly gather his senses. On 22 May, his brother arrived too late at the hospital. Forrestal had been found early that morning on a third-floor roof below a sixteenth-floor open kitchen window, which was across the hall from his hospital room.

A murky conspiracy is considered by many in Forrestal's death. A government inquiry board hastily and simply stated that the cause of death was suicide, and this was announced without a formal autopsy, investigation, or inquest. Forrestal was a major person in-the-know about Roswell, he favored release of the Roswell secret to the public, and he confided to friends that he was constantly under surveillance. Could it be that he was one of the first casualties involved with secrecy of the UFO question? There is reason to give this additional thought, which will be discussed in Chapter 17.

General Nathan Twining

There is much to consider about General Twining's involvement with Roswell. He was "point man" as Commander of AMC, and he maintained control over the intelligence areas of T-2 and T-3 at Wright Field. He was involved with delivery, receipt, and storage of the Roswell disk, associated debris, and unearthly bodies at Wright Field, which was the original intended destination for the Roswell disk mentioned in AP bulletins at the time.

Twining must have received word of multiple crash sites from either Vandenberg or McMullen, and he immediately made his way to areas in New Mexico where the objects crashed. He was intimately involved with directing and managing logistics of recovery, but it was without attracting the kind of attention that Blanchard or Ramey received. By default, he was the primary coordinating player in the flying object recovery effort. Most of the evidence would come under his control as caretaker of crash debris and body remains. This would be his responsibility for the foreseeable future, at least until a more permanent and secure site could later be established. Because of the ultra-high security attached to this, he also had responsibility for maintaining secure conduits of communication so that specific higher levels could keep informed without involving associates or certain people above and below him.

It is suspected, however, that this was not a role particularly suited to him, especially since some of his people (who were not in-the-know) were directed to investigate the UFO question. This would threaten security of the secret, and it is quite likely that this untenable situation was reason for him to initially campaign for release of the secret at highest levels. Further evidence for this will be forthcoming in following chapters.

President Harry Truman

From what can be determined, there is no record of President Truman's activities in regard to Roswell, and no specific information in the Truman Library. Researchers point out that anything generated in written form would have been considered highly classified and not likely found in memoirs. Most communication was in person between Symington, Forrestal, and generals who were involved.

Researchers, however, have found anecdotal comments involving Truman. His good friend General Doolittle, after his investigation of "Foo Fighter" and "Ghost Rockets," once told Truman that the objects were "most likely of extraterrestrial origin." Air Force Colonel Robert Landry, an aid to Truman for four and a half years, was instructed to give regular oral reports to Truman on flying object incidents considered to have "strategic threatening implications." Rogene Ramey, the wife of General Ramey, stated that Truman and his wife visited socially at their home on several occasions in later years at Fort Worth, Texas. She reported that her husband was very upset and embarrassed about lying to the public, and using the weather balloon excuse (in regard to the Roswell incident). In consideration of this, one might speculate that the lie and deception regarding Roswell was protected and honored at the highest level of government. General Ramey was undoubtedly nervous about his career and reputation if his lie were uncovered, and he and Truman may have struck a bond together.

In Truman's public statements, he always made it quite clear that he had no interest in flying objects, but research demonstrates that the UFO question was a very serious matter to him, and he took a very decisive role in it until his presidency ended.

✨

On 8 July 1947, after General Ramey took away public excitement regarding the captured disk at Roswell, there was continued concern and deliberation in Washington, D.C., but it soon simmered down with knowledge that a docile press and unconcerned public was under control.

A tough road loomed ahead, however, particularly for those controlling and maintaining the secret, and especially with the government about to undergo organizational change for an extended period, which will be outlined shortly. The secret of Roswell would need continuous protection, but how would this be coordinated with additional flying object sightings reported by credible people? Also, there were many in the general military who were not in-the-know, and seriously concerned about the intrusive flying objects making periodic appearances. How would those witnesses be handled, and what would be done about the mysterious flying objects that would continue to make an appearance?

CHAPTER TWELVE

Knocking on Truth's Door

July - August 1947

National Security Reorganization

In the beginning of the post-WWII era, there was a sense that the United States was top dog among nations, and positioned as the greatest power in the world. Great things could be accomplished with its scientific, industrial, economic, civilian, and military might. This perception, however, was quickly replaced by apprehension of the Soviet threat. In touting their prowess and contribution in defeating Hitler and the Germans, Soviet leadership figuratively flexed their muscles, which posed a challenge to the United States. They never outwardly recognized that the United States supplied them with a great quantity of strategic material and supplies for the war against Germany, and also helped divert German forces to the Western front, which enabled the Soviets to share in a WWII victory.

Joseph Stalin consolidated his leadership after the death of Vladimir Lenin in 1924, and he became a brutal dictator embracing communism. For Soviet leadership, communism provided a subservient society, and the concept of democracy and capitalism threatened control over their people. Their vision was a world dominated by communism, and this was made abundantly clear when Stalin announced in February 1946 that communism and capitalism could not coexist. Their expansionist ideas became very evident when they forcibly implemented their leftist ideas on Eastern Europe, and forcibly incorporated those countries into their vision and sphere of influence.

Soviet belligerence, and a menacing attitude toward the West, became a trade mark of its totalitarian regime, and the United States took notice for good reason.

Freedom loving people everywhere feared that the Soviets might develop the atomic bomb, with capability for delivering it. The thought of it became a nightmare.

The Cold War began to brew, and this served to emphasize a need for the United States to overhaul and strengthen foreign policy coordination, with better and more effective intelligence gathering, increased communication within governmental agencies, and greater civilian control over military services.

This was the reason for implementation of the National Security Act of 1947 that was signed by President Truman on 26 July. Within a few months, it would ultimately result in creation of the National Security Council (NSC), with a new Central Intelligence Agency (CIA) reporting to it. The War Department, consisting of AAF and Navy components, would become the National Military Establishment, with the Air arm of the Army becoming a separate and equal component with the Army and Navy. Each of the three services would interface with a new Department of Defense, and each would have representation in the NSC with the right to appeal directly to the president. The Department of Defense would not have direct authority over the military services, but would exercise "general direction and control" and serve as a coordinating organization for budgets and other matters common to the services. This would later be recognized as very difficult thing to manage.

This modernization and reorganization effort would become a huge endeavor, and it would take time to develop before becoming reality. It would require continued refinement, but it was hoped that government complexity would be reduced, and redundant and autonomous areas minimized, especially within intelligence organizations. A smoother running military was anticipated, although the Army and new Air Force would need to redefine its separate roles. Thrown in with this was the Roswell incident, occurring earlier in the month, and it was just another situation that high government officials needed to contend with.

AAF's Counter Intelligence Corp (CIC), which was active in flying object investigations, would also be involved in this transformation. Captain Sheridan Cavitt, who initially investigated the Roswell incident with Major Jesse Marcell, would become a part of this. Under the War Department, the CIC was a separate division, but it would migrate into the new Department of Defense. This meant that certain CIC units attached to the Air arm of AAF would, in time, make a transition and come under direct control of the Secretary of the Air Force, and eventually become the Air Force Office of Special Investigations (AFOSI).

Air Force Office of Intelligence (AFOIN)

Because the AAF was scheduled to split apart, it meant that there would be efforts to reorganize and streamline operations to some extent. Some divisions, branches, and offices were operated under independent authority, policy making, and directional planning, and there was disorganization and also redundancy. Such was the case in divisions under the Air arm of AAF Intelligence (AFOIN-A2), which provided specific intelligence expertise. It was somewhat analogous to the Army-G2 intelligence counterpart.

Within AFOIN, as in any military hierarchy, an institutional protocol of authority and responsibility existed, but this was not always managed properly. Authority was delegated in order to accomplish tasks, but responsibility would normally remain with those who delegated and ultimately coordinated policy. Those who delegated without insuring completion of goals, however, were not immune to failures in that regard, and this was reflected to some extent in operations between AFOIN's two divisions, especially in regard to the UFO question after separation of the AAF into Army and Air Force components.

Prior to Roswell, significant sighting reports of flying objects were collected regularly by AFOIN, which was under leadership of Assistant Chief of Air Staff Major General George McDonald. He, like Vandenberg, reported to Chief of Staff General Carl Spaatz. Reporting to McDonald was his executive officer and assistant, Brigadier General George Schulgen, who was head of the Office of Intelligence Requirements (AFOIR), which was a division within AFOIN. Another significant division was the Office of Air Intelligence (AFOAI), commanded by Colonel James Olive, which was primarily concerned with national security, air superiority, and defense.

Office of Intelligence Requirements (AFOIR)

It was General Schulgen who was most involved with the UFO question. His AFOIR division was the clearinghouse for analysis of sighting reports, and he was under constant pressure from those above in the Pentagon to find out what the flying objects were all about. Sightings were increasing in number, and becoming a concern to pilots and other military personal, but Schulgen and the vast majority of those within AFOIN, and specifically AFOIR, were not privy to what happened at Roswell.

Of the five branches within AFOIR, one was the Collections Branch (AFOIR-CO) headed by Colonel Robert Taylor, who was assisted in the office by Colonel Frank Dunn. Reporting to Taylor was Lieutenant Colonel George Garrett, whose job was to review, study, and analyze sighting reports. For many months before Roswell, Garrett was reviewing reports from AAF sources and others. The Roswell incident, however, never came across his desk, or reports of the incident were discounted because of "exposure" by General Ramey. He may have been aware of Roswell from a few brief news reports, but with the many reports coming in, only the more significant showing merit would be reviewed by him.

Other significant sightings also occurred during the time of Roswell. A sighting occurred on 7 and 8 July near Muroc Field (now Edwards AFB), and also on 9 to 11 July near Harmon Field in a remote area of Newfoundland. These particular reports received much attention, although additional significant sightings were reported at several other locations as well. All the sightings involved very credible people, but many of the sightings were not immediately known to Garrett, and it took days and sometimes weeks for the reports to come across his desk. Such was the speed of routine information transfer between various commands to AAF Headquarters.

On 9 July, after significant sighting incidents caught Schulgen's attention, he attempted to obtain further help by reaching out to Special Agent S. Wesley Reynolds, who was a liaison contact with the Federal Bureau of Investigation (FBI). Schulgen wanted assurance that reported sightings originated from credible individuals, and not from Soviet agents or someone with communist sympathies. He also wanted assurance that the sightings were authentic. His primary reason for reaching out to the FBI was because of awareness that psychological warfare was on the Soviet agenda, and this needed to be looked into. His request for assistance was answered on 30 July with a policy statement from the FBI stating that it "agreed to cooperate in the investigation of flying discs." That agreement, however, was short-lived when the FBI later intercepted an internal letter from Air Defense Command (ADC) Headquarters dated 3 September, which stated:

> The services of the FBI were enlisted in order to relieve the numbered Air Forces of the task of tracking down all the many instances which turned out to be ash can covers, toilet seats, and whatnot.

On 24 September, FBI Director J. Edgar Hoover, put an end to FBI involvement with a curt letter to AFOIN chief General McDonald that stated:

> I cannot permit the personnel and time of this organization to be dissipated in this manner.

From that point on, the FBI would remain in the background on the UFO question, but it would still maintain very serious interest in the matter.

Technical Intelligence (T-2)

In his investigation of incidents, Colonel Garrett would often obtain assistance from Technical Intelligence (T-2) at Wright Field. This organization was involved in analysis of foreign technology, and located within the "Patterson" area of the base, which was before the base was renamed Wright-Patterson in January 1948. Garrett dealt specifically with an engineer named Alfred Loedding, who was part of a unique group within a Special Projects branch of T-2, but he was not in-the-know on Roswell. Loedding was on loan from T-3, a Technical Research and Development Engineering organization located apart from T-2, and separate from the Patterson area. He was particularly knowledgeable about low aspect ratio aircraft research and design, and he was assigned to interface with Garrett as a liaison to AFOIR at the Pentagon. Becoming deeply involved in technical analysis of reported flying object sightings, Loedding was able to provide periodic and valuable help to Garrett in resolving many significant reports received at AFOIR.

Headed by Colonel Howard "Mack" McCoy, T-2 was composed of many branches with hundreds of people, and this was the place where Roswell material was delivered and stashed away. There can be no doubt this material was under strict lock and key, supposedly in Hangar 18 (some say Hangar 5), or the Building 18 complex. It was probably protected by a small, select, top secret unit within T-2, and composed of people in-the-know with ultra-high security clearance.

One can speculate about the number of people at T-2 or T-3 who may have known about Roswell. Many must have heard news announcements during the few hours on 8 July, which stated that remnants of a saucer were being shipped to Wright Field. Rumor and speculation must have been rampant on base, especially in T-2. Everyone involved with receipt and handling of recovered Roswell material was sworn to strict secrecy, but it's not specifically known who was in-the-know at Wright Field, except for General Chidlaw and General Twining. If any were involved with receipt and handling of Roswell material, and also assigned to undertake flying object investigation under Loedding, which was not likely, it would present a very difficult situation for them. Their honest, accurate, and professional analysis of flying object sighting reports would potentially confirm Truth, which would conflict with the requirement to maintain the secret of Roswell.

When one reviews Colonel McCoy's interface to his team of "unknowing" scientists and engineers in the investigation of the UFO question, and also the way he managed and backed his Special Projects branch, it's easy to consider it doubtful that McCoy ever knew of Roswell. One can debate this, but it's highly likely that McCoy was very aware of Roswell, and that he had access to recovered material. After all, he was Commander of T-2 and he must have been in-the-know, just like his boss General Twining. Also, a January 1947 JRDB roster lists him as an Army staff member of the JRDB Planning Division. He was one of eight staff reporting to its chief, Dr. David Langmuir. Both Langmuir, and Ralph Clark of the Programs Division, reported to Lloyd Berkner, Secretary and Chairman of the JRDB Executive Council. All these people can be considered in-the-know on Roswell, and it was a very convenient situation for the JRDB to have Colonel McCoy on its staff.

McCoy's participation with Twining in receipt and storage of Roswell material must be considered very likely, including the fact they both desired that Truth be revealed. If McCoy and a few in the Special Projects branch knew the Truth, they were in the untenable position of investigating Truth while maintaining the secret. This was not an enviable position, and it gave plenty of reason for them to desire that Truth be revealed and made known publicly.

It will be recalled that Twining previously made a sudden, unplanned, and unannounced trip to New Mexico in the middle of heavy action at Roswell, and also elsewhere in the general area, and he was there for at least four days, and probably many more. Ostensibly, he went for a pre-scheduled bombing course he was to take, but he also wrote a letter to the Boeing Aircraft Company in Seattle, apologizing for cancelling a previously arranged 16 July visit because of "a very important and sudden matter that developed." He was at Alamogordo Field (now Holloman AFB) on 7 July, and then at Kirtland Field on 8 July where some important and very sensitive crash material was taken. His deputy, General Benjamin Chidlaw, was designated to eventually receive material at Wright Field, where both Twining and Chidlaw managed hundreds of people at AMC. The sudden and important development that interrupted Twining's normal schedule may have continued for quite some time, and McCoy must have been heavily associated with that at Wright Field.

Senator Barry Goldwater

As an aside to this in regard to where Roswell material was stashed, Barry Goldwater, a five-term United States Senator from Arizona for thirty years (1953 - 1965, 1969 - 1987), made an interesting revelation on 1 October 1994 in a CNN TV interview where he stated the following:

> *I think that at Wright-Patterson Field if you could get into certain places you would find out what the Air Force and the government knows about UFOs. I called Curtis Lemay and I said, "General, ah, I know we have a room at Wright-Patterson where you put all this secret stuff. Could I go in there?" I've never heard him get mad, but he got madder than hell at me, cussed me out, and said, "Don't ever ask me that question again!"*

It was strange that Goldwater was cussed out for his simple question instead of being told there was no "room" with "secret stuff." The subject referred to by Goldwater was clear to LeMay, and he responded with a curt decisive response to Goldwater, which was an implicit admission that Wright-Patterson did contain the "secret stuff" that Goldwater was referring to. One would think that Goldwater might have had enough authority and clearance to ask the question of LeMay, whom he referred to as his "good friend." Other than being a senator, Goldwater was Chairman of the Senate Intelligence Committee, a 1964 presidential contender with Lyndon Johnson, and a retired command pilot in the Air Force holding the rank of major general.

Brigadier General Arthur Exon

Another person worth mentioning in this regard is General Arthur Exon, who was a lieutenant colonel when stationed at Wright Field during the Roswell incident. He was also stationed at the Pentagon as a full colonel from 1955 to 1960, in the position of Deputy for Procurement and Production, and also as Deputy Chief of Staff for Material. He was also Base Commander at Wright-Patterson AFB from August 1964 to January 1966, and was promoted to brigadier general while there. Although not directly involved with the Roswell incident, he went on record stating he knew of the Roswell crash when stationed at Wright Field, and was privy to conversations and scuttlebutt during the times he was stationed at both the Pentagon and Wright-Patterson. Exon was aware of crash debris arriving at Wright Field, and he knew of those who investigated and tested debris. He flew over the Roswell crash sites a few months after the incident, and observed them from the air. They were visually identifiable because of land tracked up around the sites from obvious and extensive activity. Exon knew that bodies had been removed from the crash site, that General Ramey fabricated a cover-up story, and that Colonel Blanchard took leave of absence to supervise total removal of crash site debris. Other than Brigadier General Thomas Dubose, Exon is another general to confirm the Roswell incident.

Suspicion Regarding "Topside"

About the middle of July, or a bit later, Lieutenant Colonel Garrett of AFOIR was deeply involved in investigation and review of sighting incidents, and also coordinating with Loedding at T-2. It was quickly realized by both of them at this time that pressure from above to produce answers on the flying objects suddenly disappeared. At the beginning of July, high level officials were exerting tremendous pressure upon General Schulgen, and the Collections Branch, to investigate the many significant sightings made by credible observers. This pressure, however, evaporated quickly with a sudden lack of interest by "topside," the term given to those at higher levels. This was apparent because many significant reports were still coming in, but the reports were without any requests from above for answers.

Garrett and Schulgen, along with FBI agent Reynolds, and also Loedding, began to suspect that lack of interest was because topside now knew the reason for the mysterious objects. They suspected that the United States, or possibly the Soviets, might be involved with secret programs or projects that they were not yet aware of. But, they could not imagine the Soviets achieving such advanced performance demonstrated by the objects, or of Soviets taking the risk in having flying objects cavort over United States territory. With no information on Roswell, or any reason to know or understand that the incident at Roswell involved recovery of a crashed object, there was no way to realize that the sudden attention by topside toward Roswell was the reason for cessation of pressure from above.

Officials at high levels, and in-the-know on Roswell, became "star struck" with their new found discovery. At the same time, they were frightened and apprehensive about their finding, and they were determined that the secret must remain hidden. For them, reason to apply further pressure on AFOIR became unnecessary. In the midst of a desperate and continuing cover-up, they failed to consider that serious investigative activity, and concern, remained in ranks below. The mysterious flying objects were still making appearances, and frantic attempts to discover what they were still remained a priority for those with continuing responsibility for investigation.

There was a reversal of interest. High officials were now attempting to suppress Truth, while those in lower ranks were increasing their concern, and noting significant sightings reported by credible people. It was another conundrum that higher levels needed to deal with. Confusion reigned at top levels, where many questions needed to be deftly handled and finessed. They were urging the media to suppress everything regarding flying objects, while others at lower levels of the military were in hot pursuit of the UFO question, which was due to incidents and incredible sightings reported by reliable and credible people.

Harmon Field Incident

The Muroc Field and Harmon Field incidents were most disturbing. When the Harmon Field incident finally came to Schulgen's attention, about the middle of July, he contacted Colonel McCoy at T-2 and asked him to immediately send investigators to Newfoundland, and then return directly to AAF headquarters with an assessment report. AFOIR was certain that officials at top levels knew what was going on, and they wanted to quickly get to the bottom of it all and not waste further time. Without specific confirmation from above, they would need to press on with serious investigation to obtain answers.

McCoy sent a team that included his top assistant, Chief of Intelligence Analysis Colonel William Clingerman. Upon receiving Clingerman's report, and after learning more of the Muroc incident, those in AFOIR and the Collections Branch became certain that something was going on, which was "very real and not fictitious." Stronger measures were now needed to obtain resolution of the matter.

From significant sighting reports received, including those of the Muroc and Harmon Fields sightings, Garrett conducted a comprehensive review of about eighteen of the incidents occurring from 17 May to 12 July. In that review, he was able to identify certain common characteristics about the objects that gave him great cause for concern. He and others involved in the review concluded that the objects were absolutely real, and he surmised that the objects were either of United States or Soviet origin. There was also a third option they were leaning toward, but to state that the objects might be from another world was best left as a final option. The notion that these particular sightings were figments of the imagination, or misidentification of known or unknown natural sources, was discarded. Of course, the vast majority of other sightings would fall into that category, especially where data was totally deficient, or sources not credible

Although AFOIR's investigation was purposely directed toward possible United States or Soviet involvement, it was difficult to believe that observed behavior of the objects could be produced by Soviet technology, let alone American technology. The objects were too technically advanced in their observed maneuvers. But the nature of the incidents sparked very serious concern, and intense study would continue.

Preliminary Estimate - Objects Are Real

On 30 July, Garrett produced a secret Preliminary Intelligence Estimate (IE) titled "Flying Discs," which provided a review of the significant recent sightings studied. Conclusions of the Estimate stated:

> *From detailed study of reports selected for their impression of veracity and reliability, several conclusions have been formed:*
> *(a) This "flying saucer" situation is not all imaginary or seeing too much in some natural phenomenon. Something is really flying around.*
> *(b) Lack of topside inquiries, when compared to the prompt and demanding inquiries that have originated topside upon former events, gives more than ordinary weight to the possibility that this is a domestic project, about which the President, etc. know.*
> *(c) Whatever the objects are, this much can be said of their physical appearance:*

1. The surface of these objects is metallic, indicating a metallic skin, at least.

2. When a trail is observed, it is lightly colored ...

3. As to shape, all observations state that the object is circular or at least elliptical, flat on the bottom and slightly domed on top. The size estimates place it somewhat near the size of a C-54 or a Constellation.

4. Some reports describe two tabs, located at the rear and symmetrical about the axis of flight motion.

5. Flights have been reported; from three to nine of them, flying in formation on each other, with speeds always above 300 knots.

6. The disks oscillate laterally while flying along, which could be snaking.

It is truly significant that, in the same month as the Roswell incident, AFOIR would come out with an IE stating that flying objects are real. This comprehensive study of a number of specific incidents provided a conclusion that complemented the Truth of Roswell.

On the same date, Garrett also formulated a letter under General Schulgen's signature, which was sent to General Nathan Twining, requesting assistance on what was now considered a very serious matter. Included in the letter was Garrett's IE. It was hoped that there would be some indication from Twining that further investigation on flying objects might not be necessary, especially if the sightings involved a top secret U.S. project. The thinking was that Twining would probably have knowledge of such a project, since he was in charge of AMC and the Intelligence Divisions at Wright Field, which specialized in examining and back-engineering captured or recovered aeronautical technology. With Garrett working closely with Loedding at T-2, it was understood that T-2 was the place where recovered flying objects would likely end up. Little did Garret, Taylor, and Schulgen realize that Twining had everything to do with flying objects, and that much of the Roswell wreckage was under his care. But Twining was not about to reveal the secret of Roswell, even if he wanted to.

T-2 Becomes TID

Upon receiving the memo from Schulgen, Twining's response to Schulgen would be delayed to allow continued study by his organization. Twining began to reorganize T-2, and in August the organization was renamed the Intelligence Department of AMC, or Technical Intelligence Division (TID). This was a large organization with many branches involved in assessment of foreign technology, and foreign capabilities regarding air and space armaments. The few people comprising the Special Projects branch, and involved with studying the UFO question, were only a very small part of the large organization.

As July turned into August, both of Garrett's bosses (Schulgen and Taylor) sent inquiries to contacts in other military services involved with research and development. They were probing and asking if anyone was aware of a secret project that might fit the mold and pattern observed with flying objects. Their efforts, however, were not productive, and they received only negative replies.

An FBI memorandum of 19 August, titled "Flying Disks," revealed information on current thinking within AFOIR, and it stated:

> *Lieutenant Colonel Garrett of Air Forces Intelligence, expressed the possibility that flying discs were, in fact, a very highly classified experiment of the Army or Navy...Colonel Garrett indicated confidentially that Mr. C. Carroll, who is a scientist attached to the Air Forces Intelligence, was of the same opinion...*

> *Colonel Garrett stated that...we have reported sightings of unknown objects over the United States, and the "high brass" appeared to be totally unconcerned. He indicated this led him to believe that they knew enough about these objects to express no concern. Colonel Garrett pointed out further that the objects in question have been seen by many individuals who are what he termed "trained observers," such as airline pilots...He stated the above has led him to come to the conclusion that there were objects seen which somebody in the government knows all about...*

> *Mr Reynolds subsequently discussed this matter with Colonel L. R. Forney of the Intelligence Division of the War Department. Colonel Forney indicated to Mr. Reynolds that he has the assurance of General Chamberlin and General Todd that the Army is conducting no experimentations with anything which could possibly be mistaken for a flying disc...*

> *Colonel Garrett indicated to Mr. Reynolds that he had pointed out his beliefs to General Schulgen and had mentioned the possibility of an embarrassing situation arising between the Air Forces Intelligence and the FBI. General Schulgen agreed with Colonel Garrett that*

a memorandum would be prepared for the signature of General McDonald, A2, to General LeMay, who is in charge of Research and Development in the Air Corps. Colonel Garrett indicated that this memorandum will set forth the characteristics of the objects seen by various reliable individuals. The memorandum will then request General LeMay to indicate whether or not any experiments are being undertaken by the Air Forces which could possibly be connected with any of the observed phenomena.

The FBI wanted assurance that it was not wasting resources and money by chasing down a secret government-sponsored program that would end up embarrassing both the FBI and AAF Intelligence. And this was before FBI Director J. Edgar Hoover abruptly canceled his cooperation with AAF Intelligence on 24 September.

Garrett subsequently wrote a letter to General LeMay, dated 22 August, under the signature of Colonel Taylor, and with approval of General Schulgen. LeMay, of course, was also in-the-know on Roswell. It will be recalled that he directed his deputy, General Laurence Craigie, to survey the crash scene at Roswell. Garrett's letter stated:

1. From a detailed study of certain reported observations on the flying saucers, selected for their veracity and reliability, it is apparent that several aspects of their appearance have a common pattern.
2. Before pursuing its investigation of these objects any further, this office requests assurance that no research project of the Army Air Force, at present being test-flown, has the following characteristics and that it may therefore be assumed that recent flying saucer "mystery" is not of United States origin ...

A week later, a resounding "No" was received from LeMay who replied:

A complete survey of research activities discloses that the Army Air Forces has no project with the characteristics similar to those which have been associated with the Flying Discs.

The result of AFOIR's efforts to find out if the United States was somehow involved with flying objects was an announcement on 5 September by General Schulgen that provided a repeat of the response from LeMay. They found through their queries that no one would claim knowledge, or take responsibility for a

United States project associated with flying objects. This was somewhat baffling to AFOIR. Credible sightings were still occurring, but higher officials were still not applying pressure for answers, and they still continued to exhibit no interest at all, which now left two other potential options available. One option was that flying objects were of foreign origin, and the other was that they were extraterrestrial.

ഗ്ര

From a more general perspective, the above really does provide one with indirect confirmation of the happening at Roswell. Prior to Roswell, much concern was generated by Air Force officials at high levels, and also at lower levels, about incredible sightings from credible people. But no concern was demonstrated by high officials after the Roswell incident, although sightings continued with increased concern by others in the military. Those who were knocking on the door of Truth were very concerned, and determined, to find answers to the UFO question.

One might surmise that a vast majority within Air Force Intelligence didn't have a clue about Roswell, and were left out of the loop, but one might equally surmise that many people in the intelligence community, especially in AFOIR, were very concerned about the UFO question, and were interpreting the situation as a very "real" problem. To them, Soviet involvement was the next most likely answer, although a very strong suspicion lingered about the other possibility. They were very concerned about specific incidents they investigated, and they believed the objects were definitely real, but they needed to have conclusive determination of origin. Strong suspicion that officials at high levels knew the origin of flying objects provided frustration, and it became a quandary they didn't quite know how to handle.

Within the military chain of command today, there are relatively few, if any, who know the whole story. Those who witness an event are quickly told to zip lips. Those who have a "need to know," in their particular job specialization, are privy to very little compared to the whole aspect of it. Any real evidence that becomes available is placed under lock and key. Those who know it all are trapped with their knowledge, and tied to it in such a way they cannot escape.

As will be seen, this aspect of continuing to keep the "carrot" of Truth just out of reach provides frustration to those motivated to go after it, and to those whose job requires them to investigate flying objects. It becomes a balancing act that could tip either way, depending on those who have control. With continued active interest by lower levels in the AAF, could that control by higher officials be threatened as time goes on, and as Truth becomes more difficult to contain?

CHAPTER THIRTEEN

Balancing Truth

August - December 1947

"UFO Question Reviewed at AMC"

In early August 1947, while General Schulgen awaited an answer to his letter, General Twining ordered his staff to undertake a complete review of the UFO question. Participants included Chief of TID/T-2 Colonel Howard McCoy, Chief of T-3 Major General Alden Crawford, Chief of T-3 Research and Development General Franklin Carroll, Chief of T-3 Power-Plant Laboratory Colonel Russell Minty, Chief of T-3 Aircraft Laboratory Colonel C. K. Moore, and Commander of the Air Force Institute of Technology Brigadier General Edgar Sorenson. They were provided information on flying object incidents that were collected by AFOIR, and it included Garrett's Intelligence Estimate (IE).

Other than McCoy, it's possible that none of the above were in-the-know on Roswell. But it's also quite possible that some were. The delivery of Roswell material, and its storage at Wright Field, was a big deal prior to this, especially to those involved, and this activity must have filtered around to others on base who were not directly involved, such as Lieutenant Colonel Exon.

It is worthwhile to consider Twining's motivation and possible reason for educating his staff on investigations made by AFOIR, which received help from Alfred Loedding in TID's Special Projects branch. Twining must have been struck by the IE from Garrett, which then caused him to ponder the letter from Schulgen. Twining realized that the UFO question was now a serious concern to the Directorate of Intelligence, and that Garrett's investigation was on track to discover Truth.

It is noted that most of the staff selected by Twining to participate in the review were people from his T-3 organization, with only McCoy representing TID/T-2. The TID organization had previously received Roswell material, which was isolated in a small unit under tight security, but exactly how that was done, or possibly managed by a few in McCoy's immediate staff, is not known. Even if some of his staff were not in-the-know, they would not participate in Twining's review, as it would have been considered unnecessary. But it was necessary for all others in Twining's organization to come to a consensus about the potential reality of flying objects, and this would help facilitate an answer to General Schulgen.

Securing of Roswell material at AMC created multiple problems for Twining. The material needed to be completely isolated on base and hidden away, including a small specialized unit responsible for its study, analysis, and security. It became problematic how to effectively do this, while also providing necessary resources and personnel. The other side to this is that Twining was also responsible for keeping track of current activity of flying objects, and helping to determine potential impact by the objects on national security.

Twining must have realized that the monitoring of current flying object activity was necessary, which was a task initiated prior to Roswell in AFOIR, and also by McCoy's Special Projects branch. He also knew that this process could not be effectively accomplished by those in TID who were involved with the secret of Roswell. They could not be allowed to investigate their own secret. One could speculate that Twining must have been thinking he needed to get his entire organization together on the notion that flying objects were "real," and then he would eventually be able to bring the UFO question out into the open. His first priority was to get all of his staff in consensus on reality of flying objects, especially those who were not "officially" in-the-know, and then proceed from there. After a review by his staff, Twining would then summon a meeting between them, and render an "opinion," which would then constitute a response to General Schulgen.

Government Changes Take Effect

On 18 September, as a result of the National Security Act, the AAF was divided into separate services. General Spaatz became chief of staff of the Air Force and Stuart Symington became its secretary. Also, Roscoe Hillenkoetter became head of the new CIA on that date, and on 19 September John Kenney became secretary of the Navy. With resignation of both Howard Peterson and Robert Patterson of the War Department, which was now named the Department of Defense, James Forrestal took over as secretary of defense on 17 September. He became top dog in the military arena, but he had little to no control over individual service branches. He, however, was privileged to have control of a mountain of classified information, including the UFO question.

Deliberation on Control Group

At this time, in the middle of September, Vandenberg and other high officials realized it was time to put plans in place to more strictly control the UFO question, and to insure future security of recovered material. Although the JRDB was already involved with this, and the caretakers at TID were currently maintaining physical security, the JRDB was about to be replaced by a new organization. It was now time to determine, plan, and formally establish the requirements for a "controlling organization" that would manage and oversee the UFO question, and the secret of Roswell. It would be necessary to establish leadership policies and procedures, along with administrative functions to handle personnel assignments and other tasks. This was crucial for controlling and protecting the monumental secret of Roswell and other recoveries. A high level group would be required to provide a stable and controlled administrative link to a lower level where recovered items were located. The lower level needed to have a command and administrative infrastructure involved with providing storage, inspection, and diagnostic analysis for recovered evidence. Also required, at some point in time, was a more secure and remote location for facilities. All of this needed to be "compartmentalized," which would provide the greatest possible protection.

President Truman, who returned on 19 September from a twenty-day diplomatic trip to Brazil, must have become informed and actively involved with organizing the "Group." Also, the following people were probably much involved in consultation and collective decision making: Secretary of Defense James Forrestal, Secretary of the Air Force Stuart Symington, Air Force Deputy Chief of Staff General Hoyt Vandenberg, Commander of Air Material Command General Nathan Twining, Commander of Air Force Research and Development General Curtis LeMay, CIA Director Roscoe Hillenkoetter, and Chairman of the JRDB Dr. Vannevar Bush. This list is not complete, and it's likely that certain others were also involved.

It was critically important at this time to insure integrity and security of the UFO question, for however long it would take. Significant personnel changes were about to take place, and there would be no end to controlling the situation.

"Real and Not Visionary or Fictitious"

After sufficient time was given to Twining's staff to review information previously provided to them, Colonel McCoy conducted a meeting with them to render an opinion on the UFO question. On 23 September, he then provided a response to Schulgen in a secret three-page letter titled "AMC Opinion Concerning

Flying Discs." This significant document, signed by Twining, indicated agreement with Garrett's IE assessment. It stated in part that:

> *The phenomenon is something real and not visionary or fictitious...*
> *There are objects probably approximating the shape of a disc, of such*
> *appreciable size as to appear to be as large as man-made aircraft... There*
> *is a possibility that some of the incidents may be caused by natural*
> *phenomena, such as meteors... The reported operating characteristics*
> *such as extreme rates of climb, maneuverability (particularly in roll),*
> *and action which must be considered evasive when sighted or contacted*
> *by friendly aircraft and radar, lend belief to the possibility that some of*
> *the objects are controlled either manually, automatically, or remotely.*

It also stated:

> *...Due consideration must be given the following:—The lack of*
> *physical evidence in the shape of crash recovered exhibits which would*
> *undeniably prove the existence of these subjects—The possibility that*
> *some foreign nation has a form of propulsion possibly nuclear, which is*
> *outside of our domestic knowledge.*

One cannot overlook the significance of this communication from General Twining. It is here that a general of the Air Force, who was in-the-know about Roswell, actually admits reality of flying objects, and that they were not visions or figments of imagination. He described in detail the characteristics associated with the objects, including the fact they were "circular or elliptical in shape, flat on the bottom, and domed on top." And he alluded to the fact that they were intelligently controlled. This vision of the objects suggested an extraterrestrial answer, and it was a risky addition to his letter. He, however, was substantially motivated to provide this vision, which will be further discussed.

Twining also had a balancing act to perform, and he needed to carefully balance the Truth and secret of Roswell with an ongoing investigation of flying object activity. On one hand, he needed to set up and manage an investigation into flying objects to gain information on their activity, intentions, and origin. On the other hand, he needed to ensure that investigation of the objects did not compromise the ultimate secret. His letter confirmed that the objects were quite real, as determined by those under his command who studied several recent flying object incidents. But he also suggested that a foreign/domestic source might be involved, which would need to be resolved.

In his response to Schulgen, Twining requested in his letter that the Air Force:

...issue a directive assigning a priority, security classification and Code Name for a detailed study of this matter...

His letter then closed with the following:

Awaiting a special directive AMC will continue the investigation within its current resources in order to more closely define the nature of the phenomena. Detailed Essential Elements of Information will be formulated immediately for transmittal through channels.

Twining requested that the Air Force set up a special project to study the UFO question in-depth, and that TID be restructured and reorganized with a budget to finance investigation of flying objects. It was within TID that a specialized unit was previously set up by Colonel McCoy, in the Special Projects branch, to look into the UFO question, but the unit was operating without designated funding.

Of special note is that Twining's letter also stated that an Essential Elements of Information (EEI) would be provided and distributed. This was to be a detailed account regarding possible characteristics of construction, and flight capabilities of the flying objects. The premise was that some of the objects were of possible Soviet origin developed from recovered German plans involving the Horton brothers Flying Wing, which may have been produced with assistance of Soviet captured WWII German scientists and engineers. It was necessary, in the investigation of flying objects, to account for possible Soviet involvement, without necessarily affirming the extraterrestrial possibility. The EEI was to be given to various commands around the world, including intelligence offices and agencies, which would aid in searching for and identifying Soviet involvement with the objects.

TID and its Special Projects branch, which worked closely with AFOIN, and especially AFOIR at Air Force Headquarters, was supervised and administered by Twining's AMC. Reorganization and restructuring of Colonel McCoy's TID to investigate flying objects was already in progress prior to Twining's formal letter to Schulgen, but Twining's letter served as a formalized request to gain authorization to activate the investigative unit and secure operating funds. Garrett was already transferring his documentation of incidents and reports to Loedding, and he was continuing to pass down new information after initial review. Loedding was also meeting periodically at the Pentagon with AFOIN, which included General Schulgen, Colonel Taylor, Colonel Garrett, Dr. Charles Carroll, and others.

TID's intention to steer investigation of flying objects in the direction of

Soviet involvement, meant that a more concentrated analysis and critical look at technical aspects of the sighted and reported objects would be taken, which TID was equipped to handle. It also meant that the national security aspect would receive more focus and attention, which would be more in line with AFOAI's mission, while leaving AFOIR less involved to some extent.

Discussions at higher levels regarding setup of the specialized investigative unit must have been rigorous. Efforts by AFOIR to bring the UFO question into the open, along with stating that they were real, served to provide notice of a concerted effort to get to the bottom of the UFO question. Higher levels saw advantages to an investigative unit, and recognized that they also had a continuing need to learn more about the objects. It was also necessary so that the public, and general military, would have a central place to report incidents, which would provide assurance to them that reported sightings or encounters would be processed and investigated. The investigative effort, however, would need to be carefully monitored and controlled so that the protected secret would not spin out of control. There needed to be assurance that intelligence people at AFOIN, and investigators in the Special Projects branch at TID, would not come to an extraterrestrial conclusion.

When one takes a closer look at Twining's letter to Schulgen, it leads one to speculate that Twining must have sought additional critical discussions at higher levels. It must be understood that Twining and McCoy were caught in the middle of a very difficult situation, and it was something they knew would be very difficult to handle. TID's primary mission was to obtain and analyze technical intelligence, and the investigation of flying objects fell into that category, which higher officials seriously depended on. This meant, however, that Twining and McCoy would be involved with protecting the magnificent secret, while others under them in their Special Projects branch would be attempting to investigate the secret. This equated to an unrealistic endeavor, and it provided great motivation for Twining and McCoy to have Truth revealed.

Because of the untenable situation faced by Twining, one is tempted to speculate that he probably lobbied for release of the Roswell secret before sending his letter to Schulgen. Having no success with that, Twining made sure that his letter to Schulgen stressed that flying objects were definitely very "real, and not visionary or fictitious," and this was backed up by his immediate staff who previously issued their "opinion." Disclosure of this "reality," however, may not have set well with some who were in-the-know, even though his letter expressed intent to identify "foreign" origin and involvement. Those in-the-know knew that flying objects were not of Soviet origin, and Twining's letter did nothing to dispel the extraterrestrial idea. This probably caused considerable concern at higher levels, especially with the knowledge that Twining was partial to revealing Truth.

Those few in-the-know at AMC wanted financing, a budget, and an official project for their investigative effort, but it's likely they also wanted higher levels to recognize, and understand, that Truth might be revealed with such an effort. Investigation of the reality, identity, and origin of flying objects would surely lead in that direction. This gave impetus to those who were in-the-know and involved with the investigative effort at TID to have Truth revealed sooner than later. This aspect is an important key toward a deeper understanding of the Air Force's continuing effort to cover up Truth.

Establishment of RDB

On 30 September, Dr. Vannevar Bush was appointed by Truman to take charge of a newly created Research and Development Board (RDB). It directly replaced the JRDB, which was still involved with the Roswell incident and helping to manage its security and debris analysis. The JRDB was originally conceived by the secretaries of War and Navy, and then organized and directed by Bush under military control. The new RDB was designed to have a chairman appointed by the president, with approval by the Senate and military service representatives. Its purpose was to advise the secretary of defense on scientific development and research in regard to national security.

It is interesting that Truman took his time to finally name Bush as RDB chairman, but Secretary of Defense James Forrestal managed to convince Truman to give Bush the position. Forrestal realized it was critically important to keep Bush and members of the previous JRDB intact at this particular time, which was also when Truman was beginning to learn more of the facts regarding Roswell. Truman hesitated naming Bush as head the new RDB because Bush strongly advocated creation of a different organization, which he referred to as the National Science Foundation (NSF). This particular organization, according to Bush's vision, would be completely autonomous, and something akin to the previous JRDB. Bush envisioned that the NSF Board would consist of top scientists and civilian administrators, with an executive director appointed from within the organization. This did not agree with Truman's view, and he was wary of Bush's attempt to gain control of such an organization without it being responsible to the president, or to Congress. There needed to be accountability to the highest levels in government because of critical secrets such as Roswell, which it would be involved with.

The new RDB came under administration of Secretary of Defense Forrestal, and its board members included the following: General Jacob Devers, General Anthony McAuliffe, Vice Admiral Earle Mills, Vice Admiral John Price, General Joseph McNarney, and General Laurence Craigie. Under Vannevar Bush, the RDB would continue to manage and "control" the secret of Roswell.

Reassignment of Twining and McNarney

From the beginning of October and into November, major personnel changes would take place impacting those involved with the UFO question. On 13 October, Twining left AMC at Wright Field for a new assignment as commander in chief of the Alaska Air Command (ALCOM) at Fort Richardson, Alaska. General Joseph McNarney replaced him on 14 October. McNarney was on the Board of the new RDB with General Laurence Craigie, chaired by Vannevar Bush, and it appeared that the UFO question was still in good hands with McNarney because the magnificent secret he was now in charge of was not secret to him. Benjamin Chidlaw, who was now promoted to lieutenant general, became McNarney's deputy commanding general at AMC.

One might wonder why Twining was sent to Alaska, and whether he ruffled feathers of higher-ups in Washington, D.C. This was possible because of his letter to Schulgen stating that flying objects were real, and also the fact that he was in-the-know on Roswell and favored release of the Truth. That would be enough to concern a number of people, especially those in the Control Group. When considering Twining's position in AMC, however, and his high esteem within Air Force Headquarters, it's likely he remained in good standing. In fact, he would later be posted to the military's highest position in the Department of Defense.

Twining's assignment to Alaska was probably due to the fact that the Soviet Union was a serious concern. Someone with Twining's ability was needed to manage buildup of air defenses in the Alaska region, and Twining was known to the Soviets for ending WWII by dropping the atomic bomb on Japan. It must be mentioned, however, that Twining had been in charge of the Twentieth Air Force and the 509th Composite Bomb Group for only four days when the bomb was dropped. Also, the bombing action had been approved by President Truman.

Twining replaced Lieutenant General Howard Craig, who was then sent to the Pentagon as deputy chief of staff for Material. Brigadier General Joseph Atkinson, who was the commanding general of ALCOM, continued to remain in his position to serve under Twining. Due to the fact that Twining was located away from close scrutiny, it's tempting to contemplate that he was occasionally assigned to other temporary duties related to Roswell.

Reassignment of General LeMay

Also in October, General LeMay received orders to command U.S. Air Forces in Europe; this was a major assignment for him. His transfer would take place by the end of the year, which was in response to an increased threat from the Soviet Union. LeMay was known for an overbearing and brash behavior, a demand for perfection in performance, and a preference for overwhelming air bombardment to defeat an aggressor. This was his trademark in WWII air campaigns over Europe and Japan. As it turned out, his new job would put him face-to-face with Soviet forces before the middle of 1948, during the Soviet blockade of Berlin. It was then that Soviet forces suddenly put a stranglehold on the city in an attempt to add it to their dominion. LeMay subsequently led a massive effort to save the beleaguered and isolated city with a huge airlift of supplies.

LeMay's involvement with the UFO question was about to come to an end, and his deputy, General Craigie, was in line to replace him as the new director of Research and Development at Air Force Headquarters. Craigie would report to General Craig of Material Command, who was recently replaced by General Twining at ALCOM.

Reassignment of Colonel Harold Watson

On 17 October, another move involved Colonel Harold Watson. According to his official Air Force records, he was assigned chief of the Strategic Vulnerability Branch of AFOAI, but it appears that his duties may have remained elsewhere on a temporary basis. His official Air Force biography is quite confusing, and erroneously indicates he was assigned in October (some sources indicate 1 October), as chief of Air Technical Intelligence Center, which would become the future name for TID at Wright Field. But this was the current position of Colonel McCoy, who would remain in that position for another year and a half. Watson would not arrive to take charge of that organization until July 1949. This provides fodder for speculation about why Watson's official biography is incorrect. Was it intended to divert attention from what he was actually doing?

Watson's official records also indicate he was previously assigned as "Advisor to Under-Secretary of War" Howard Peterson in June 1947. Peterson was in-the-know on the Roswell incident, and because of later circumstances regarding Watson, one can easily speculate that Watson was involved with investigation and securing of Roswell crash evidence before "arriving" at his Strategic Vulnerability job. Even then, however, he may have continued with a temporary duty assignment in that regard. Watson was in a perfect position, at the opportune time, to independently advise and help with the recovery at Roswell. He previously worked for Twining, and it would seem reasonable that he was with Twining in managing the Roswell recovery effort.

Previously, Watson was instrumental in leading an effort to collect captured German air technology upon defeat of Germany in WWII, and he became very successful, and famous, in that post-war endeavor. He was at the forefront of discovering, locating, and removing sensitive material to secure locations, and shipping it to Wright Field where he was based until August 1946.

Watson's exceptional talent in managing recovery and analysis of foreign technology was significant. He was the primary architect in establishment of T-2 at Wright Field, which subsequently evolved into the National Air and Space Intelligence Center (NASIC). He is recognized as the "progenitor" of NASIC and is memorialized as a pioneer in its Foreign Material Exploitation Squadron, which is denoted by "Watson's Legacy" emblazoned on its insignia. Watson was the foremost expert in the analysis of foreign technology, its recovery, and securing of evidence. And he was undoubtedly the one most capable of managing the recovery at Roswell and getting evidence shipped to Wright Field. Further speculation on this, however, is left for others to resolve and confirm about his role at Roswell. Much more, however, will come to light on Colonel Watson in following chapters, where he becomes a major player in a blatant and overwhelming cover-up of the UFO question.

Soviet Origin Probed for Flying Objects

Although there is some question whether Colonel Watson was active in the Strategic Vulnerability Branch at AFOAI, the Soviet threat was a primary consideration in that organization. One of its responsibilities was to identify potential Soviet targets in case of hostility in the future. With the Cold War in its infancy, it was necessary to formulate strategic nuclear contingency plans. This included matching B-29 bombers stationed at new Strategic Air Command (SAC) bases with specific Soviet targets. These would be neutralized in the event of war. Given the concern about a possible Soviet connection with flying objects, especially with many objects sighted near military installations around New Mexico and other areas, it was necessary to determine whether the flying objects originated in the Soviet Union, and exactly where in that country. Those in-the-know about the Roswell incident, however, were not particularly concerned about possible Soviet involvement with the objects.

On 28 October, Garrett wrote a secret memo under Schulgen's signature titled "Intelligence Requirements on Flying Saucer Type Aircraft." It included a "Draft Collection Memorandum," and an "Intelligence Requirements" document. This memo provided instructions for the copy and distribution of the included documents to Counter Intelligence Commands around the world, especially in Europe. The Collection Memorandum stated the following:

> An alleged "Flying Saucer" type aircraft or object in flight, approximately the shape of a disc has been reported by many observers from widely scattered places...This object has been reported by many competent observers, including USAF rated officers. Sightings have been made

from the ground as well as the air...

Commonly reported features that are very significant and which may aid in the investigation are as follows: Relatively flat bottom with extreme light-reflecting ability...Absence of sound...Extreme maneuverability and apparent ability to almost hover...A plan form approximating that of an oval or disc with a dome shape on the top surface...The ability to quickly disappear by high speed or by complete disintegration... The ability to suddenly appear without warning as if from extremely high altitude...The ability to group together very quickly in a tight formation when more than one aircraft are together...Evasive action ability...

The first sightings in the U.S. were reported around the middle of May. The last reported sighting took place in Toronto, Canada, 14 September. The greatest activity in the U.S. was during the last week of June and the first week of July...

For the purpose of analysis and evaluation of the so-called "flying saucer" phenomenon, the object sighted is being assumed to be a manned aircraft, of Russian origin, and based on the perspective thinking and actual accomplishments of the Germans...

The purpose of the document was to discretely inform intelligence agencies, especially CIC units in Europe, of the situation with the UFO question, and to promote gathering of information on Soviet aircraft created from original plans and experiments conducted by the Germans, specifically the Horton brothers. Particular interest was directed toward the flying wing design that fell into Russian hands, which was possibly modified into a disc-shaped craft. It was desired that investigation be carried out to locate German plans, aircraft specialists, and test pilots who could provide information on such a vehicle.

The Intelligence Requirements document, referenced by and included with the Collection Memorandum, contained specifics on suspected technical engineering aspects regarding control, construction, architectural arrangement, landing gear, and power plant of the discs. With direction from McCoy at TID, and help from Dr. Charles Carroll at Air Force Headquarters, Alfred Loedding authored the document, since he was experienced in studying and modeling disc-type aircraft. Created previously on 6 October, this document was apparently the EEI referenced in Twining's 23 September letter. Both the EEI and the Collections Memorandum were hand-carried and delivered to the European Command by McCoy's former

Special Projects manager, Lieutenant Colonel Malcolm Seashore, who worked under Colonel Clingerman in TID.

It is particularly notable that this release of information to intelligence operatives was a strong attempt to determine possible Soviet involvement with flying objects. It was done in spite of observed and described actions of the objects, which gave them an unearthly quality and put them far beyond known or imagined capability of human technology, or principles of physics. While providing a conventional engineering breakdown in the EEI, it must have seemed a bit ludicrous to describe an unearthly-type object that could hover, quickly appear from high altitude without warning, evade pursuit, and quickly disappear or fade from view. Loedding, Carroll, Garrett, and others, were quite aware of this disparity, but it was necessary to eliminate the Soviet factor by providing this technical document, which could only be stated in specific terms of human capability and technology.

AFOIR and TID were quite aware that high level brass continued to exhibit no concern about encounters and sightings by reliable and competent witnesses. But it was determined, without tangible evidence, that flying objects were quite real, and not of U.S. origin. One might wonder what AFOIR or TID would conclude when finally establishing that the Soviets were not involved with flying objects. No other domestic alternative would be available for their origin, and only one other possibility would remain.

General Cabell Replaces General Schulgen

In November, General Schulgen exits the picture, and General Charles Cabell enters as AFOIR's new chief. Cabell previously arrived at Air Force Headquarters in August, and served as acting deputy to the director of Joint Staff. In that position, one might infer that he heard and knew all. But this was not necessarily the case, because the UFO question was a very high priority secret, and only available for discussion between a few who were in control and in-the-know. In his new position, one of Cabell's requirements was to continue where Schulgen left off in the attempt to get to the bottom of what flying objects were all about.

In early December, General LeMay of Air Force Research and Development was about to depart for his new assignment overseas, and he sent a request to AFOIN Director General McDonald to obtain current status on the analysis of flying objects. Apparently, this was done in reference to Twining's previous written request to authorize and establish a flying object investigative project, and LeMay needed current status to pass on to his successor.

The task of responding back to LeMay with a new Estimate on this was given to Chief of AFOAI Colonel Olive, who then coordinated a status review with his Chief of Offensive Air Lieutenant Colonel Jack Thomas. The completed Estimate corresponded with views of Garrett in AFOIR, and this is quite significant because AFOAI would never again align so closely to the views of AFOIR. AFOAI's ideas regarding the UFO question would later be revised with future changes in AFOAI leadership.

On 18 December, AFOAI responded back to Research and Development with its Estimate, which included signatures by AFOIR's new chief General Cabell, and also his Chief of Collections Colonel Taylor. For Cabell, this was a strong introduction to an important area in which he would soon be fully immersed. He was now quite privy to the progressive and coordinated views of several people in both AFOIR and AFOAI.

Project SIGN Approved

On 22 December, a memorandum by General Cabell to Research and Development agreed with General Twining's previous request that a project be established for detailed study of flying objects. Cabell's concurrence was based on Twining's 23 September letter stating that flying objects were real, which indicated that some objects approximated the shape of a disk, and some displayed characteristics of evasive tactics. The concurrence was required from Cabell because the Directorate of Intelligence and AFOIR were the benefactors of such a project, but the project would be established and run by TID with approval by Research and Development. To Cabell, it was imperative that an official project be approved, which was the desire of AFOIR.

On 30 December, General Craigie, the new director of Research and Development, issued an order establishing Project SIGN, which was the answer to Twining's request for implementation of a special project with funding. This authorization, however, was probably not decided without a great amount of discussion between Generals Craigie, LeMay, Vandenberg, and perhaps others in the Control Group. The order stated that the purpose of the project was to:

> *...collect, collate, evaluate and distribute to interested government agencies and contractors all information concerning sightings and phenomena in the atmosphere which can be construed to be of concern to the national security.*

It is very interesting how this was stated. Rather than specifying that the origin and identity of flying objects needed to be resolved, the primary goal and purpose of the new project was to collect, collate, evaluate, and distribute information about flying objects. It seemed to suggest that there was no desire to analyze and determine origin and identity. The wording is very strange, and it would appear to derive from a collective agreement by a Control Group attempting to protect its secret. It was not in line with what should be the inclination of serious investigators who were unaware of the identity and origin of flying objects, which would be a priority to determine. One would expect that the stated purpose of such a project would be much more explicit in this regard, although higher officials and the Control Group would be reluctant to propose that the objects be identified. The stated purpose was a naïve and unassuming lack of intelligent foresight, which would not serve the general military, or those critically concerned with resolving the sightings and encounters.

❧❧

When one examines the reality of events in 1947, it becomes clear that many people, especially those in the military, were aware that something truly mysterious was happening in skies overhead. Those in the Air Force, with expertise and responsibility for identifying the mystery, were convinced that flying objects were real. There was no evidence of U.S. origin for the objects, but a slim possibility of Soviet origin was still on the table. That left one other remaining possibility. From now on, those who would become involved with Project SIGN would be looking for any "sign" pointing to identity and origin of the objects.

There was a great amount of intense interest generated within the Air Force about flying objects, although the public didn't have a clue about this interest. Those tasked with investigating the objects, were struck by their unearthly nature, as demonstrated by fantastic maneuvers and seemingly intelligent actions. If any TID investigators were in-the-know about Roswell, other than their bosses (McNarney, Chidlaw, and McCoy), they were in for a difficult time with their new project.

In exploring the history of government interaction on the UFO question, it is found that various factions in the Pentagon evolve to take sides at this particular time. This was because Truth became a tightly held, and highly-classified secret, which was maintained by those in-the-know. Some wanted to proclaim that the Soviets must be involved, especially if the United States was not responsible for the objects. Some chose to exhibit overt skepticism, because they refused to consider the extraterrestrial alternative, or they were in-the-know and working to deflect Truth. Others in-the-know believed that Truth should be revealed. Still others were unaware of the Truth, but they knew with absolute certainty that flying objects exhibited unearthly traits not possible with current human technology. As mentioned previously, they were the ones knocking on Truth's door.

Everyone was interested in flying objects. The public was transfixed with sightings, and the military establishment was wary of potential serious threats posed by the objects. Those in-the-know, and in the Control Group, were extremely nervous about the secret they harbored, but they needed more answers. The situation could not be ignored, but there was a delicate balance to maintain. With Project SIGN officially approved to look for Truth, could that delicate balance be maintained without Truth being revealed?

PART FOUR

Project SIGN

֍֎

Within a month following the Roswell incident, in July 1947, independent Air Force investigators, who were not in-the-know about Roswell, determined that flying objects were "real." A couple months later, General Nathan Twining, who was in-the-know about Roswell, stated that flying objects were "real, and not visionary or fictitious." But what exactly were the objects up to? In order to find out, the Air Force set up an investigative study called Project SIGN.

Considering that there was a Control Group maintaining the secret of Roswell, one might wonder why Project SIGN would be allowed. Was it a plan to actually unmask flying objects, or was there some other motivation? Perhaps there was a combination of interests in this regard.

Many in the general military were anxious to get to the bottom of the UFO question, because sightings and encounters by creditable people were raising concern. Unit commanders, and others, wanted answers on how to deal with the objects.

There were also those in-the-know who didn't have all the answers. They wanted to know where flying objects originated, and whether the objects had hostile intent, or posed a threat to national security. The Control Group, however, knew the objects were not of this earth, but it was a scary proposition to release such information to an unprepared public. Also, it was unwise for the public to find out that the Air Force, or military, was powerless to deal with the objects. It would also be unwise for adversaries to know that the U.S. possessed alien spaceships containing ultra high technology.

But there were also those in-the-know who were conflicted with keeping their secret. The gut feeling was that the Roswell incident must be revealed, and the public informed. In a democratic society, it would serve no purpose to withhold such information. Sooner or later, Truth would spread and be revealed with possible repercussions. Those most conflicted, however, would be those in-the-know and masquerading in the investigation of Truth.

To prevent revelation of Truth, one might consider that constraints would be imposed on Project SIGN. However, the fact that flying objects seemed to be illusive, may have suggested to those in control that no actual danger existed in allowing investigators to gather information, and interview witnesses. Such information would be helpful to the Control Group. But if an object crashed or was forcibly put down, it would present problems if quick action was unable to surround the object and protect it from SIGN investigators, or a curious public. It would bring back reminders of the Roswell incident.

From all aspects, it appeared to be a worthwhile endeavor for Project SIGN to take investigative action. It would allow the Control Group to keep tabs on flying objects, and gain information on their activity. And there was little chance that an investigative group would proclaim an extraterrestrial answer without physical evidence in hand. The project would also serve as a point of contact for the general military, or public, for reporting of sightings and encounters.

It is difficult to get into the mind of active Air Force elements during this period of time, or the Control Group from so long ago, but one can examine and speculate based on actions and activities recorded in documents from the past. The following provides a scenario from those documents, which shows very active interest by the general Air Force, by Project SIGN investigators fighting for Truth, and by others who must have been in-the-know and preventing revelation of the magnificent secret.

CHAPTER FOURTEEN

Finding Truth

Project SIGN: January - July 1948

Concerned Military

At the close of 1947, General Craigie authorized Project SIGN to obtain information for "government agencies and contractors" on airborne phenomena that could present a threat to national security. This was a rather strange objective, but motivation for the project was to learn as much as possible about the mysterious flying objects, especially for those who were in-the-know on Roswell. The unstated objective was to provide a "public front" to the UFO question for both the general military and public. General Twining recognized that an investigative project was needed, which was based on Schulgen's communication with him. No doubt General LeMay, General Craigie, and General Vandenberg became aware of this and recognized they also had a need; so the project was born.

Interest for the project came primarily from those in the Air Force who were not in-the-know, but with a concern that UFOs were real and posed a possible threat. Significant numbers of sightings, since the spring of 1947, caused much interest and talk in the public sector, but ranking officers in the military were even more anxious for serious answers. There were many sightings reported near military bases by competent and credible people, and much concern was building that the flying objects might be hostile, even if they were not of Soviet origin. From a military perspective, this was a valid consideration, since the military's role was to protect the country from all existing threats. No potential threat could be ignored, and it was therefore necessary to conduct an intensive investigation of the flying objects being sighted and reported.

Project SIGN Accepted

On 3 January, project SIGN was accepted by TID, and it was internally identified as Project HT-304, but it was not yet officially confirmed as a functioning project. Under direction of General McNarney in AMC, the project was given to Colonel Howard McCoy's TID, and he buried it within the Material Command Intelligence Analysis Special Projects Branch (MCIAXO), which was commanded by Major Raymond Llewellyn. Llewellyn reported to Miles Goll, who was chief of the Technical Analysis Division, and he reported to Colonel Clingerman, who was chief of the Intelligence Analysis Division under McCoy.

Major Llewellyn assigned Project SIGN to Captain Robert Sneider, who took charge of the project team designated MCIAXO-3. Alfred Loedding became director of the team, which was composed of about a dozen technical engineers. The unit was located within barbed-wire enclosed Quonset huts at Wright-Patterson AFB (the new name for the base). As previously mentioned, Loedding was already working with Colonel Garrett at Air Force Headquarters on all flying object incident reports passed to him.

Due to many sightings taking place, it was hoped that Project SIGN would conduct investigations, serve as a clearing house for information, and provide further answers about flying objects. Those assigned to work on the project understood the restricted nature of their work, but that didn't mean they understood they might be prevented from proclaiming an extraterrestrial conclusion with their project, which was a conclusion they were already leaning toward.

Mantell Incident

Much military interest was famously demonstrated on 7 January 1948 when Kentucky Air National Guard Captain Thomas Mantell was notified by a tower operator from Godman Field near Fort Knox, Kentucky, of a strange object overhead. Mantell, who was a very experienced and decorated pilot flying his F-51 fighter (known as a P-51 Mustang during the war), soon obtained visual contact and began to pursue the object. As he climbed higher, over 15 thousand feet, he reported:

> *It appears to be a metallic object or possibly a reflection of the sun from a metallic object, and it is of tremendous size.*

He later radioed:

> *I'm still climbing, the object is still above and ahead of me moving at about my speed or faster. I'm going to close in for a better look.*

Unfortunately, that was the last heard from Mantell. His crashed plane was later recovered from a farmer's field. With no oxygen on board, he evidently succumbed from hypoxia at high altitude, and then he met his fate. Being a very experienced pilot, and having an understanding of altitude limitations, it is quite surprising he became so enthralled with his pursuit that he chanced to get "a better look" at the object, which was "of tremendous size."

This incident drew the attention of Alfred Loedding and his team of TID investigators. After spending much time eliminating other possible explanations, the "extraterrestrial" idea took hold. Due to the fact that they were not yet ready to publicly proclaim their honest conviction, they used "Planet Venus" as a euphemism to describe what Mantell was chasing, but they were quite convinced of what they considered to be the only answer.

On 31 January, in response to publicity about the Mantell incident, the *East Oregonian* newspaper in Pendleton, Oregon, published an article with the title "Flying Disks Book Declared Closed." It mentioned the following in regard to a statement by the RDB:

> *A spokesman said the board experts dismiss the flying saucers as a mirage induced by mass self-hypnosis. The scientists declare that the discs were nothing more than optical illusions and say that no evidence has ever been found to show that the saucers were either man-made or products of nature.*

It is surprising that an RDB spokesman, possibly Vannevar Bush or maybe General Joseph McNarney, felt compelled to provide input with official authoritative expertise, especially with absurd verbiage about "mass self-hypnosis" and "optical illusions." It makes sense that the RDB wanted to quell further public concern about flying objects, although it might have made more sense to declare that Mantell was "chasing" a high altitude research balloon.

Loedding and his team were working diligently to find additional answers to the mysterious flying objects, and the possibility of an unearthly origin kept their attention, especially after the Mantell incident. Previously, by the end of December 1947, Garrett and Loedding were directing serious attention toward the extraterrestrial answer, but now they were mostly convinced. Certain cases provided Loedding's investigators with overwhelming evidence of something unearthly about flying object sightings. For them, an "Extraterrestrial Hypothesis" loomed as a final answer, but it was still necessary to look into Soviet involvement. Others were still questioning the Soviet possibility, and they would continue to conduct further investigations to insure their unearthly conclusion was solid.

Soviet Involvement Extremely Remote

Aside from the fact that he was in-the-know, and wanted Truth revealed, Colonel McCoy knew his investigative team needed to cover all bases in their quest to find an answer to flying objects. This needed to be done before settling on the extraterrestrial answer. Previously, in August, General LeMay provided an emphatic "no" to possible U.S. responsibility for flying objects, and his implication was that no U.S. "black projects" were involved. McCoy was also fairly comfortable with the fact that a foreign domestic origin for the objects was addressed, because nothing of substance was received in regard to the EEI and Collection Memorandum previously sent out.

Nonetheless, McCoy still needed conclusive confirmation that no Soviet/German connection was involved in the adaptation of a flying wing design by the German Horton brothers. McCoy's experience in recovering German technology at the end of WWII, in association with his good friend Colonel Harold Watson, influenced this concern. Other people previously involved with McCoy in the recovery effort in Germany were now connected with him at Wright-Patterson. They included Miles Goll, Colonel Malcolm Seashore, and Colonel John O'Mara. Colonel Albert Deyarmond would later join them in May 1948, and then General Donald Putt. McCoy knew that a foreign domestic origin for the objects must be totally ruled out, especially in minds of others. It would then be possible to concentrate more on the extraterrestrial hypothesis, which could lead to Truth of the UFO question.

Just prior to serious work commencing on Project SIGN, McCoy received an Army Intelligence Division memorandum dated 21 January, which included an Army Intelligence Requirements document. It provided information on Russian development of "Unconventional Aircraft," and it described an update to the Horton flying wing concept, but this information did not equate to the ultra-high technology required to produce demonstrated maneuvers observed in sightings of reported flying objects. This confirmed that it was highly improbable that Soviets were responsible for the mysterious object sightings. It was also hard to imagine any Soviet country becoming so technically capable, especially when it was known that the Russians were reverse-engineering U.S. hardware, such as the B-29 bomber.

The Army Intelligence Division memorandum satisfied McCoy that TID should be able to proceed beyond the idea that flying objects were foreign incursions. He now had a document proving that Soviet involvement with flying objects was a false assumption. But he would also find, however, that the Army memorandum would not suffice for some people. They were still reluctant to abandon the notion of Soviet involvement, and McCoy would need to figure some other way to totally eliminate the Soviet possibility, perhaps by obtaining tangible evidence.

Project SIGN Confirmed

Project SIGN was implemented on 26 January, but it was not officially recognized as an active functioning project until 11 February when Air Force Technical Instruction TI-2185 confirmed project guidelines, procedures, and instructions, which was primarily put together by Loedding. The project was given a "2A priority" level, the second highest, and a "restricted" security classification, which was the lowest.

The usual practice of the military was to provide an additional code name for such a project, which could then be referred to in the public arena without revealing entrenched secrets. In this case, Project SIGN became widely known as "Project Saucer" to the general public and press. Informally and unofficially, however, Project SIGN was already in place beginning in late 1946 and early 1947 when the AAF first took an interest in the UFO question.

Also on 11 February, the RDB Board was convened, possibly to recognize and discuss the new project. In attendance was General McNarney of AMC, General Laurence Craigie, General Jacob Devers, Admiral Earle Mills, Admiral John Price, General Henry Arnold, and RDB Chairman Vannevar Bush. One might easily imagine conversations taking place at this meeting discussing the fact that SIGN investigators were flirting with the extraterrestrial answer.

As previously mentioned, serious work was already in progress in TID's Special Projects branch by the later part of January, which included review of incident reports sent down by Garrett. On 27 February, however, it was ordered that military sources send sighting reports directly to TID. This directive, from AFOIR Director General Cabell was relayed and distributed by Garrett to all commands. Its intent was to speed up the investigative process. It also allowed Project SIGN to be selective about which sightings were worthwhile to send up to Air Force Headquarters.

Tangible Evidence Needed

On 17 and 18 March, an Air Force Scientific Advisory Board (AFSAB) conference headed by Dr. Theodore Von Karman was convened. It was attended by Colonel Robert Taylor, chief of AFOIR's Collections Branch, and Colonel McCoy of TID. At the conference, McCoy commented:

> *We have a new project–Project SIGN–which may surprise you as a development from the so-called mass hysteria of the past Summer when we had all the unidentified flying objects or discs. This can't be laughed off. We have over 300 reports which haven't been publicized in the papers from very competent personnel, in many instances–men as*

capable as Dr. K. D. Wood and practically all Air Force, Airline people
with broad experience. We are running down every report. I can't even
tell you how much we would give to have one of those crash in an area
so that we could recover what ever they are.

This short comment from McCoy to AFSAB is significant for several reasons. He referred to "hysteria" during the summer of 1947, and "all the unidentified flying objects or discs," as reason for implementing Project SIGN. In speaking to a group of such distinguished scientists—a group where each would be tempted to dismiss such nonsense rather than look foolish in front of the another—McCoy quickly added that this was no laughing matter. He went on to state that more than three hundred unpublicized recent sightings were witnessed by very competent people with broad experience in aeronautics. Among them was Dr. Karl Dawson Wood, a well-known and greatly respected aeronautical/aerospace engineer and educator. He once worked at the National Bureau of Standards in Washington, D.C. and was now the current department head for aerospace engineering at Colorado University. It is doubtful that McCoy would have dropped this particular name if he was not certain of his facts. Moreover, he indicated he wished he might have tangible evidence from a flying object crash site.

Based on what McCoy related to AFSAB, one might consider he was not personally aware of Roswell, and was seriously toying with the extraterrestrial idea, especially since he wished an object might crash. But McCoy was definitely in-the-know, due to being on the staff of the former JRDB, being in charge of T-2 during Roswell (which was now TID), and also being associated with Generals Twining, McNarney, and Chidlaw. His talk to the AFSAB was simply a way for him to introduce the subject to a group of influential scientists, with whom the Air Force often consulted, and it provided an opportunity for McCoy to prepare the scientists for an extraterrestrial awakening.

The AFSAB still exists today, and "provides independent advice on matters of science and technology relating to the Air Force mission." It had its beginning in 1944 under the leadership of Dr. Von Karman. Subsequent leaders would include General James "Jimmy" Doolittle and Lieutenant General Donald Putt. Many of its scientific members were leading scientists and influential technological thinkers, but many did not have a security clearance or a need-to-know on many matters. Most meetings consisted of selective and generalized topics geared to the particular audience. In light of existing secrecy on the UFO question, discussion of recovered extraterrestrial evidence would not be expected to be one of those topics, even if a flying object did crash and drop into the lap of Colonel McCoy. He knew this, but he also knew it was necessary to get the AFSAB in touch with the UFO question.

Although McCoy suggested to the AFSAB that he hoped an object might crash, he actually attempted to facilitate the possibility. He previously sent a letter to Vandenberg asking that fighter planes be stationed around the country in an effort to engage flying objects, and take one down. In answer to him, a rather curt letter, dated 3 March, was sent by General Samuel Anderson, director of Plans and Operations at Air Force Headquarters, to Director of Air Force Intelligence General McDonald. It stated the following:

> *The proposal of Air Materiel Command for stationing fighter aircraft at all bases on a continuous alert status is not considered feasible... It is recommended that Air Materiel Command's responsibility for "collecting" information on unusual phenomena in the atmosphere be confined to the establishment of direct channels for the receipt of such information.*

On 17 March, the same day McCoy spoke to the AFSAB, Colonel Garrett of AFOIR was directed to write a letter on behalf of General McDonald to McCoy's boss General McNarney. It stated:

> *Reference is made to Colonel McCoy's informal proposal that certain fighter aircraft be maintained on a continuous alert status, within the Zone of the Interior, to aid in gathering information on flying Discs... This proposal is considered unfeasible...*

Colonel McCoy decided that it was necessary to recover one of the flying objects for evidence, even if one had to be shot down. To him, the reasoning was quite simple. If there was no U.S. involvement with the objects, and Soviet involvement was extremely doubtful, no harm would result in shooting one down, and it would put the UFO question into the open for revelation of Truth to everyone. This put Vandenberg, and those in the Control Group, on notice that McCoy was determined to recover an object, which they were not anxious to have happen.

McCoy knew the objects were not Soviet craft, and the response to him essentially confirmed that there was no high level concern about that either, but there were still a few minds that couldn't be changed in that regard. If an object was shot down, it would solve the problem. National security, however, was a concern in regard to flying objects, and shooting one down would not resolve whether the objects possessed menacing intentions or not. This dichotomy for McCoy must have been frustrating, as well as learning that recovery of "another" object would not be allowed.

Incident at Aztec, New Mexico

If TID investigators were not sheltered from top secret sighting events, they and McCoy may have had an object within their grasp, but remains were taken to Los Alamos. On 25 March, near Aztec, New Mexico, oil field workers discovered an object, or objects, sitting relatively intact atop a mesa off Heart Canyon Road, which was brought to their attention by a fire in the local area. The craft were supposedly disabled by Air Force high-powered radar located not far away. Although this incident continues to be controversial, researchers discovered years later that people in the local area were quite familiar with the incident. The incident was confirmed by eyewitnesses, and testimony was obtained from several individuals. Today, a plaque at the crash site honors discovery of this incident, and it states:

> On or about this site, on March 25th, 1948, a spacecraft of origins unknown crashed landed on this mesa. The 767th ANW radar base, in nearby El Vado, New Mexico tracked the errant landing to this site. High security recovery operation took approximately two weeks with all remains being taken to Los Alamos, New Mexico for scientific study and evaluation by some of the world's leading scientists. The recovery of this craft by the U.S. government military was one of the most secretive recoveries of a spacecraft with origins unknown since the similar recovery in Roswell, New Mexico eight months earlier. Sadly, all occupants, as many as sixteen, died as the result of this crash, making full disclosure of both purpose and origination all but impossible.

Flying Object Evidence Available

It is ironic that investigative efforts for Project SIGN were taking place at Wright-Patterson AFB, where real tangible evidence of the Roswell crash was stashed away. In reality, shooting down a flying object was not necessary, although it would provide tangible evidence for those not privy to Roswell, and it would smooth the way for disclosure of Truth. Other than McNarney, Chidlaw, Exon, Twining, and McCoy, who were in-the-know at Wright-Patterson, there must have been a select few at TID who were previously involved with receipt, handling, and storage of Roswell crash debris. It would seem clear, however, that Project SIGN investigators were oblivious to Roswell. But it is highly likely that a few others in TID were tightly bound to the magnificent secret.

It is not known how long Roswell evidence was kept at Wright-Patterson, or if it was ever relocated, but there must have been plans to substantially update security of the facility where it was housed. Roswell evidence was too valuable to remain in such a congested and vulnerable location. The facility was probably updated at some point in time by placing evidence underground, or evidence was transported to a more secure and remote location designed for it. This is conjecture, but it's also tempting to believe that Colonel Watson had a hand in creating better security for crash evidence while "assigned" to Air Force Headquarters, and until he left there in July 1949.

If any Project SIGN investigators were in-the-know on Roswell, and sworn to secrecy, they were tasked with playing a very difficult game of pretending to discover Truth, while also determining whether the objects were a threat to national security. Either way, as investigators, they first needed to determine if the objects were real, or a misidentification of man-made objects, or natural phenomena. If the objects were real, they needed to determine if they were U.S. or Soviet origin. If Soviet, then the presumption would be that the objects were a threat to national security. If the objects were neither U.S. nor Soviet origin, investigators were then left with a final option, and a confrontation with Truth. Whether or not they knew the Truth, they were still inspired with a conscientious and concerted effort to find Truth, which was their general expectation and personal goal. Of the investigators involved with Project SIGN, the only ones possibly coming under suspicion of being in-the-know are Albert Deyarmond and Lawrence Truettner. Their names will come up later, but there is no definite evidence available to verify whether they knew of Roswell or not.

For those in control, and in-the-know, it was convenient to have Air Force investigators look into the UFO question. It would satisfy those in the military who were concerned about potential menace of flying objects. Those in control, however, were not too keen in having Truth proclaimed by investigators who might happen upon physical evidence, or actually figure it out. It was preferable to maintain the secret, but also have incidents tracked and reported so that an overall view of what was taking place could be obtained. To the Control Group, the real nature of these objects was still a mystery, even with available Roswell evidence. They still needed to know whether the objects were hostile, what their intentions were, and where they came from.

First Project SIGN Report

On 23 April, Project SIGN produced its first report, which was sent to Air Force Chief of Staff Vandenberg, with attention to Air Force Director of Intelligence General McDonald. It was signed by Colonel McCoy, and it provided a tabulation of sighting incidents received up to February 1948. Of the incidents listed, one was from December 1946, and others from June 1947 up through succeeding months, but the vast majority were from June and July 1947. Included were the Kenneth Arnold incident, Harmon Field incident, Muroc Field incident, and

Mantell incident. Altogether, a hundred incidents were documented with multiple witness sightings for many.

Quite noticeable in the SIGN Report was the classification of witnesses. Forty of them were Air Force personnel, including ten Captains, five Majors and four Colonels. Also included were forty pilots, nine police officers, an astronomer, a lieutenant governor, judge, air traffic controller, radar operator, and others employed in professional government and non-government positions. None were considered uneducated or of dubious character, and it is likely that the listing was taken from a selection of the three hundred reports from competent people that McCoy previously referred to in his talk with the AFSAB.

Most important, it is very likely that this first Project SIGN report caught the attention of Vandenberg, primarily because the reported incidents did not provide prosaic explanations for flying objects. Many descriptions and details of the incidents gave the objects an unearthly quality. The significance of this is that Vandenberg may have suspected, or decided, that project investigators might potentially reveal Truth at some point in time, especially with McCoy's expressed desire to shoot down an object. Vandenberg may have been reminded of this by General McNarney, McCoy's boss, and it's likely this provided Vandenberg with an incentive to install a "road block" at Air Force Headquarters to counter any damage that might be presented by McCoy and his investigators. Whether this was Vandenberg's intent or not, it was not long before Major Aaron "Jere" Boggs was assigned to the Defensive Air Branch of AFOAI, and tasked with that responsibility, or so it seemed. That is the way it appeared based on Boggs' efforts to follow, and on his subsequent reputation as AFOAI's flying object specialist. But he would also have help in that regard from additional new assignments of personnel to AFOAI.

New Assignments

On 30 April, General Vandenberg took over from Air Force Chief of Staff General Carl Spaatz, who retired on the same date. At age forty-nine, Vandenberg became the Air Force's top dog. He came into a position of tremendous responsibility, which required management of many secrets, including the UFO question. General Muir Fairchild assumed Vandenberg's old position as Air Force deputy chief of staff.

On 15 May, General Cabell officially replaced General George McDonald as head of AFOIN, and Cabell now reported to Vandenberg. Cabell's replacement at AFOIR was Colonel Walter Agee, and he would take a lesser role in regard to the UFO question, but Cabell would now employ Colonel John Schweizer to help with that. Shweizer previously arrived in April, and was now Cabell's executive officer. Also, both AFOIR and AFOAI were now under Cabell's control, but those two organizations were about to take slightly different positions on the UFO question, and AFOAI would assume more of the lead. This would not help the situation for Cabell, who was dependent on them and TID to resolve the UFO question.

Also in May, Colonel Brooke Allen entered the picture at the same time Cabell came into his new position. As stated in his Air Force biography, Colonel Allen took charge, "for the most part," as chief of the Defensive Air Branch of AFOAI. This was a rename of the Strategic Vulnerability Branch, which would later become Air Targets Division in March 1950.

Of special note is that this assignment of Allen conflicts with official Air Force records for Colonel Watson, previously noted in Chapter 13, which indicated that Watson was chief of the Strategic Vulnerability Branch. Speculation has already been mentioned about Watson's possible involvement in other activities for the Air Force in regard to Roswell, and one wonders exactly where Watson fits into the scheme of things. Watson's actual activities during this specific time is quite vague, with no special mention of him in historical records,

Loedding Receives Excellent Rating

Albert Loedding and his team became very busy in the investigation of reported sightings. He often communicated with Garrett at Air Force Headquarters, while also serving as TID's special consultant to Division Chief Miles Goll, and others. On 5 May, Goll produced an "Efficiency Report" on Loedding and gave him an "Excellent" rating. This report mentioned several accomplishments for him, "which were of outstanding credit to the Division." It stated that Loedding acted in many technical matters that "required broad engineering background as well as accurate concepts of intelligence [the field of intelligence]," and "required initiative action dealing with high officials of this command and also with higher authority in Headquarters, USAF." In seven more months, this excellent rating would become noteworthy, especially with Loedding's subsequent and unfortunate downfall.

Miles Goll, who was third in command under McCoy and Clingerman, was the civilian chief of the Technical Analysis Division (MCIAT), and he was in a unique position to closely monitor Project SIGN activity. He possessed much knowledge of what was going on, often signing for his boss Clingerman. Most important, his later activity, early in 1953, would implicate him as a possible agent of the CIA.

Conflicting Intelligence Agencies

In June, a government task force was organized to examine the country's national security and intelligence systems. Ferdinand Eberstadt, a close friend of Defense Secretary Forrestal, was assigned to head it up. One purpose of the task force was to examine the lack of coordination and cooperation within the community of intelligence agencies, and also the considerably poor relations between them. These agencies included the CIA, FBI, Atomic Energy Commission (AEC), intelligence agencies of the separate military services, and also the intelligence section of the State Department. Differing intelligence estimates from each agency suggested that each was submitting, in one way or another, subjective and biased estimates.

Providing intelligence estimates, or an "estimate of the situation" (EOTS), was the usual practice of such agencies, which was a method used to advise and inform on intelligence findings. This activity was intrinsic to an agency's operation, and a basis to their formation and foundation. The CIA was the nation's foremost intelligence agency, and it was reporting to the National Security Council and the president. It was also considered the most infringed upon by interference from other agencies. All were interfering with each other, however, in seeking information, and they were often investigating one another rather than cooperating in sharing information.

The primary informant and instigator who prompted examination of this problem was Ralph Clark of the RDB staff. He was on the staff of the JRDB at the time of Roswell, but he was also later associated with what would become CIA's Office of Scientific Intelligence (CIA-OSI), which will be discussed in Chapter 17.

Gathering scientific intelligence was only one area where interference between agencies was prevalent. With so many agencies involved with similar agendas, and investigating the same thing, none trusted the other in regard to what each considered its own domain in certain areas. And this certainly applied to the UFO question. For example: The RDB did not want other intelligence agencies involved with secrets they were carefully monitoring and protecting, but AFOIN and TID were attempting to uncover those secrets, as were the FBI and CIA. This competition and duel responsibility in various areas still exists today, and intelligence agencies still continue to overlap and conflict with one another, which results in interference and duplication of effort. With a new government task force involved at this time to sort out these problems, the UFO question and associated evidence may have been carefully but discretely considered by the Control Group for future migration from the RDB to the CIA. With Vannevar Bush in charge of the RDB, he was seriously concerned with securing the Truth and secret of Roswell.

Chiles-Whitted Incident

In early morning hours of 24 July, the Chiles-Whitted incident occurred, which set intelligence agencies, the public, and the press abuzz. Two commercial pilots flying an Eastern Airlines Douglas DC-3, from Mobile to Montgomery, Alabama, needed to take immediate evasive action to prevent collision with a flying object, which quickly swept by them. In getting a very close look, they described the object as one hundred-feet-long, wingless, and cylindrical, with two rows of square lighted windows. It had a blue glow on its underside, a fifty-foot-long orange-red glowing tail, and it was traveling up to seven hundred miles per hour heading directly toward them. After passing by on the right side, it pulled up into a steep climb through broken scattered clouds. A passenger, Clarence McKelvie, also saw it and provided confirmation. Eastern Airlines then released information about it, which created national news headlines.

The *Seattle Daily Times* on that date displayed an article on its front page with a heading that said, "Wingless Air Craft 'Seen,' Spouts Flame." The article stated:

> *Two Eastern Air Lines pilots reported today that their plane last night passed a tremendous, wingless aircraft that shot a 40-foot flame out its back end and traveled between 500 and 700 miles an hour. Capt. Clarence Shipe Chiles and his co-pilot, John B. Whitted, told of seeing the fantastic airship and said it looked like a B-29 fuselage "blown up after four times." The aircraft was brilliantly lighted and had two decks of big square windows, they asserted. It passed within several hundred feet of the pilots' DC3 and then shot up in the clouds, they said.*

Everyone took notice of the Chiles-Whitted incident. For the first time, the public and press learned that professional and credible witnesses viewed a flying object up close. The incident became widely known and caused a great deal of attention. Talk of the incident quickly spread. Within a few days, General Cabell came under intense pressure from the general military, certain elements of the Air Force, and especially the press. There was an immediate demand get to the bottom of what was going on, and General Cabell quickly contacted Colonel McCoy about the incident. He wanted an immediate investigation started.

Personal interviews were then conducted by Loedding, Albert Deyarmond, and their boss Llewellyn. During the investigation, Loedding and his team noted other similar sightings, especially a nearly identical object sighted a few days earlier over the Netherlands. This and several other major incidents convinced these investigators that they were dealing with an unearthly vehicle that was not man-made. The ultra-high technology necessary for the observed flight characteristics of this and other objects left them with no other option. They even went so far as to determine when the orbits of Mars or Venus might place those planets in close vicinity to Earth, which might provide predictions of future flying object activity. The only thing missing for a definitive conclusion on the mysterious object was some sort of physical evidence. This incident made a deep impression on Loedding and his investigators, and they became more convinced than ever of the extraterrestrial answer. They then set out to prepare a report of the incident.

Dr. Allen Hynek Enters Picture

Dr. Hynek also became involved with the Chiles-Whitted investigation and stated:

> *If the pilots had reported accurately on what they had seen, no astronomical explanation was remotely plausible.*

But Hynek did suggest that they may have seen an "extraordinary meteor," although he could not justify his conclusion.

It was just before this that TID contracted with Ohio State University (OSU) for astronomical services for Project SIGN, and to obtain identification of meteors, planetoids, and associated phenomena. Hynek, who worked at OSU as a physics and astronomy professor, was chosen for this, and he was recommended by astronomer Dr. Paul Herget, from Cincinnati Observatory, who was an expert on asteroids, and a leading authority on orbital mechanics.

This was Hynek's first introduction to many years of investigation for the Air Force, where he primarily reviewed incident summary sheets on the sightings, and then added comment. He would later state that he initially enjoyed his job as a "respected scientific debunker," which was a major part of his duty. It was much later, however, that he attempted to be more objective in his scientific analysis. His earlier work then became a great source of embarrassment to him.

Dr. Hynek came to work for Project SIGN in the spring of 1948, and he commented on his selection for the job as follows:

> *One day three men, and they weren't dressed in black, came over to see me from Wright Patterson Air Force Base in nearby Dayton. They started out by talking about the weather, as I remember, and this and that, and then finally one of them asked me what I thought about flying saucers. I told them I thought they were a lot of junk and nonsense and that seemed to please them, so they got down to business. They said they needed some astronomical consultation because it was their job to find out what these flying saucer stories were all about.*

AFOAI Leadership Change

Even though Project SIGN was underway in TID, Cabell was losing patience, and he needed to have definitive answers, especially in light of the Chiles-Whitted incident. On 27 July, he asked AFOAI to "determine the origin of flying objects and their purpose in invading U.S. airspace." If there was a foreign element involved with flying objects, AFOAI was the place to get this settled. This was primarily because the Office of Air Estimates, and the Office of Defensive Air, would have responsible interest in this area in regard to national security.

At this time, however, AFOAI was gaining new leadership in some important positions. General Ernest Moore arrived to replace General James Olive as chief of AFOAI. Colonel Lester Harris became Moore's executive officer, and he also became chief of Defensive Air. This was after switching jobs with Colonel Brooke Allen, who then became the new chief of Air Estimates. Both Allen and Moore knew each other, having previously served together on the Military Staff Committee at the United Nations, which was an office of the Joint Chiefs of Staff. This was also a time when the Berlin Airlift was coming into full operation, which began about a month earlier under command of General LeMay. In their new positions, both Moore and Allen would become valuable assets in providing assistance to LeMay.

૭∾ഛ

Albert Loedding and his team of Project SIGN investigators began their new project with much enthusiasm. They were anxious to find answers to the mysterious flying objects, and identify what they were. Colonel McCoy, however, was stressed about protecting the Truth of Roswell, while also attempting to investigate flying objects with a team of investigators who were not in-the-know. He had one team attempting to investigate the UFO question, but had another team that was sworn to protect the "secret." It was not a good situation to be in. When his investigative team became convinced of the extraterrestrial answer, he attempted to hasten revelation of Truth by making a personal request to General Vandenberg to have a flying object shot down, which would provide tangible evidence and assist in revealing the Truth. But his request was bluntly denied.

At the end of July, Project SIGN investigators were convinced that flying objects were real, and that they were something not of this Earth. The Chiles-Whitted incident served to cement the idea for them, and this was after they produced their first Project SIGN Report.

Most SIGN investigators had been involved with the UFO question for more than a year, but now they were seriously contemplating creation of a preliminary review of their findings. It is recalled that a year previous, on 30 July 1947, Colonel George Garrett in AFOIR produced a Preliminary Intelligence Estimate, which essentially stated that flying objects were real. Then, on 23 September 1947, General Twining, who was in-the-know, confirmed it in a letter to AFOIR Director General Schulgen. Project SIGN investigators were now ready to confirm it again, and even more clearly, although no tangible evidence was available to them.

The big question now facing Project SIGN investigators was whether to release a report of their findings, look for additional confirmation and delay their report, or continue with further investigation and hope to obtain tangible evidence.

In the background, and behind the scenes, the Control Group and those in-the-know were protecting their monumental secret. But were they keeping a close eye on Project SIGN? Would they allow Project SIGN to confirm the extraterrestrial possibility?

CHAPTER FIFTEEN

Signs of Trouble

Project SIGN: August - November 1948

Project SIGN Verifies Roswell

On 5 August, TID's investigation of the Chiles-Whitted incident was documented in a preliminary Estimate of the Situation (EOTS). Alfred Loedding and his team created their report after completing a thorough investigation, which included many interviews and much research. The high-profile incident created intense interest, and General Cabell was anxious to have an immediate answer. The EOTS was then sent directly to Cabell, and it announced that flying objects were interplanetary.

It was about six months prior to this that Project SIGN was authorized, but Loedding and his investigators were involved with serious investigation of flying objects for more than a year. It started prior to the Roswell incident because of demands originally placed on them to find out what was going on with the objects. With continued sightings by credible people, which were followed by on-site investigation of many significant incidents, especially within the last few months, they were totally confident in stating that flying objects were not of this Earth. They knew that much work remained to further substantiate and solidify their argument, but it was still decided that their best estimate of the situation was appropriate for release, even if tangible evidence was not available.

It's highly significant that independent Air Force investigators came to an extraterrestrial answer based on the remarkable incidents they studied. They had indirectly, and unknowingly, verified the incident at Roswell, which was a very unique circumstance.

Extraterrestrial Hypothesis

TID and AFOIR were convinced beyond a reasonable doubt. Their substantial investigative effort provided no other choice than to proclaim an extraterrestrial answer to the UFO question. Release of the EOTS indicated their confidence, but they also knew their decision on the matter could be a tough sell, especially if others not familiar with their engineering and scientific expertise saw things differently. There was a potential for the pendulum to swing against them. But they also knew their professional expertise and integrity was highly regarded, and they were quite certain they could defend their comprehensive analysis. They would continue to investigate and study new sightings, and also review previous sightings to insure nothing was left uncovered. Their confidence was high, but the lack of tangible evidence required that they carefully proceed with continued investigative efforts.

It now becomes necessary to elaborate on the existing confusion regarding the EOTS document. The situation is problematic because the document is significant to the history of Project SIGN, but no known copy of it currently exists. The reason for this will come to light later, but this confusing matter centers on the date of its initial release. Some researchers believe the document, or a more complete one, was produced in late September or early October. Others believe it was produced later, but prior to 12 November 1948. Allen Hynek confirmed, years later, that the document did exist. And others, including Major Donald Keyhoe, Major Dewey Fournet, and Captain Edward Ruppelt (of Project BLUE BOOK fame) viewed the document and indicated it was created on 5 August, which was the time TID investigators became totally convinced of the extraterrestrial answer. The named individuals have considerable credibility, and the 5 August date is likely correct.

TID's work on the important document commenced directly after investigating the Chiles-Whitted incident. Within a day or two after the incident, Cabell experienced intense pressure to find out what was behind the sighting, and TID experienced even more pressure. After conducting an intensive investigation, and interviewing those they considered competent witnesses, TID investigators were left with no doubt about the extraterrestrial answer. Because of the great interest generated, and Cabell's desire for an answer, it was decided to finalize and release the special report they were working on, and they were quite confident in doing so.

Their report was actually a preliminary appraisal, or an estimate of what might be forthcoming in a final Project SIGN report. Presumably, with no realistic astronomical answer possible for the Chiles-Whitted incident, or any other viable possibility, Loedding and his team felt obligated and compelled to state what seemed to be a rational, although very shocking conclusion in their EOTS. Their announcement was referred to as the Extraterrestrial Hypothesis (ETH).

Prepared by Loedding's team, it was unprecedented that Air Force investigators were willing to address the UFO question with an EOTS containing the extraterrestrial answer, and also with a suggestion that flying objects were potentially hostile, which attracted additional attention. As mentioned, however, this was not intended to be a final report for Project SIGN, but it was a status report and notice of current work in progress, which would indicate a possible or probable conclusion in a final Project SIGN report.

Captain Sneider, the leader of Project SIGN, was ultimately responsible for the EOTS document. Loedding was the author, but others on his highly-qualified team provided write-up and editorial work. Final approval, with confident and confirming signatures, was provided by Colonel Clingerman and Colonel McCoy. The document was then bound in a black legal-size cover, stamped "top secret," and sent to General Cabell at Air Force Headquarters.

"Tactics" and "Probability" to be Studied

It did not take long for Cabell to react to the EOTS, although he may have been aware of it prior to its release. He must have discussed it with certain people in his organization, and possibly with Vandenberg. A day after its release, and a week after his previous notice to AFOAI to "determine the origin of flying objects and their purpose in invading U.S. airspace," Cabell sent a memo on 6 August to AFOAI directing that a study of flying objects be specifically conducted by the Office of Defensive Air, which was the unit where Major Jere Boggs was recently assigned. The memo also requested AFOAI to:

> *examine the pattern of tactics of reported flying saucers and develop conclusions as to their probability.*

Cabell wanted to determine if national security was threatened by the objects, and one might consider that he sent his request to AFOAI to get that organization aligned with TID and AFOIR. It's easy to assume that Cabell wanted his complete organization involved, and to have a consensus on the matter. He must have been struck by the conclusion in the EOTS, but it's also possible it confirmed his own suspicion. He was aware of the extraterrestrial possibility due to his previous close association with AFOIR, and close familiarity with flying object incidents. He knew where AFOIR stood on the UFO question, but AFOAI had not been involved since the middle of December 1947, when the views of each of the AFOIN branches coincided.

This request to AFOAI, however, also provides one with reason to further pause, reflect, and review this situation. One can envision that Cabell, while interacting with Vandenberg, must have had serious discussions with him regarding the extraterrestrial conclusion of the EOTS, and also with the unearthly aspects of flying objects. Cabell likely expressed his own ideas about that, and Vandenberg probably recognized strong views expressed in that regard. This would suggest to Vandenberg that there was a serious challenge to the secret and Truth of Roswell. And this was in addition to Colonel McCoy's request in March to shoot down an object. And it was also in addition to the SIGN Report released in April suggesting that many competent witnesses had sighted objects of an unearthly nature.

There is no record of a meeting between Cabell and Vandenberg on this, but one might consider it possible that Vandenberg pressured Cabell to involve AFOAI and its Defensive Air Branch. The determination of "tactics" and "probability" of flying objects was an existing priority of Vandenberg and the Control Group, and it was now Cabell's specific request to AFOAI. With new leadership in AFOAI, and with the particular assignment of Major Boggs, it provided Vandenberg with a possible solution to counter the situation that now existed.

Another consideration one could make at this particular time is that Cabell was no longer in the dark on the UFO question, and finally aware of Truth. This, however, was not necessarily the case due to Cabell's later and further concern in looking for Truth. It's quite likely he was still in the dark.

The Truth Game

The thing that needs to be kept in mind in this situation is that a game was being played between those in-the-know about Roswell, and those who didn't know. Those not in-the-know were being used as pawns in obtaining additional information on flying objects, which was desperately needed. But there were also those in-the-know who did not agree with the game being played, and believed that the UFO question should be revealed. It was a contentious situation that would continue, and it would muddy the waters at this particular time with this confusing factor. Those who did not believe in playing the game were obligated to maintain the secret, while also motivated to having it revealed.

In this regard, and a case in point, involves Chief of TID Colonel McCoy. He was required to maintain the secret of Roswell, but he was also attempting to manage a staff that was not in-the-know, and charged with investigating the reality, origin, tactics, and probability of flying objects. This meant he had great incentive to have the UFO question revealed, which also reflected the position of

General Twining and a number of others. McCoy's only recourse was to support and back his people, let them do their job, and let them declare the answer to the UFO question if it came to that. His boss, General Joseph McNarney, probably understood the unmanageable situation McCoy was presented with.

AFOAI and the Office of Naval Intelligence

On 11 August, Cabell received an "interim" response from AFOAI regarding his 29 July letter and 6 August memo, and it outlined initial progress and specific analysis procedures that would be taken in the AFOAI study. With new leadership in the AFOAI organization, now headed by General Ernest Moore, there was special determination to take on the requested study, but also with an awareness of the direction Project SIGN was headed. Assigned to monitor and assist with creation of the study, and make sure it would be acceptable, was Chief of Planning Estimates Lieutenant Colonel Robert Smith. Planning Estimates was a section of Air Estimates headed by Colonel Brooke Allen.

Responsibility for conducting the AFOAI study was assigned to Major Jere Boggs, who was AFOAI's "flying object specialist" in Defensive Air. When considering Cabell's 6 August memo, which directed that the study be specifically conducted by Defensive Air, it's quite possible that Boggs was previously under consideration for the new study.

Special assistance and consultation for the study was made available to Boggs by the Office of Naval Intelligence (ONI). The fact that ONI was involved in this effort is somewhat strange since cooperation between services was a bit strained at the time, but there is a hint later on that elements in ONI were in-the-know. Just how ONI specifically became involved is not clear, but Air Estimates and Defensive Air would work closely together to produce the directed study in close cooperation with ONI.

General Putt Replaces General Craigie

In September, Brigadier General Donald Putt took over General Craigie's job as director of Research and Development at Air Force Headquarters. General Putt had been chief of AMC's Engineering Division at Wright-Patterson AFB since the previous month of May, and this is where he may have been clued in about Roswell, probably by his very good friend Colonel McCoy.

Craigie was then transferred to Wright-Patterson and became head of the Air Force Institute of Technology (AFIT). This is where Craigie's involvement with the UFO question might have come to an end, except he was now in the area where Roswell material was stored. This is significant because Craigie was a board member of the RDB, and he was a Roswell participant with firsthand knowledge of the incident. The RDB now had two board members at Wright-Patterson, which included General McNarney.

The Gorman Incident

On 1 October, another major flying object incident occurred. Captain Edward Ruppelt (of BLUE BOOK fame) described it as a "duel of death with a UFO." This involved a F-51 piloted by Second Lieutenant George Gorman of the North Dakota Air National Guard out of Fargo. The encounter turned into a twisting dog fight with a "lighted object" he was pursuing. At one point, the object passed within a few feet of his plane's canopy before the "light" climbed quickly upward and disappeared. Later, Gorman described his incident to TID investigators and stated:

> *I had the distinct impression that its maneuvers were controlled by thought or reason.*

Again, TID investigators were given additional reinforcement for their EOTS, and the extraterrestrial answer. There was no possible way to equate the incident to an overt challenge from a foreign country, or contradict credible witnesses and ground observers they interviewed. Loedding and his team of investigators were totally unable to reconcile the flying object to any kind of earthly domestic origin.

U.S. Involvement? - No

On 7 October, Colonel McCoy felt obligated, on behalf of his investigators, to write again to the Army, Navy, and the CIA to determine their possible involvement with flying objects. Previously, General LeMay had responded for the Air Force with an emphatic "No." This would now be a final last ditch effort to determine if a national covert operation might be involved to account for object sightings. McCoy stated in his letter:

> *This Headquarters is currently engaged in an intelligence investigation of all reported unidentified aerial phenomena. To date, no concrete evidence as to the exact identity of any of the reported objects has been received. Similarly, the origin of the so-called "flying disks" remains obscure. The possibility exists that some are of domestic origin, i.e., unrecognized configurations of some of our latest aeronautical attainments, or that they are objects not readily recognized by the public—test vehicles in various stages of development, etc.*

His letter went on to request a response, and to forward details of possible U.S. domestic origin for the objects. But his letter drew no positive reply, and this was a definite confirmation to TID investigators that a U.S. or foreign origin for flying objects did not exist. No earthly possibility could account for the objects.

Foreign Involvement? "Unwise to Overlook"

On 11 October, Chief of Air Estimates Colonel Brooke Allen wrote a letter to General Cabell providing an update to AFOAI's study. It stated:

> *An exhausted study was made of all information pertinent to the subject in this Division and the Intelligence Division of Air Material Command. Opinions of both aeronautical engineers and well-qualified intelligence specialists have been solicited in an endeavor to consider all possible aspects of the question...It must be accepted that some type of flying objects have been observed, although their identification and origin are not discernible. In the interest of national defense it would be unwise to overlook the possibility that some of these objects may be of foreign origin.*

AFOAI indicated that flying objects were real, but good judgment dictated a need to be on guard for a foreign presence. This was a confirmation that the extraterrestrial answer was not going to be considered, or admitted to by AFOAI. It's study encompassed a review of some two hundred and ten incidents that were supplied by TID. These consisted of some of the best available, with most reported by very credible witnesses, including a substantial number of military personnel. It is significant that a foreign origin was the center of attention in AFOAI's review, although it must have been difficult for them to imagine that the Soviets would chance to play games in the middle of United States territory.

TID investigators already ruled out a foreign origin for flying objects, and were making one last effort to rule out covert U.S. involvement. Within AFOAI, however, one might consider that they were a bit paranoid toward possible Soviet involvement, which would be understandable considering the nature of their responsibility. One might also consider that there was no additional "realistic" answer they could provide on the matter.

When looking back at AFOAI's turnaround in attitude, within six months after changeover in new personnel, one might consider that AFOAI's study was somewhat suspicious. Colonel Allen gave a measure of respect in his letter to those in TID by mentioning that flying objects were real, but it was countered with the suggestion that flying objects could be of foreign origin, and it was "unwise to overlook the possibility." Of course, the foreign possibility was seriously and previously considered by TID, and it is obvious that AFOAI took an independent position without consulting with AFOIR or TID. It was also done without interviewing sighting witnesses or conducting any kind of professional technical investigation.

Allen's letter did not resolve Cabell's original request for a solution to the UFO question, or of "tactics" or "probability," as he would have liked. But it probably satisfied higher officials, and it left Cabell with much contention in his organization, and also with TID. The Control Group now had an ally in Cabell's AFOAI organization.

RDB Leadership Change

Serving just two years as chairman of the RDB, Dr. Vannevar Bush resigned his position on 15 October. He didn't have the same power and influence he previously enjoyed with the JRDB, and one can speculate he was not satisfied. Also, he must have been concerned with developments in TID regarding the EOTS, and safeguarding of the secret. Perhaps that is why he remained on the RDB Oversight Committee, which still gave him much leverage. Another individual with him on the Oversight Committee was Dr. Howard Robertson. He would later play a very significant role in debunking the UFO question, in January 1953 with the "Robertson Panel," and changing the operations of Project BLUE BOOK

Secretary of Defense Forrestal appointed Dr. Karl Compton to replace Bush, and this indicated that chairmanship of the RDB had reverted to an internally selected position with the prospect for consistent leadership, which Bush preferred when he originally proposed the NSF. Compton was currently the president of MIT, and he was previously involved with Bush in the ultra-secret Manhattan Project in development of the atomic bomb.

As former head of the RDB, one can entertain the idea that Vannevar Bush may have created a special autonomous group within the organization to oversee the business of unearthly flying objects, and also the secret of Roswell. Just how that Control Group may have evolved is buried within the cover-up. It undoubtedly contained top scientists and civilian administrators, along with certain high-ranking military in its initial makeup. It likely included Lloyd Berkner and Ralph Clark, who were in-the-know, and originally involved with the JRDB. Much more will come to light about this in Chapter 26, where details are provided about a top secret Canadian government memo, which implicates Bush as being in charge of an ultra-secret group within the RDB, and involved with flying objects. There can be no doubt that control of the UFO question migrated from the former JRDB to the RDB, and that many who were previously involved in the JRDB were now in control within the RDB.

TID Put On Notice

At the end of October, the AFOAI study requested by Cabell was nearing its final form, and was under review. Cabell probably understood that the study satisfied the desires of Vandenberg, and that it would be necessary to put TID and Project SIGN in line with AFOAI's study, or have TID provide proof of origin for flying objects.

On 3 November, Cabell wrote to Lieutenant General McNarney of AMC in reference to Project SIGN, and tersely stated the following:

> *...the conclusion appears inescapable that some type of object has been observed. Identification and the origin of the objects is not discernible to this Headquarters. It is imperative, therefore, that efforts to determine whether these objects are of domestic or foreign origin must be increased until conclusive evidence is obtained. The needs of national defense require such evidence in order that appropriate counter-measures may be taken.*

Included in the letter, Cabell also mentioned that the public needed to be properly informed on matters because of their increased impatience. Obviously, Cabell himself was frustrated and impatient. Answers he now needed, however, were not in the direction TID was headed. His previous request to AFOAI was to determine "tactics" and "probability" for the "saucers," but now his focus was on "national defense" and security. The objects must now be determined to be either U.S. or foreign origin. Otherwise, "conclusive evidence" needed to be provided for another origin. This was Cabell's signal that no other conclusion on the matter was acceptable without verifiable evidence, which was also Vandenberg's demand. He then closed by asking AMC for "conclusions to date and recommendations as to information to be given to the press."

His letter was written to inform McNarney of the immediate importance of TID conforming to a "domestic or foreign origin." Part of this was that Cabell was finding it increasingly difficult to contain or satisfy the press, and he was aware that higher levels would not take kindly to the extraterrestrial alternative. He was probably unaware of being manipulated by Vandenberg and others who were in-the-know, but he now needed a turn around from TID. He was also unaware that McNarney was in-the-know, and this put McNarney in a corner. One could speculate that McNarney was advising others about this situation, including Vandenberg, but McNarney also wanted Truth revealed, and preferred that Air Force Headquarters just deal with the situation.

Cabell understood the nature of incidents investigated by TID, and also the dilemma facing them and AFOIR. But he was now looking for a specific answer from TID, which would help put AFOAI and TID on the same page, especially with AFOAI's study about to close. Cabell was probably heavily involved with Vandenberg on this, but we do not know what kind of specific response or feedback Cabell was receiving from him. One might suspect that Cabell was possibly aware of Truth at this time, but either way, his job was not made any easier, and he still had a major conflict in his organization to deal with.

USAFE Item 14 Document

Within AFOIN, two sides were pitted against each other. AFOIR was sided with TID and convinced of an unearthly origin for flying objects, but AFOAI was not convinced. AFOIR was aware of additional evidence accumulating on the extraterrestrial answer, and the idea was making more sense as time progressed. Sighting reports continued to be received from reliable sources, and it was deemed important to prepare the entire intelligence community on this, including Cabell, AFOAI, and others.

On 4 November, the day after Cabell's memo to McNarney, AFOIR sent out a document titled "USAFE Item 14," which included information received in a cable from United States Air Force Europe (USAFE). In part, the top secret document stated the following:

> *For some time we have been concerned by the recurring reports on flying saucers... They have been reported by so many sources and from such a variety of places that we are convinced that they cannot be disregarded and must be explained on some basis which is perhaps slightly beyond the scope of our present intelligence thinking. This question was put to the Swedes. Their answer was that some reliable and fully technically qualified people have reached the conclusion that "these phenomena are obviously the result of a high technical skill which cannot be credited to any presently known culture on earth." They are therefore assuming that these objects originate from some previously unknown or unidentified technology, possibly outside the earth.*

This document specifically referenced the fact that flying objects "cannot be credited to any presently known culture on earth," and it ended by stating "we are inclined not to discredit entirely this somewhat spectacular theory, meantime keeping an open mind on the subject." In AFOIR, this USAFE document presented an irresistible opportunity to confront those opposing the extraterrestrial answer, and help confirm TID's investigative efforts. It was forwarded to all AFOAI branches, and also to the CIA and AMC. But AFOIR may not have been aware of Cabell's letter sent to McNarney the day before, which must have resulted in an interesting reaction from McNarney and others.

Prior to its release, the document must have received considerable debate within AFOIR, and there must have been great anticipation on how it would be accepted. Upon its release, one can imagine much conversation taking place within AFOIN, and the conversation was probably centered on the mounting conflict regarding the UFO question.

TID's Response - Extraterrestrial Possibility

On 8 November, Colonel McCoy responded to Cabell's 3 November communication to General McNarney, with a long, detailed statement drafted by McCoy's longtime friend and investigator, Albert Deyarmond. It described many of the significant flying object incidents, but also acknowledged that "conclusive evidence" could not be provided on flying objects until such time that physical, tangible evidence became available. It stated:

> *Although explanation of many incidents can be obtained from the investigations described above there remains a certain number of reports for which no reasonable everyday explanation is available. So far no physical evidence of the unidentified sightings has been obtained ...The possibility that the reported objects are vehicles from another planet has not been ignored. However, tangible evidence to support conclusions about such a possibility are completely lacking.*

The letter ended with the following:

> *It is not considered advisable to present to the press information on those objects which we cannot yet identify or about which we cannot yet identify conclusions. In the event that they insist on some kind of statement, it is suggested that they be informed that many of the objects sighted have been identified as weather balloons or astral bodies, and that investigation is being pursued to determine reasonable explanations for the others.*

To deal with the press on difficult flying object cases, McCoy advised that release of information on some cases should be avoided, especially where it was impossible to provide an earthly explanation. Instead, he suggested telling the press that the vast majority of cases had easy answers, and that reasonable answers were being studied for others. This would be a standard response the Air Force would provide for years to come, including the statement "insufficient information available." In providing this advice, McCoy was probably "biting his tongue," but it is clear that McCoy was more or less resigned to the fact that tangible evidence was the only way that Truth might be revealed. He was also aware of the fact that he would not be allowed to obtain tangible evidence, as previously made clear to him.

With this letter from McCoy, Cabell was again informed that the extraterrestrial answer must be considered. Cabell was being advised of this, but also informed that tangible evidence to support the idea of extraterrestrial, or foreign origin, was not available. This was not the response Cabell was looking for, and it served to seal the aggravation developing between opposing views in AFOIN. Cabell was now beset with chaos in his organization, and he was disappointed with AMC.

TID - Confident and Convinced

It could be said that submittal of the EOTS document by TID was a naïve and one-sided hasty action, and that it was foolish not to retract it when AFOAI began supporting the foreign domestic origin of flying objects. But TID investigators knew they had accomplished their homework on the UFO question, which was an investigative process they spent about a year and a half on, and they were not about to give in to those who did not have a real grasp of the situation. They were aware of opposing views, but they were beyond any reconsideration, mostly because of a strong conviction based on intensive investigation and analysis, which included on-site investigation of incidents and personal interviews with credible witnesses.

TID was a completely separate organization from AFOIN. It was simply an unattached field investigative unit belonging to AMC, and utilized by AFOIN. TID was probably concerned about AFOAI having a different mind-set, but at least the AFOIR Division in the Directorate of Intelligence was partial to TID's side. AFOIR recognized TID's technical analysis capability, which was backed up by a strong reputation due to highly skilled and competent engineers and scientists. TID investigators were of high-caliber, with a proven track record of superior scientific and technical excellence.

TID was absolutely convinced of their in-depth analysis of sightings investigated, and they believed they adequately prepared the intelligence community for their view of the situation. It is also significant that Colonel Clingerman and Colonel McCoy were satisfied enough to sign off on the EOTS, although they were also motivated to do so.

Previously, when the EOTS reached the Pentagon, it was no surprise to AFOIR because Colonel Garrett and others were prepared with advance knowledge, knew what it contained, and knew it was coming. It's possible, however, that Garrett and others in AFOIR underestimated the intense debate that would come from it. It's also possible that Cabell was satisfied that the objects were properly investigated, researched, and studied by TID, and also by AFOIR, but it was disconcerting that AFOAI was only concerned with a foreign domestic reality. But this "reality" was also without direct evidence. Cabell was more than aware of the serious conflict he now faced within his AFOIN organization, but he was now out of options, and TID was not willing to help resolve the situation.

EOTS Rejected

Cabell was distressed by the fact he was unable to obtain a consensus on the UFO question, or to turn around the extraterrestrial answer from TID, especially after the 4 November memo distributed by AFOIR, and the 8 November letter from Colonel McCoy. There seemed to be nothing further he could do about the situation except to send the problem, and the EOTS, up to his boss.

When General Vandenberg eventually received the EOTS, and reviewed it, he very quickly rejected it "for lack of physical evidence." Of course, due to the chain of events since Roswell, one could easily predict this would be Vandenberg's reaction. It's quite likely, however, that Cabell previously reviewed or discussed the document with Vandenberg, and also indicated that his staff was heavily conflicted. Vandenberg may have simply informed Cabell to pass the document on up, and he would take care of it.

This did not resolve Cabell's problem with his organization, but it certainly put AFOIR and TID on notice that their position on the UFO question was not acceptable. Both were very disappointed. More than that, those in TID were now compelled to justify their position. They could not let their hard work and reputation be swept away so easily. Cabell, however, now understood he needed to bring people together, to hash it out, and have a meeting of minds.

Meeting at National Bureau of Standards

As a result of chaos brought on by the ETOS, a hastily arranged meeting took place on 12 November at the National Bureau of Standards (NBS) in Washington, D.C. to discuss the situation, and the EOTS. One can imagine that it was a classic confrontational meeting. Project SIGN staff was flown in, represented by Captain Sneider, Alfred Loedding, and also by SIGN's primary investigators, who were Lawrence Truettner and Albert Deyarmond. They were pitted against representatives from AFOAI and Major Boggs of Defensive Air. AFOIR was also there, along with Dr. George Valley of MIT, who represented the Air Force Scientific Advisory Board.

This was a very crucial meeting for TID and Project SIGN. They needed to present a strong convincing argument for their EOTS, and take a demanding confident position. They were the ultimate experts, and they possessed the ammunition needed, except for tangible evidence. Debate raged, but AFOAI controlled the winning hand in the form of Vandenberg's rejection of the EOTS.

An interesting aside to this meeting at NBS is that it was personally hosted by NBS Director Dr. Edward Condon. Twenty-one years later, in 1969, he and Robert Low (mentioned in Chapter 5) were instrumental in putting an end to the Air Force's "overt" investigation of flying objects.

Also, immediately prior to the meeting, Condon endured many months of grueling and intensive examination by the House Un-American Affairs Committee, which accused Condon of communist affiliation, and of having close association with Soviet spies and diplomats. He was rescued by many prominent fellow scientists who stood up for him, as well as President Truman, and he avoided serious consequences, except to his name and reputation. He was seriously challenged on this because of his highly questionable connections, and close involvement with the nation's highest secrets and technology. The FBI collected an extensive dossier on him, and he was continuously monitored until 1 December when a memo from the FBI stated that "technical surveillance was discontinued." This FBI memo came from Guy Hottel, who was in charge of the Washington, D.C. Field Office, and he would later author another exclusive and mysterious memo detailing reality of flying objects, which will be reviewed in Chapter 23.

TID Discredited

Of course, Vandenberg's immediate action in slapping down the EOTS did not sit well with those in TID who had worked very diligently and carefully, and with systematic analysis of the UFO question. The same goes for those in the chain of command who previously approved the Estimate. A through professional analysis of flying object incidents convinced them beyond doubt they had the correct answer. TID's skills in engineering and scientific technical analysis were previously without question, but now the team of quality professional technicians, engineers, scientists and experts were disheartened, disappointed, and demoralized. The great significance of this is that the Air Force would never again employ such a team of high-caliber investigators to confirm Truth of the UFO question. This situation provides testament to the reality of Roswell, and that there was Truth to what the Air Force and others were attempting to cover-up.

$$\wp\!\!\sim\!\!\wp$$

It is an amazing picture to behold. The Special Projects branch of TID, which consisted of talented, professional, and dedicated people, was a very valuable component of the U.S. Air Force. And it was summarily discredited without recourse. One sees a situation where many leading Air Force scientists and engineers, whose primary job was to analyze foreign technology, came to a consensus in their analysis of the UFO question. Their certain conclusion, without physical evidence in hand, was that there must be an extraterrestrial presence based on demonstrated and observed advanced technology of the flying objects.

Their conclusion was based on the notion that no nation on Earth could possibly develop and produce the objects observed. The objects appeared to be intelligently controlled, while exhibiting flight characteristics that no known technology could possibly duplicate. The objects managed to avoid and evade encounters by pilots attempting to pursue them, and they were visually observed by ground witnesses and confirmed by radar. Many incidents were observed by multiple people at the same time, and the witnesses were totally credible, technically articulate, and beyond reproach. These were incidents that caught particular notice, compared to a majority determined to be misidentifications of man-made objects or natural phenomena.

At this point in time, toward the end of November 1948, one thing was very clear. The Air Force was taking the UFO question very seriously. It was not just those harboring the secret of Roswell, and other recoveries, it was also those in the general Air Force and military who were continuing to witness, report, and investigate incredible antics of the flying objects. It cannot be emphasized enough that there were many in the Air Force who were not privy to the Truth of Roswell, but they knew the objects were very real, with attributes putting them in the realm of not being from this Earth. At the same time, it was still very difficult to grasp such unearthly reality. For many, however, there was no other alternative than to accept the extraterrestrial answer. And this became a great concern to the Control Group.

CHAPTER SIXTEEN

Project SIGN Terminated

Project SIGN: December 1948

Green Fireballs Appear

In late November 1948, a new phenomenon entered the skies that would soon capture attention, especially from government observers. At first, little notice was paid to what was described as speeding green balls of light seen in the northern New Mexico area, and traveling in a relatively low, flat trajectory, which was quite unlike ordinary meteors. They were likened to bright illuminated flares with a very distinctive green color. They would suddenly catch one's attention as they quickly flashed across the sky, and then they would quickly fade and blink out. TID investigators became aware of the sightings, but they chalked them up to possible flares or unusual meteors.

In early December, a flurry of fireballs was reported. On 5 December, a crew of an Air Force C-47 flying at eighteen thousand feet encountered what appeared to be a huge green meteor just east of Albuquerque, New Mexico. It suddenly appeared in front of them, low on the horizon, and it appeared to climb slightly upward and then level out. This was the second one they had seen after viewing one just minutes before near Las Vegas, New Mexico.

On the same date, pilots of a Pioneer Airlines DC-3, Flight 63, sighted a similar object about a half hour later. Upon landing at Albuquerque and describing the encounter, they said the Green Fireball approached them head-on, and they were required to quickly bank in a tight turn to avoid collision.

Reports of these fireballs were unique to the New Mexico area, except for one sighted near Hanford Nuclear Reservation in Washington State. They continued to appear regularly in the month of December, and the Air Force and Army took particular notice because of their proximity to sensitive military bases and nuclear facilities. Due to security concerns, meteor expert Dr. Lincoln LaPaz was contacted by intelligence officers at Kirtland AFB to investigate the matter. Over a period of several days, LaPaz discovered that dozens of people on the ground observed the objects, and he calculated that eight different fireballs were seen during that time. In a further study of the objects, Dr. LaPaz began to doubt that the Green Fireballs were meteors. Their nature didn't fit the mold, especially with most of them localized to the New Mexico area. If they were meteors, it would be expected that the objects would be observed worldwide due to Earth's rotation. This was a mystery, and it was not about to end. The objects would continue to be seen for many more months to come. But what were they? Were they a natural phenomena, unearthly flying objects, or man-made devices?

EOTS Destroyed

While some in the Air Force and a few others were learning of Green Fireballs, the public was not generally aware of them. But the media was making frequent inquiries to Air Force Headquarters and pressing for release of information on other flying objects sighted by the public. There was general public recognition that many credible people were becoming involved with sightings, and this did not equate with flippant remarks often made by the Air Force stating that flying objects did not exist. This awareness and concern was enough to prompt additional questions to the Air Force on what the objects were, or what was being done about them. A side effect to this was that it provided confirmation to General Vandenberg that the EOTS was a dangerous report to have around, or to be available for circulation, and Vandenberg quickly ordered that all copies be burned and destroyed.

Vandenberg realized that allowing the top secret EOTS to remain in Air Force files would, sooner or later, jeopardize the ultimate secret maintained by the Control Group. It could lead to possible revelation that the United States possessed recovered extraterrestrial vehicles and bodies. Vandenberg did not have a choice in the matter except to take quick action and destroy the EOTS. Those in Project SIGN had no choice except to acquiesce and accept a possible end to their project, and possibly their careers.

A few copies of the EOTS did escape to become souvenirs for posterity, and some of these were later seen and reviewed. Given more time, a surviving document may yet possibly reappear, perhaps now buried in government files or eventually recovered from someone's attic. If that happens, the valuable document will confirm the dedicated hard work of those in Project SIGN. Last seen in 1951, another resurrected copy could provide a trail of Truth.

AFOAI Releases Study No. 203

On 10 December, AFOAI released its long-awaited top secret report titled "Analysis of Flying Object Incidents in the United States," which was otherwise known as "Air Intelligence Division Study No. 203," or Report No. 100-203-79. This was AFOAI's final report, based on Cabell's 27 July request made after the Chiles-Whitted incident. It was AFOAI's answer to the EOTS produced by the Project SIGN team. Presented as a joint document between the Air Force Directorate of Intelligence and the Office of Naval Intelligence (ONI), this document was more docile and slightly more preferable, which meant that it was something top brass could live with. It dealt extensively on the possibility of Soviet involvement, while avoiding "extraterrestrial" reasoning, although some wording did not appear that way. In part, it stated:

IT MUST be accepted that some type of flying objects have been observed, although their identification and origin are not discernible.

AMONG THOSE incidents still not positively explained...
a. Most of the objects are a thin disk, round on top and flat on the bottom. The front half of the disk is often circular...
b. A high rate of climb as well as the apparent ability to remain motionless or hover for a considerable length of time is indicated.
c. Reported sizes have varied from that of a 25-cent piece to 250 feet in diameter, and from the size of a pursuit plane to the bulk of six B-29 airplanes.
d. Speeds have been estimated throughout the entire range from very slow or hovering to supersonic.
e. Sounds and visual trails are not normally associated with the sightings.

It is evident from the performance characteristics attributed to the unidentified objects at this time that if they are foreign, they involve efficiencies of performance which have not been realized in any operational airborne device in this country.

It is likewise impossible at this time to contain discussions of possible performance characteristics or tactics within limits of practical reason, if for no other reason than the fact that proof of the existence of a foreign development of this type would necessarily introduce considerations of new principles and means not yet considered practical possibilities in our own research and development.

It is significant that this report admits that "performance characteristics" of the objects would not be addressed because it would "introduce considerations of new principles and means not yet considered practical possibilities." With recognition that the objects seemed to possess performance characteristics not possible with current technology, it was tacit admission that highly advanced technology was likely involved with the flying objects, but there was no willingness to proceed beyond the sphere of current technical understanding. Of course, that is the same rational (not proceeding beyond "current practical reason") that limits current thinking of doubters, skeptics, and debunkers. When considering the reality of advancements in science and technology occurring during the last few years, which long ago may have seemed impractical or impossible, such an excuse would seem to be a flagrant cop-out. This convenient excuse, however, does have another side to it, because it also allows those in-the-know, and those keeping the secret, to label credible witnesses and serious investigators as "lunatics" for suggesting that flying objects might utilize far greater technology than one can currently imagine.

Those working on Project SIGN were highly-trained scientific and technical engineering professionals in the Special Projects branch of TID at Wright-Patterson. They were closely connected with the Directorate of Intelligence at Air Force Headquarters, and worked closely with AFOIR in the Office of Collections. Likewise, AFOIR dealt closely with the highly-skilled technical people at TID who were known for high standards in technical analysis. And both TID and AFOIR accepted the fact that flying objects exhibited characteristics that were far beyond any means of current human technology.

In comparison, one might consider that AFOAI, in their partnership with ONI, was working to squelch Truth. In analyzing sighting reports obtained from TID, they avoided interviewing witnesses and they ignored information they could not objectively relate to. It is quite clear that they preferred using "objective reality," and they were not interested in considering further possibilities. It appeared that they were only interested in creating their document in order to help General Vandenberg discard the extraterrestrial idea.

A Telling Scenario

What has been witnessed here is a telling scenario of internal conflict within Air Force ranks due to an Air Force cover-up. There was a mixture of internal Air Force political ramifications, and a conglomerate of views on flying objects. A few people in-the-know preferred that Truth be protected, while others believed it should be revealed. Some were not in-the-know, but convinced the objects were extraterrestrial. Some were suspicious that flying objects were demonstrating new Soviet technology, and they were concerned that these "real" objects were toying with our sensitive military installations, and with the public. There were also those who were highly critical of any notion about flying objects. But there were also professional Air Force investigators who concluded that flying objects were of extraterrestrial origin. Along with this, there was the reality of a Cold War threat that Air Force officials needed to contend with.

It's remarkable that Air Force intelligence, as a whole, exhibited much concern about flying objects. Concern within AFOIR, AFOAI, and TID was generated because of many sightings by the public, and also by credible military personnel. Without considering those who were in-the-know, it's very significant that the general Air Force took it all very seriously, especially those charged with flying object investigation. Investigators exhibited intense concern, and they were the first to confirm that flying objects were quite real. They then confirmed, except for AFOAI, that flying objects were extraterrestrial.

Project GRUDGE Proclaimed

On 16 December, amid all the turmoil regarding the UFO question, Air Force Research and Development Chief General Donald Putt at Air Force Headquarters announced that the code name of "SIGN" would be changed to "GRUDGE." This action, and the name chosen, can be considered amusing, but it was also a signal that big changes would be in store. It would later be confirmed, by Edward Ruppelt (of Project BLUE BOOK fame), that the GRUDGE name "did have significance, a lot of it." No doubt, it reflected considerable internal conflict at Air Force Headquarters.

Originally confirmed by General Twining, in his 23 September 1947 letter, that flying objects were "real and not visionary," flying objects continued to be observed in sightings by competent people, especially military personnel. Accordingly, Project SIGN was created to monitor their ongoing presence, while also keeping tabs on what seemed to be their harassing nature. It was necessary to determine whether they posed a hostile threat, and what their intentions might be. This project provided the general military with a place to report sightings, but it also provided the Control Group with needed information on exactly what the objects were up to.

The unadvertised purpose of Project SIGN was to act as a public relations front, and it became known to the public and media as Project Saucer. The project, however, was not allowed to provide reality to the extraterrestrial idea. Those at the top, and in-the-know about Roswell, could not allow release of the secret, but it was necessary to determine what the objects were all about, including their tactics, purpose, nature, and origin. Serious and continuous investigation would be required, but only to a certain extent. The Control Group needed to obtain information on the flying objects, but Truth of them would need to be protected, especially from project investigators and the public. This wasn't working out, however, and now it was time to rename the project and begin again.

Project SIGN Historically Significant

With rejection of Project SIGN's "Estimate of the Situation," the fate of Project SIGN was sealed before its completion. It stands out, however, as the only time the Air Force took a realistic approach and true accounting of the UFO question. It's also the only time that a substantial number of very bright and highly respected civilian scientists and engineers, including a number of highly-qualified Air Force personnel, became involved to such a great extent. They were highly-regarded specialists looking into foreign technology matters, and they carefully weighed all evidence available to them from various incidents and encounters. They took into account many witnesses who were confirmed to be respectable, responsible, reputable, competent, and credible beyond reproach. Investigators carefully weighed all evidence, and then came to an honest and objective "Estimate" that they were duty bound to provide.

Hynek stated:

> ...it was called Project Sign, and some of the personnel at least were taking the problem quite seriously. At the same time a big split was occurring in the Air Force between two schools of thought. The serious school prepared an estimation of the situation which they sent to General Vandenberg, but the other side eventually won out and the serious ones were shipped off to other places. The negatives won the day, in other words.

Without doubt, Project SIGN would have been a historical breakwater of huge significance if project investigators had been provided special access to physical evidence, and also access to people who were involved in the incidents at Roswell and other places. SIGN investigators could have been considered heroes if Vandenberg accepted the Estimate, and then let the project continue toward eventual discovery of Truth. Instead, the Air Force elected to press forward in managing a huge and highly classified cover-up of the UFO question. Project GRUDGE was not yet official, but it would soon take over with a great amount of intimidation in the air. There would also be additional caution on the part of those in-the-know to insure that any continued investigation would not get out of control.

Even if you take Roswell out of the picture, well-respected professional Air Force investigators from Project SIGN still came to a final conclusion that sightings and incidents involving flying objects, reported by credible people in 1947 and 1948, were of unearthly origin. It is truly significant that the quality, capability, and talent of those investigators would never be matched or duplicated again in subsequent and continued investigation of flying objects. When one contemplates this even further, it becomes evident that Project SIGN, as well as the EOTS, was a very significant piece of U.S. history. Project SIGN, however, was abruptly terminated by the Air Force, and has been kept away from general public knowledge since that time.

ఞఎ

One might wonder, would General Vandenberg's quick and sudden disposal of Project SIGN provide a possible glimpse into his thinking or reasoning on the UFO question? Was he trying to eliminate the possibility of the Air Force looking silly in accepting extraterrestrial reality without tangible evidence? Was he thinking the public would not receive news calmly regarding an extraterrestrial answer, which might lead to possible panic? Was he trying to protect highly valuable technology and information obtained from recovery of a flying object? Was he trying to avoid acknowledgement that the Air Force was powerless to deal with flying objects? Was he trying to maintain a cover-up about something that only a select few in government, or the Control Group, had knowledge of? Was he attempting to protect people in-the-know? Perhaps the answer to all those questions is "Yes," but he was also attempting to satisfy desires of higher authority, and the Control Group.

AAF recovery of a crashed unearthly object containing occupants in July 1947, along with subsequent release of the EOTS by independent Air Force investigators, should be considered significant history in regard to the UFO question. Additional perspective will come to light on that history as Air Force involvement with the UFO question is looked into further. It will be observed, however, that protection of the closely-held secret, which resulted in the cover-up of Roswell and the demise of Project SIGN, portends the same for succeeding projects as dictated by circumstances and events impacting the secret of the UFO question.

Investigators in Project SIGN, who were involved in the first official U.S. Air Force project for investigation of flying objects, came to a significant conclusion regarding the UFO question, but what would happen to them? Would they continue to remain in their jobs? Would they transfer to the new Project GRUDGE? How would investigations take place in the new project? How would those in control of the UFO question proceed to insure continued security of the magnificent secret?

The most important thing to note about all this is that the UFO question was a very serious concern to the Air Force at this particular time, and it played a serious role in disrupting and causing interference in Air Force operations. It was quite costly in a number of respects, and it was simply because the Truth of Roswell, and the UFO question, needed protection from general public knowledge.

CHAPTER SEVENTEEN

SIGN's Last Gasp

Project SIGN: January - February 1949

CIA's Office of Scientific Intelligence

On 1 January 1949, the CIA's new Office of Scientific Intelligence was established with help from Lloyd Berkner. This is where many high priority and sensitive projects would be based. The CIA had been continuously involved with the UFO question since the time of Roswell, and it was closely monitoring and collecting information from TID. The UFO question was now one of those high priority CIA projects that this organization would closely monitor and help manage behind the scenes, but in coordination with the RDB.

It will be recalled that Berkner was secretary/chairman of the Executive Council of the JRDB under Vannevar Bush during the time of Roswell, and he was also in the same position in the RDB. He was the one who was effectively managing the large organization. As a distinguished scientist, and one known for having exceptional managerial capability, he was well-connected within the military, industrial, and educational/scientific communities. He was particularly noted for having important influential connections between the science of technology and the clandestine world of the CIA, and he was a man who could be called upon by both the CIA and high officials within the military to get difficult things accomplished. He possessed superior expertise to effectively deal with huge technical problems involving current and future military goals, and he knew how to set up and compartmentalize areas so that work could be accomplished securely and independently. There is no doubt that Air Force Chief of Staff Vandenberg, and several who were in-the-know, knew this man quite well, and depended on his services.

Berkner was associated with the Control Group, which resided within the RDB. He was closely involved with maintaining the secret of Roswell and relying heavily on the CIA to provide intelligence information. But the RDB was a relatively unstable organization at this particular time, and it was a difficult place to keep a monumental secret secure. It was probably the same for many other scientific and technological efforts involving "black projects" supported by the RDB. Because of this, it's quite likely that the CIA's OSI was under consideration as a potential future home for managing and protecting Truth of the extraterrestrial secret.

Dr. Donald Menzel

In regard to Berkner and his many contacts in various scientific fields, one of his acquaintances was Dr. Donald Menzel, who was a renowned astrophysicist from Harvard University. Menzel was a greatly admired and respected public figure, and his fame and talent were well-known. His intellect and character were unsurpassed, and this placed him as a top authority in celestial and astronomical matters. In addition to this, and unknown to the public, he led a double life, which now involved a parallel intelligence career with the CIA. During WWII, he had been an expert cryptanalyst, managed a section of a Naval security group, chaired a committee under the Joint Chiefs of Staff, and maintained a top secret "ultra" security clearance. He and Dr. Vannevar Bush came to know each other quite well when both worked under the Joint Chiefs of Staff.

Menzel's double life was an unknown factor to the public, which allowed him to become heavily involved with securing the UFO question and managing a great amount of debunking that went with it. He is mentioned here because both he and Berkner will be mentioned in succeeding chapters as they intercede to covertly control the UFO question.

Press Involvement - Sidney Shalett

One can speculate that turmoil regarding the UFO question at Air Force Headquarters was leaking out, and that the press may have been getting wind of something going on. This is understandable when considering rejection of the EOTS by Air Force Chief of Staff General Hoyt Vandenberg, and then his ordered destruction of the document. Also, the situation occurring at the joint meeting held at the National Bureau of Standards on 12 November 1948, and then release of "AFOAI Study 203" didn't help either. There must have been much talk slipping out, which may have caught the attention of many. Secretary of Defense Forrestal was probably one of them. Others would include Dr. Vannevar Bush of the RDB Oversight Committee, and Roscoe Hillenkoetter of the CIA.

Another of those whose attention was possibly attracted by talk leaking from Air Force Headquarters was Sidney Shalett, who was a writer for the *Saturday Evening Post*. At this particular time, he was given a grand opportunity to look into Air Force handling of the UFO question.

Details on how he became specifically authorized for this opportunity are very interesting. What is known provides fodder for intrigue. One would suspect, however, due to recent turmoil involving the UFO question at Air Force Headquarters, that neither Cabell nor Air Force Secretary Symington, or Vandenberg in particular, would want Shalett's involvement. It was not an appropriate time.

Contributing to this, however, was Deputy Public Relations Director Stephan Leo, who was in charge of a large office under Air Force Secretary Symington at Air Force Headquarters. He was a highly respected former newsman, who had direct access to many high level brass at the Pentagon, including Secretary of Defense General Forrestal. Leo was also a close friend of Shalett's, and this served to facilitate a meeting between Forrestal and Shalett. It led to Shalett receiving permission by Forrestal to look into the Air Force's investigation of the UFO question.

Previously, Forrestal planned to ease out of office during the month of November 1948, but only if Truman was defeated for reelection. Even though Truman was not defeated, things began to go sour for Forrestal. Pressure was being directed at him from all directions, with no apparent support from Truman on anything. A big reason for this is that Forrestal, and his department of defense, possessed no leverage with military services. All Forrestal could do was attempt to make agreements and suggestions on courses of action with the equally powerful secretaries of each service, and this put him at great disadvantage with headaches that couldn't be resolved. As previously mentioned in Chapter 14, his inability to effectively coordinate between government agencies is one of the reasons that Forrestal's friend, Ferdinand Eberstadt, formed a task force in June 1948 to help resolve this, and this study was still ongoing.

On 11 January, Forrestal submitted his offer of resignation to Truman with the idea that he would "ease out" of office within six months after his resignation was accepted. Truman, however, asked Forrestal to stay on.

Forrestal was instrumental in putting together the Control Group, but he was also partial to having Truth of the UFO question revealed, which Project SIGN had the potential of doing before the hatchet fell on it. In late January, with authorization from Forrestal, Stephan Leo was able to finesse special access for Shalett, who would work on his story over the next two months. Harry Haberer, from Air Force Public Relations, was directed to shepherd Shalett, and provide special access for his investigation, including opening files held at TID. The files were immediately made available for Shalett's inspection, and the only restriction placed on Shalett was that he could have no information from foreign sources, or anything top secret, but he was free to interview people made available to him. This was a considerable coup for Shalett, and it leaves one with thoughts about Forrestal's motivation for giving Shalett authorization.

Green Fireballs Attract Attention

Previously, in late November, and during December 1948, there had been concern about the Green Fireballs, which were now quite prevalent in southwestern skies. The fireballs were particularly centered near sensitive outlying areas surrounding Albuquerque and Kirtland AFB in New Mexico. Many reports were coming in to the Air Force Directorate of Intelligence and also TID, and the general feeling among certain elements in the military was that the objects were potentially dangerous and threatening, with no assurance they weren't. The Green Fireballs seemed to have a malicious nature to them, and they didn't act like meteors. Some people were concerned they were of Soviet origin. On 13 January, an Army G-2 intelligence memo stated the following:

> *Agencies in New Mexico are greatly concerned over these phenomena. They are of the opinion that some foreign power is making 'sensing shots' with some super-stratospheric devise* [sic] *designed to be self-disentergrating* [sic]. *They also believe that when the devise* [sic] *is perfected for accuracy, the disentegrating* [sic] *factor will be eliminated in favor of a warhead.*

On 31 January, an FBI memo stated:

> *… This matter is considered top secret by Intelligence Officers of both the Army and the Air Force.*

> *…various sightings of unexplained phenomena have been reported in the vicinity of the A.E.C. Installation at Los Alamos, New Mexico, where these phenomena now appear to be concentrated. During December 1948 on the 5th, 6th, 7th, 8th, 11th, 13th, 14th, 20th and 25th sightings of unexplained phenomena were made near Los Alamos by Special Agents of the Office of Special Investigation; Air Line Pilots; Military Pilots, Los Alamos Security Inspectors, and private citizens.*

> *…Night-time sightings have taken the form of lights usually described as brilliant green, similar to a green traffic signal or green neon light…*

It is noted that no debris has ever been known to be located anywhere resulting from the unexplained phenomena.

…When observed they seem to be in level flight at a height of six to ten miles and thus traveling on a tangent to the earth's surface.

Also, on 31 January, a message from the commander of Kirtland AFB to USAF Chief of Staff Vandenberg stated:

Reference previous reports, subject: unknown, aerial phenomena… Sighting of identical object reported…by aprx 30 people. Estimate at least 100 total sightings. AEC, AFSWP, 4th Army, local commanders perturbed by implications of phenomena. Sighting reported from El Paso, Albuquerque, Alamogordo, Roswell, Socorro, and other locations. All appear to be same object at different points in trajectory.

Dr. Lincoln LaPaz, who was mentioned in the last chapter, was a mathematician and meteor expert teaching at the University of New Mexico, and he became heavily involved with investigation of the fireballs. Coincidentally, he was once a colleague of Alan Hynek during WWII in development of the military's proximity fuse. LaPaz also saw a fireball, as did his wife. After extensive study and interviews with many people, he wrote in his notes that he was "now convinced the various 'green flare' incidents reported to the O.S.I. are not meteoric in nature." He believed they were man-made flying missiles of some kind. His reference to the Air Force OSI is interesting, because nearly all sightings of Green Fireballs were investigated by that organization. But the reports it sent to TID were simply filed away.

SIGN's Last Day

On 11 February, the last day of Project SIGN, there was no concluding final report for the project, but a secret Technical Report (No. F-TR-2274-IA) titled "Unidentified Aerial Objects Project Sign" did get produced by TID. It was authored by Lawrence Truettner and Albert Deyarmond. This was their final assignment on the project, and it was the last in a series of periodic technical reports for Project SIGN. It was approved under the names of Colonel McCoy and Colonel Clingerman, but Clingerman signed for his boss McCoy, and Colonel Miles Goll signed for his boss Clingerman.

This report would keep the status of flying objects officially unidentified rather than extraterrestrial, although the extraterrestrial idea was not discarded, and it received a fair amount of comment and discussion. This was probably accomplished with a bit of contemptuous intent by Truettner and Deyarmond in response to AFOAI's Study No. 203, and rejection of the EOTS.

In the SIGN Technical Report, the extraterrestrial idea was not totally dispelled, and some of this was displayed in Appendix C of the report, which was a "discussion" by Dr. George Valley of MIT, who was on the Air Force Scientific Advisory Board. The discussion was titled "Some Considerations Affecting the Interpretation of Reports of Unidentified Flying Objects," and one section stated the following:

> *If there is an extraterrestrial civilization which can make such objects as are reported then it is most probable that its development is far in advance of ours. This argument can be supported on probability arguments alone without recourse to astronomical hypotheses...Such a civilization might observe that on Earth we now have atomic bombs and are fast developing rockets. In view of the past history of mankind, they should be alarmed. We should, therefore, expect at this time above all to behold such visitations.*

It is interesting to note that Valley mentioned that we should expect extraterrestrial visitations because of our alarming past history in development of atomic weapons. This observation has attracted much attention since then, which seems to be substantiated by the fact that so many sightings and encounters have been associated with nuclear installations, and also witnessed by missileers at Minot AFB and other places with nuclear involvement.

Also, a letter by Dr. James Lipp of the RAND Corporation was included as Appendix D, which was a copy of a letter addressed to General Donald Putt. It was dated 13 December, which was three days prior to Putt's announcement that Project SIGN would become Project GRUDGE. The letter was a response to Putt, who previously sent a letter to Dr. Lipp on 18 November regarding the "Flying Object Problem." Lipp's response addressed the "likelihood of a visit from other worlds as an engineering problem." It was primarily a discussion of chances of life existing elsewhere in space, and the potential of receiving visits from Venus or Mars. The last part of his letter stated:

> *Only one motive can be assigned; that the space men are "feeling out" our defenses without wanting to be belligerent. If so, they must have been satisfied long ago that we can't catch them. It seems fruitless for them to keep repeating the same experiment...Although visits from outer space are believed to be possible, they are believed to be*

very improbable. In particular, the actions attributed to the "flying objects" reported during 1947 and 1948 seem inconsistent with the requirements for space travel.

Truettner and Deyarmond also included a summary in their report, where some very interesting connotations can be read between the lines. Part of the summary stated:

The possibility that some of the incidents may represent technical developments far in advance of knowledge available to engineers and scientists of this country has been considered. No facts are available to personnel of this command that will permit an objective assessment of this possibility. All information so far presented on the possible existence of space ships from another planet or of aircraft propelled by an advanced type of atomic power plant have been largely conjecture. Based on experience with nuclear power plant research in this country, the existence on Earth of such engines of small enough size and weight to have powered the objects described is highly improbable.

In this statement, the two remaining analysts on Project SIGN were able to indirectly state their extraterrestrial conclusion. They indicated that some incidents were analyzed where highly-advanced technical aspects of the unknown objects, which were like "spaceships from another planet," were considered to be far beyond knowledge available to our scientists and engineers, and that it was highly unlikely that any technology on "Earth" exists to power such objects. This was another way for those investigators of Project SIGN to indicate that a domestic origin was not an option for some of the sightings. However, the last paragraph of the report stated that the possibility of the aerial objects being visitors from another planet, "will not be further explored...pending elimination of all other solutions or definite proof of the nature-of-these-objects."

The recommendation of this last SIGN Technical Report stated:

...future activity should be carried on at the minimum level necessary to complete investigations in progress...special project status should be terminated when it was determined that sightings no longer represented a threat to the nation's security.

Dr. Hynek was not specifically challenged or confronted in the demise of Project SIGN, although Project GRUDGE would never find a use for him. Hynek did note, however, that the change at TID "was rapid." Civilians Deyarmond and Truettner stayed at TID, although they were separated from the Special Projects branch. Alfred Loedding was located elsewhere by the end of 1948. In disgust, he was forced by "treatment" he was receiving to resign from Wright-Patterson in February 1951. Colonel Clingerman was replaced by Colonel A. J. Hemstreet. Colonel McCoy remained, although he was involved with other things, including the magnificent secret, but he disappeared from the spotlight. Other people who remained, still connected to the Branch, were Lieutenant Howard Smith, George Towles, and Miles Goll. Both McCoy and Goll were in the regrettable position of having to remove the valuable people in the Special Projects branch.

৵৶

With completion of the SIGN Technical Report, the project came to a screeching halt, and this was the desire of General Vandenberg. The tipping point for the project was that it was permissible to state that flying objects were "real," but not permissible to state they were "extraterrestrial." This same "rule" would hold for Project GRUDGE, but it would now be understood by its personnel.

The most important thing about this situation is that it was still necessary to find out whether the nation's security was threatened by flying objects. For the Control Group, Project SIGN was an effort to learn more about the objects. For those not in-the-know, however, there was great concern by the general military and Air Force about the potential threat posed by the flying objects, which were occasionally observed and encountered. Additionally, many in the public sector were also concerned. They wanted continued investigation of the flying objects, and a place to report sightings and encounters.

At this particular time, high officials in the Air Force were without options. It was still necessary to learn more about, and track, the activity of flying objects. The Control Group was relying on the Air Force to do this. But it would be a contorted scenario of investigation and denial, while attempting to protect the magnificent secret, and nothing would change in that regard. This scenario is all the evidence one needs to realize that the Air Force was actively involved in covering-up Truth of the UFO question.

The Air Force and Control Group turned away a chance to reveal Truth to the world with Project SIGN. Their choice was to forsake the opportunity and press ahead with further contorted gyrations of investigating and denying, and tossing innocent people aside in the process.

How could they not do otherwise? Would it be a good thing to admit they were powerless to confront the objects, which might then spread fear throughout the land?

PART FIVE

Project GRUDGE

ೞ๑

A quick, inglorious end was reflected in the demise of Project SIGN, all because professional Air Force investigators became convinced of an extraterrestrial answer to the UFO question. The investigators became expendable, not because high Air Force officials thought they had become looney, but because Truth of the UFO question needed to be kept secret. This was the legacy of Project SIGN. It was a very disappointing time, if not a confusing one for SIGN investigators. If they had been in-the-know, and revealed the secret of Roswell, they would have faced grave consequences. But they were not in-the-know, and they were still ostracized.

It was a "twisted" effort by high Air Force officials to discover what was behind the mysterious flying objects, but the effort could not be abandoned. Further information was needed on the objects by the Control Group. At the same time, many in the general military were witnessing sightings and encounters, and this was generating considerable concern. So the investigation would continue with Project GRUDGE, but also with new ground rules.

Remarkably, Project SIGN was just a small taste of things to come for Project GRUDGE. The Air Force would eventually find itself tied in knots, and with its hands tied behind its back on the UFO question. It could not compete with the many sightings by competent witnesses, or journalists who were intent on writing articles about Air Force investigative involvement, or Air Force duplicity on the matter. Flying objects were not going to disappear and fly away to another place in the universe. Air Force attempts to point out that flying objects were merely figments of the imagination, or fantasies gone wild, would not help the situation. It simply created intense anxiety in sighting witnesses and others. Through it all, the Control Group would closely watch behind the scenes, which would be revealed and confirmed in official documents discovered years later.

Project GRUDGE was to become a very contentious period in annals of U.S.

Air Force cover-up history, with periods of relative calm interspersed with periods of very intense frustration within Air Force Headquarters. Paradoxically, which is difficult to contemplate and imagine, it was a problem due to what the Air Force previously proclaimed as "real" flying objects, but then relegated to the status of "nonexistent" objects.

The Air Force would, however, end Project GRUDGE on a somewhat positive note. It would begin proactive investigation of flying objects without rendering an "opinion," or a suggestion of Truth. It was similar to a "no ask, no tell policy," which was in vogue on a different subject years later.

CHAPTER EIGHTEEN

New Project - New Problems

Project GRUDGE: February - April 1949

Project GRUDGE Takes Over

Rejection of the "Estimate of the Situation" by General Vandenberg signaled demise of Project SIGN. The EOTS concluded that analysis of flying objects, even without tangible evidence, were determined to be of extraterrestrial origin. The project was initially set up because of Air Force concern about the objects, but there was also a Control Group that needed to keep track of the objects, and to determine what they were up to. Although the general military was experiencing frequent sightings and encounters, the public was mostly unaware of Air Force involvement, except when Project Saucer was sometimes referred to by the press. The Air Force was not about to ignore the UFO question, although it would never again allow serious investigation by professional Air Force technical analysts, engineers, and scientists.

Chief of Research and Development General Donald Putt at Air Force Headquarters had announced on 16 December 1948 that the code name of SIGN would be changed to GRUDGE, and on 12 February 1949 it became official. Project SIGN then disintegrated into an inglorious and ignominious end. It is not known if Putt received orders from Vandenberg to make necessary changes, but it must have been very obvious to Putt that the existing project was in limbo, and faced a brick wall.

Because such projects were created under auspices of the Air Force Research and Development organization, the name change was purposely made by General Putt, since Project SIGN was essentially defunct for most project personnel. It was necessary to now protect possible compromise of the on-going project with a new name. It is amusing and quite appropriate that the name "GRUDGE" was chosen, when considering the conflict that existed at Air Force Headquarters. Allen Hynek commented on the name change as follows:

> *My final report...was issued after Project Sign had somewhat mysteriously been transformed into Project Grudge, on February 11, 1949.*

Personnel on Project SIGN, who were still around in the Special Projects unit, and leaning toward the extraterrestrial idea, were left with two choices. They either needed to do an about face on the UFO question, or get out of the program altogether. Either way, their careers were not in a favorable position. If they maintained the extraterrestrial idea, they could not survive in their jobs. If they discarded the extraterrestrial idea, their reputation was still tarnished. As a result, nearly all SIGN personnel disappeared from the project, probably by 12 February. Those individuals who held their ground were quickly gone, and those who contemplated the extraterrestrial idea were soon replaced.

It was apparent, as Project GRUDGE continued in place of Project SIGN, that few of the original people remained. In fact, the primary investigative talent at TID was represented by Lieutenant Howard Smith, and he possessed little experience and was receiving little help. The basic premise now was that investigation of sightings would be based on pure simple objective science, and the flying objects would be categorized as misidentification of known aircraft, or occurrences of natural phenomena. If any investigation was to be accomplished, this was the modus operandi for Project GRUDGE. In other words, extraterrestrial flying objects could not and did not exist, and they would not exist after Project GRUDGE was finished. If an object could not be identified, it would be listed as "unknown," and it would remain unknown without a suggestion of it being extraterrestrial.

Shalett Creates Concern

In February and into early April, Sidney Shalett of the *Saturday Evening Post* proceeded to spend considerable effort in his review of Air Force investigation into flying objects. General Cabell was not happy about Shalett receiving access to Project GRUDGE, but Cabell was in a cumbersome position, and not able to do much about it. He did initiate a request to rescind approval, but he didn't get satisfaction, and this must have frustrated him greatly. As head of the Directorate of Intelligence, and charged with resolution of the UFO question, Cabell thought he should have absolute control in the matter, and this may have been Vandenberg's thinking also, but Cabell obviously didn't have control. This must have been a great concern to many others also, including the Control Group. In fact, it may have been terrifying to think that Shalett might find out about the top secret EOTS, which was supposedly "destroyed," although a few copies were still floating around. It would be disastrous for Shalett to take on the view of Project SIGN.

It was hoped Shalett would adopt current thinking of high officials, and take on the mind-set of Project GRUDGE, but there was no guarantee. All the Air Force could hope for was to convince Shalett that flying objects had logical and mundane answers. They would continue to monitor his questions, his prevailing attitude, and attempt to deduce his demeanor on the subject.

With access to TID's secret documents, sighting reports, and investigations, it's likely Shalett reviewed the secret SIGN Technical Report. The Air Force probably understood he would be using information gained from it, and that he would also have information from about two-hundred and fifty sighting cases, including information gained from staff interviews.

Cabell knew it would be necessary to prepare a countering or clarifying document, or something to negate Shalett's anticipated article upon its publication in late April. To facilitate this, the Air Force would take an option to review a preliminary draft of his article.

Further Changes at AMC

On 15 February, Chief of AMC General Joseph McNarney took on a new job at Air Force Headquarters, and was replaced by AMC Deputy Commander General Benjamin Chidlaw. Biographies of both men indicate their transfer took place on 1 September 1949, but both reported to their new duties much earlier, which was in mid-February. McNarney became chairman of the Defense Management Committee, working for Defense Secretary Forrestal, and he became a dominant force in government acting as an interface to NSC and the CIA, which is noted in released top secret documents.

It must be considered that McNarney's transfer took place at a significant time for him, especially with all the commotion surrounding the UFO question, and also his TID organization. It would be interesting to learn more of what was happening behind the scenes in that regard, including his possible connections involving Shalett, Cabell, Vandenberg, Forrestal, and the Control Group. There must have been extreme turmoil occurring at this specific time at Air Force Headquarters, where Forrestal and McNarney were partial to revealing Truth.

Chidlaw is remembered as the one who received Roswell debris from General Clements McMullen at Wright Field in July 1947. He was now acting commander of AMC until 1 September when his new orders would become effective.

AFOIR Memorandum Number Four

Despite a realigned mind-set in Project GRUDGE, flying objects were not about to disappear. In fact, sightings of flying objects, aside from Green Fireballs, were increasing in frequency and importance, and they were generating more concern for many in the Air Force, and also the FBI. The FBI preferred to stay apart from investigation of incidents, but it was still necessary to forward any information coming to its attention.

In the Directorate of Intelligence, particularly in AFOIR, the sudden increase in flying object activity was a definite concern. The concern was always present, but now it was deemed necessary to release "Air Intelligence Requirements Memorandum Number 4," which was issued worldwide on 15 February. It was titled "Unconventional Aircraft," and signed by General Cabell. Its stated purpose was:

> *To enunciate continuing Air Force requirements for information pertaining to sightings of unconventional aircraft and unidentified flying objects, including so-called Flying Discs... To establish procedures for reporting such information.*

This memorandum specified reporting requirements and information desired in the transmittal of flying object sighting reports. But release of the memorandum is interesting for a number of reasons. It indicated that AFOIR was continuing to have an interest in investigation of flying objects, and it was not about to give this up, especially with concern increasing about the objects, and what they might be. They needed to be identified. Not only that, it reflected General Cabell's priority for continued investigation of the objects. The memo specifically indicated that stateside commands would initially report incidents and sightings to MCIAXO-3 (Project GRUDGE) at Wright-Patterson, with copies sent to the Air Force Directorate of Intelligence. Overseas units would send reports directly to the Directorate of Intelligence, where they would then be forwarded to MCIAXO-3. General Cabell was still demanding that reported sightings of flying objects be investigated, and he expected Project GRUDGE to uphold its responsibility for investigating sighting reports.

When FBI Director J. Edgar Hoover received the Air Force memorandum, he sent a memo to all FBI offices with a copy of the memorandum, and stated that no FBI investigation of flying objects would be attempted. The Air Force memorandum would be used as a reference, however, for "securing data from persons who desire to voluntarily furnish information relating to flying disks." Hoover also stated the following:

...a reliable and confidential source has advised the Bureau that flying disks are believed to be man-made missiles rather than natural phenomenon. It has also been determined that for approximately the past four years the USSR has been engaged in experimentation on an unknown type of flying disk.

This statement from Hoover relates to the fact that a number of people had the idea that Green Fireballs were some kind of missile, and the Soviets were often thought of as responsible for them, and other flying objects. But there was good reason for this because fiery Ghost Rockets, previously seen over Sweden and the Baltic in 1946, were sometimes reported circling great distances, and sometimes looping back as if under remote control. The only country with potential capability of doing that was the Soviet Union. There would now be increased attention to Green Fireballs and other objects.

Los Alamos Conference on Green Fireballs

On 16 February, a "Conference on Aerial Phenomena" was held at Los Alamos, New Mexico, due to great concern in the local area regarding sightings of Green Fireballs. Fourteen people attended, including representatives from the Army, OSI, FBI, Atomic Energy Commission, and a number of scientists, including Dr. Edward Teller and Dr. LaPaz. It was primarily an information gathering session to look at the scientific aspects of the fireballs, with discussion on how Green Fireballs differed from normal meteors, and what they might potentially be. The subject was creating serious attention, because many of the scientists and security personal associated with sensitive nuclear facilities in the area had observed the mysterious objects, and they wanted to get a handle on exactly what they were.

A FBI memorandum on 16 March referenced this meeting. It indicated that the subject was considered secret, and that "investigations concerning such matters have been given the name 'Project Grudge'." It's evident that the FBI did not have the whole story, or know the connection between Project SIGN and Project GRUDGE, but the FBI now knew of the new project.

Forrestal Fired

On 1 March, Truman abruptly asked for Forrestal's resignation. This occurred after Truman's friend, Louis Johnson, paid Truman a visit that day, which evidently prompted Truman to immediately inform Forrestal that Johnson would be taking over as secretary of Defense on the first of April. It was an immediate and abrupt turnaround for Truman, after having previously assured Forrestal he would continue to have his job. This sudden reversal leads one to further speculate on this decision.

On 28 March, Forrestal made his final farewell at the Department of Defense, which was his last public appearance. On 2 April, after suffering an "acute" mental collapse, Forrestal was admitted to Bethesda Naval Hospital, and he remained there in "isolation" for about seven weeks until his end.

One is tempted to consider a possibility that Forrestal was sacrificed for the sake of protecting the secret of Roswell. Forrestal was known for wanting Truth revealed, and also for sponsoring Shalett's investigation of Air Force activity regarding the UFO question.

Another viewpoint, however, is that the Air Force was earnestly attempting to quiet public interest in the UFO question by using Shalett. His usefulness, however, hinged on whether the Air Force could convince Shalett that flying objects were nothing to be concerned about. The one thing favorable to this is that Shalett was under escort, and he was continuously monitored in regard to information being collected. Some have connected his personal escort to the CIA.

The other side to this is that it was not a particularly good time for the Air Force to get involved with Shalett, especially with active conflict at Air Force Headquarters on the UFO question, and with Project SIGN becoming Project GRUDGE. In fact, Shalett's visit coincided with release of SIGN's Technical Report. At this particular time, it's highly unlikely that there was any incentive to employ Shalett for the purpose of quieting public interest in flying objects. But it was highly likely that people in the Control Group were very unhappy with Forrestall's authorization allowing Shalett to probe the U.S. government's most critically important secret.

Upon receipt of Shalett's draft article in early April, the Air Force's primary objection was remarks lambasting the Air Force's investigative project, indicating it was a wasted effort. It also disagreed with remarks by Shalett that inaccurately stated official Air Force policy.

Walter Winchell Creates Concern

Renewed concern was building that the Soviets were involved with flying objects, and also Green Fireballs, and this was brought to public attention by the famous newscaster Walter Winchell. On 3 April, he announced the following on his newscast:

> ...flying disks seen in this country definitely emanated from Russia... This will not be confirmed by anyone in authority in Washington at this time, but if anybody denies it the denier will be a liar. The flying saucers, never explained by anyone in authority are now definitely known to have been guided missiles shot all the way from Russia.

On 4 April, G-2 of the 4th Army sent a letter to the director of the FBI stating:

> ...*it is suggested that the Bureau may desire to arrange to have Mr. Winchell interviewed concerning the source of his information that "flying discs" emanate from Russia.*

> *It would be appreciated if the Bureau can supply any information that can be transmitted to military authorities which would clarify this matter.*

The Air Force responded to Winchell's claims on 8 April with the following:

> *To date there has been no tangible evidence which would support a theory that any of the incidents are attributable to activity of a foreign nation. On the other hand, there is no evidence to deny categorically such a possibility...However, there are some incidents reported by reliable and competent observers which are still unexplained.*

On 10 April, Winchell received a letter from the mayor's office in Hollywood, California, (later obtained by the FBI) that described a flying object incident on 8 April. It stated:

> ...*I am satisfied you are very much right in your statements and I am inclined to call a person a liar that says (there are no dis'c* [sic]*).*

On 26 April, the San Antonio FBI office notified the director of the FBI in regard to a possible interview with Winchell, and it stated:

> *For your strictly confidential information...data...was obtained in confidence from a colonel in the United States Air Material Command, who obtained his information from persons actively engaged in the investigation of this subject...No interview with Walter Winchell will be made by the Bureau concerning the source of his statements...*

Sighting by Charles Moore at White Sands

Flying objects, other than Green Fireballs, were still being noticed. On 24 April, Charles Moore and his four General Mills technicians at White Sands Proving Ground in New Mexico reported that a large disk-like object was observed through their sighting instruments. The object was circling and maneuvering near their high altitude research balloon they were monitoring. Moore, an expert associated with Project Skyhook, and previously the head of Project Mogul for the Air Force, said the object traveled at speeds of several miles per second, made sharp turns, and accelerated very quickly out of sight.

Later, astronomer Dr. Donald Menzel confidently confirmed that Moore sighted a mirage of his balloon in the sky. Moore later wrote:

> ...What I saw was not a mirage; it was a craft with highly unusual performance. It was not a balloon...I am cynical about Dr. Menzel and his approach to science.

Being an accomplished physicist, engineer, and meteorologist, Moore was not happy with Menzel. And sometime after that he told Allen Hynek he was disgusted with the Air Force for ignoring his sighting report. Of course, Moore was not aware that Project GRUDGE was essentially defunct, and he was not aware of the background of Dr. Menzel, which was mentioned in the last chapter.

AFSAB Takes Notice of Green Fireballs

Before Moore's sighting at White Sands, General Cabell was aware that little activity was taking place in Project GRUDGE. He was further advised of this on 25 April when Colonel Doyle Rees, commander of the 17th AFOSI District, contacted Air Force Headquarters regarding a significant lack of response, or interest, from Project GRUDGE on reports sent to it, particularly the reports on Green Fireballs. Project GRUDGE was not interested, and considered the fireballs to be meteors, which was the easy answer.

It was then that General Cabell contacted Theodore Von Karman, who was board chairman of the AFSAB, and they decided to have Dr. Joseph Kaplan of the AFSAB look into possibility of setting up a comprehensive investigation of the fireballs. Project GRUDGE would be disregarded and ignored on this, since it appeared there was great apathy and disinterest within the TID organization. This may have been understandable to Cabell, although he previously made it clear to TID, in his 15 February "Air Intelligence Requirements Memorandum Number 4," that he expected continued investigation of flying object sighting reports

On 27 April, Kaplan was at Sandia Base in New Mexico reviewing reports of fireball sightings. Kaplan was accompanied by Dr. LaPaz, and representatives from AFOSI, who jointly met with Sandia staff. At the conclusion of the meeting, Kaplan indicated he would recommend a full scientific investigation, with coordination from the University of New Mexico. The next day, Kaplan, LaPaz, and representatives from AFOSI met with five AEC security people at Los Alamos who previously observed the fireballs. Afterward, Kaplan expressed great concern that this was something directly related to "defense of the United States," and that this was of "extreme importance." The implication here is that Kaplan believed that the fireballs were of foreign origin, and flying over U.S. territory,

Air Force Preliminary Response to Shalett's Article

Shortly after receiving a draft of Shalett's article in early April, an Air Force response to the article was apparently completed, but puzzlement exists on its creation, review, and approval. Previously, the Air Force deemed it necessary to bring the public up-to-date prior to Shalett's publication in the *Saturday Evening Post*. Supposedly, a preemptive release was intended to blunt any misrepresentation by Shalett, and also provide and promote the view the Air Force wanted the public to have.

The response, dated 27 April, was a "Memorandum to the Press" titled "Project Saucer" (NO. M 26-49), and it indicated on its cover that, "The following report is a digest of preliminary studies made by the Air Material Command, Wright Field, Dayton, Ohio, on 'Flying Saucers'." It was released to the press on 29 April, which was the same day the first installment of Shalett's two-part article was published in the *Saturday Evening Post*.

This memorandum is very puzzling in the fact that one would think a different "slant" would be expressed by the Air Force in its prepared response, especially if there was the slightest attention paid to Shalett's draft. Of the twenty-two pages in the memorandum, fourteen were dedicated to specific sighting cases, including several classics. Another four and a half pages contained topics that were titled: "From Another Planet," "Other Star Systems," and "Space Ships." This was completely opposite to what would be expected in a preemptive "Memorandum to the Press," especially with the attitude that now existed in Project GRUDGE and the Control Group. It was incomprehensible why the Air Force would decide to respond in the manner it did. Could it be that some kind of conspiracy was involved, which resulted in a press release that bypassed an approval process? Or, was there a severe lapse of coordination or involvement by those who would have preferred different wording by the Air Force? Since it indicated it came from studies by Air Material Command, one might wonder if Air Force Public Relations simply used the SIGN Technical Report it obtained, and then formulated the response.

The Air Force "Memorandum to the Press" was unusually candid, and it essentially stated what was communicated previously in classified documents, including the SIGN Technical Report, and much of it was almost word-for-word. In addition, the press release described how TID performed its investigation of incidents. It mentioned some of the more famous and classic unsolved incidents, as well as investigations that led to misidentifications and hoaxes. The following is some of what was stated:

> *…Of course, the possibility that some of the incidents reported to AMC Intelligence may represent technical developments far in advance of knowledge available to American engineers and scientists has been widely considered. But observations based on experience with nuclear power plant research in this country label as "highly improbable" the existence on Earth of engines small enough in size and weight to have powered any of the capricious "saucers". The other obvious possibility— visitations from Mars, Venus, or distant planets attached to other star systems—is also looked upon as an almost complete impossibility.*

> *… The question marks in Project "Saucer" are not dangerous ones.*

> *…Possibilities that the saucers are foreign aircraft have also been considered.   But the reported performance of the discs is so superior to anything we have yet approached in this country that it is considered only an accidental discovery of "a degree of novelty never before achieved" could suffice to explain such devices.*

> *… The mere existence of some yet unidentified flying objects necessitates a constant vigilance on the part of Project "Saucer" personnel, and on the part of the civilian population.*

> *…Although visits from outer space are believed to be possible, they are thought to be highly improbable. In particular, detailed reports made on study of individual incidents and the overall picture of project "Saucer" point to the fact that actions attributed to the flying objects reported during the last two years are inconsistent with the requirements for space travel.*

> *… The "saucers" are not a joke. Neither are they a cause for alarm to the population. Many of the incidents already have answers, Meteors, Balloons, Falling stars. Birds in flight. Testing devices, etc. Some of them still end in question marks.*

Release of this statement by the Air Force is curious, because it does reflect the SIGN Technical Report, and it raises the idea of an extraterrestrial possibility, which was due to the "question marks" it referred to. With conflictive statements, it also provided a warning with the following:

> *...Exhaustive investigations have turned up no alarming probabilities... existence of some yet unidentified flying objects necessitates a constant vigilance on the part of Project "Saucer" personnel, and on the part of the civilian population.*

To the public, it would appear that the Air Force knew more than it was letting on, and that there was great concern about flying objects. In taking a look back at this, one cannot help but observe a major blunder taking place in the Air Force's cover-up of the UFO question. And the Control Group must have been very disturbed. It also becomes easy to speculate that there was serious intrigue involved.

Air Force Intelligence Report to JIC

On the same day that the "Memorandum to the Press" was released, the Joint Intelligence Committee (JIC) was briefed on the UFO question by General Cabell. The JIC was an intelligence organization that was set up prior to WWII, and before creation of the CIA and NSC. It specialized in estimating national strategic policy of the Soviet Union, including political, military, economic, and intelligence capabilities. Also presented to the JIC, at Cabell's briefing, was a top secret report titled "Unidentified Aerial Objects," and its purpose was stated as follows:

> *...To advise the Joint Intelligence Committee of the findings of the Directorate of Intelligence, USAF, regarding the sightings of unidentified aerial objects and the Air Force organization established for further investigation and solution of the problem.*

The report also stated in its Appendix A that:

> *...Creditable unexplained incidents which might involve the use of atomic powered craft of usual [sic] design should be considered jointly by the Atomic Energy Commission and highly competent aerodynamicists to determine the necessity for further consideration of such incidents by the National Defense Intelligence Agencies.*

> *...There are numerous reports from reliable and competent observers for which a conclusive explanation has not been made. Some of these involve descriptions which would place them in the category of new*

manifestations of probable natural phenomena but others involve
configurations and described performance which might conceivably
represent an advanced aerodynamical development. A few unexplained
incidents surpass these limits of credibility.

…It is unlikely that a foreign power would expose a superior aerial
weapon by a prolonged ineffectual penetration of the United States.

This report to the JIC was very candid, and it certainly expressed what General
Cabell knew regarding the UFO question, but it also intimated that it was "unlikely
that a foreign power" would display flying objects of "advanced aerodynamical
development" over U.S. territory. Most important, it did not reflect the tone and
substance of Shalett's published article that same day, or the attitude within Project
GRUDGE. Based on the character of the report to the JIC, one is tempted to think
that maybe General Cabell was not too concerned with the Air Force's preemptive
press release.

ശ◌ർ

Although General Cabell may have been unhappy with Shalett's investigation
into Air Force activity regarding the UFO question, it is evident that Cabell
maintained serious intent to resolve what flying objects were all about. But this is
a confusing area for researchers to diagnose because there must have been scattered
factions within Air Force Headquarters at this time. Some were in favor of releasing
the "secret," some not, and some were not aware of the secret of Roswell. General
Cabell was probably one of those not knowing, but he believed in reality of the
objects, and desperately wanted answers. This was made evident by his release of
the Intelligence Requirements Memorandum on 15 February, and his report to
the JIC on 29 April. It is also possible he was not concerned with release of the
preliminary "Memorandum to the Press" on that date. All three of these documents
were in line with his efforts to resolve the UFO question. The thing he didn't want,
however, was a public hyped on flying objects.

Working against Cabell were elements in his Directorate of Intelligence,
specifically Major Boggs and his boss Colonel Hearn in AFOAI. They were directly
interfacing with Project GRUDGE. Of course the Control Group was monitoring
the situation with help from the CIA, and this combination was enough to stifle
any investigative progress in TID.

It was an interesting time for the Air Force in regard to the UFO question.
But the situation was about to become much more contentious, due to release of
Shalett's article, and that article will be reviewed in the next chapter.

CHAPTER NINETEEN

Shalett's Article

Project GRUDGE: April - May 1949

Part One - 30 April 1949

Sidney Shalett's first installment of his article titled "What You Can Believe About Flying Saucers" was published in the 30 April 1949 issue of the *Saturday Evening Post,* which was released on 29 April. It was quite complete in its introduction to the mystery of flying objects, although its theme was total satire. He labeled it the "Great Flying Saucer Scare." The article covered most of the classic sightings, and Shalett provided his "reasoning" for what he thought the flying objects were.

When reading the article, one has little doubt that Shalett spent considerable time gathering his facts, but it was quite likely he received much help from GRUDGE staff. In this regard, Edward Ruppelt later stated:

> *I think that he just wrote the story as it was told to him, told to him by Project Grudge.*

Shalett offered the following on some of the things he became aware of:

> *...Some weeks before the Mantel tragedy, the Air Force had already made the decision that the saucer business must be investigated. Gen. Hoyt S. Vandenberg, then the Vice Chief (now Chief) of Staff for the Air Force, was instrumental in making the decision. However illogical and absurd the notion of unexplainable objects soaring over America might sound, the Air Force could not afford to close its mind arbitrarily to the*

possibility that something new might be in the skies…the United States Air Force–with considerable reluctance–finally set up a special project to investigate the reported phenomena. To date, those in charge of the projects have collected reports on some 250 instances of "unidentified flying objects," and, though the scare has substantially subsided in late months, the list of incidents continues to grow…

So the Air Force set up its project, under Air Intelligence, to investigate. For some obscure security reason, it still is not permissible to mention the code name of the project, so I will call it Project Saucer. Top supervision is from Washington, but the leg work, cataloging and evaluating are done by the Technical Intelligence Division at Air Material Command Headquarters, Wright-Patterson Air Force Base, Dayton, Ohio…Project Saucer at Wright-Patterson provides a task force of intelligence officers and civilian technical experts who can jump into a plane any time and go out to investigate. In practice, the task force has not personally investigated many cases on the scene, but has left the questioning to local intelligence officers [AFOSI], who report all information back to Dayton…At Wright Field…there is a civilian technical consultant, an expert on rockets in his own right, who is convinced that the saucers are visitations either from Moscow or Mars…

The most common sources of innocent deception in the balloon field are the so-called RAWIN (radar-wind) target balloons…Queer-shaped objects leaving trails of green light were seen coming from Peenemunde, the ex-Nazi missile center now in Soviet hands…When I went to Wright Field armed only with reports of what witnesses said they had seen, the Great Flying Saucer Scare seemed reasonably mysterious to me. When I had finished my investigation in Dayton, Washington and elsewhere, the thing seemed less mysterious than odd. There are any number of logical and perfectly normal solutions by which most of the saucer sightings can be explained…

I am necessarily accepting the assurances of the highest officers of the Air Force, and those of its research and development experts, that they have nothing concealed up their sleeves. Of course, there are a lot of people, some of them quite sober citizens, who insist that there is "something funny" about the saucer business. These will probably insist that the Air Force is kidding me. But I don't think it is.

These comments from Shalett's article allow for further examination and study about what he may have been told, or not told. The article provides an indication of the amount of information, and possible sources, made available to him. He was aware that a project was initially set up by the Air Force to "investigate the reported phenomena," but his stated reason for setup of the project by the Air Force was totally incorrect. Instead of stating the Air Force knew that the objects were real (according to General Twining), and that the project was requested by General Twining, Shalett related a completely different view on the matter. He stated that, even though "the notion of unexplainable objects" was "illogical and absurd," the project was set up "with considerable reluctance" because "the Air Force could not afford to close its mind arbitrarily." Those words of Shalett contain cover-up rhetoric, with "accepting assurances of the highest officers of the Air Force."

When he visited TID and the people at Project GRUDGE, it appears that he discovered that investigators had been involved with more than one project, which is suggested by his statement that "those in charge of projects have collected 250 instances." He likely became aware of the initial Project SIGN, because he mentioned, "For some obscure security reason, it still is not permissible to mention the code name of the project." If a careful look is taken at his article, it is realized that he obtained much of his information from the secret SIGN Technical Report.

It is also interesting to note that Shalett encountered a "civilian technical consultant" at Wright-Patterson who was "convinced that the saucers are visitations." Could this person have been Alfred Loedding, Albert Deyarmond, Lawrence Truettner, or some other holdover individual who was previously in the Special Projects branch at TID?

Another thing to note is that Shalett was convinced that the Great Flying Saucer Scare was not "mysterious," but "odd," and he justified this by stating that there are "logical and perfectly normal solutions by which most of the saucer sightings can be explained." His mistake was that "most" sightings don't equate to "all" sightings, and it's the unexplained sightings by credible people that remain most important. In total, Shalett's first installment of his *Saturday Evening Post* article did not support the extraterrestrial idea, and he inferred throughout his article why the Air Force had a hand in influencing and contributing to public anxiety.

Part Two - 7 May 1949

The second installment of Shalett's article was published in the 7 May issue of the *Saturday Evening Post,* and it again expressed a negative tone. It was filled with a great amount of debunking, which must have come from individuals in Project GRUDGE, but also from high Air Force officials at Air Force Headquarters. This was confirmed in the following:

...men who constitute the high command of the United States Air Force do not believe in flying saucers, disks, space ships from Mars – or Russia–which citizens of the United States have been reporting with increasing frequency since the atomic age burst upon the world.

The idea that flying object reports have increased since the start of the atomic age is what modern day researchers often comment about, as did George Valley in the SIGN Technical Report.

In this second installment of the article, Shalett made sure that Air Force brass was mentioned in regard to debunking of flying objects. He identified those "men who constitute the high command" as General Carl Spaatz, General Vandenberg, General Lauris Norstad, and General Curtis LeMay. He also put TID in line with the debunking by stating:

The officers and technical experts assigned to Project Saucer–a nickname for the top-secret Air Force investigative effort–sometimes get to feeling they're living in a dream world, so utterly unfettered and mysterious are some of the reports they are assigned to investigate.

Shalett also included Colonel Howard McCoy as part of the Air Force brass, and mentioned a comment by McCoy stating that he:

...once thought he saw a disk while flying a P-51 fighter in broad daylight, but it turned out to be a glint of sunlight from the canopy of another distant P-51."

It is very interesting that McCoy would be mentioned, although he was the one in charge of TID, and this was where Shalett was spending time pursuing his investigation. But this is also the organization where McCoy confidently put his signature to the EOTS.

It seems obvious that McCoy was playing it very "coy" at this particular time, and catering to desires of Vandenberg. He really had no choice. He was still managing the magnificent secret of Roswell, and he possibly suspected he was on Vandenberg's "short list." But he must have considered himself finished with the investigation of flying objects, which was probably the direction he gave to Lieutenant Howard Smith, who was representing Project GRUDGE. But one can also contemplate that McCoy was hoping that Shalett would stumble onto the Project SIGN Technical Report, which was probably just completed, or that Shalett would run into a Project SIGN investigator.

Shalett then proceeded to list some of the more significant reports received from witnesses, including some who would be considered credible. But he also alluded to the fact that they were possibly impacted by sensory illusions and other ramifications. In addition, he stated:

> ...*One of the main solutions to the reported phenomena lies in the aeromedical field. Both the Air Force and Navy aero-medical experts have prepared volumes of research findings, spelling out in detail how vertigo, hypnosis, and other sensory illusions affect pilots traveling at high altitudes and extreme speeds...Case histories...have established that vertigo and self-hypnosis brought on by staring too long at a fixed light have caused pilots to dogfight with stars, to mistake ground lights for other aircraft, flying saucers or what not, and to have outright hallucinations about things which weren't even there... In general, they feel that when a flyer starts chasing an illuminated weather balloon or a star, and vertigo or hypnosis sets in, the pilot can come down and practically tell you how many rivets were on the nose of the Martian space ship...Another wide area through which Project Saucer investigators have had to plow is the rich, intangible field of hallucinations, hoaxes, and mass hysteria...Mass hysteria is a phenomenon that has fascinated philosophers and psychologists for ages; there is no limitation on what impressionable people will think they've seen if someone starts a sufficiently convincing rumor. Even an honest rumor will do the trick.*

To finish debunking the UFO question in the Great Flying Saucer Scare, Shalett shared an interview he conducted with Dr. Irving Langmuir, and Shalett stated:

> ...*Though we have not yet produced the rocket-to-the-moon and homemade satellite, it is small wonder that harassed humans, already suffering from atomic psychosis, have started seeing saucers and Martians. Perhaps the most outspoken foe of the flying saucer in the United States is Dr. Irving Langmuir, the distinguished scientist and Nobel Prize winner. Doctor Langmuir, associate director of General Electric's Research Laboratory at Schenectady, has spent a lifetime debunking what he calls "pathological science"—that is, untruthful scientific theories which were carelessly accepted as truthful until someone came along to prick a hole in them—and he lumps saucers in this category...Though Dr. Langmuir speaks on saucers in his*

nonofficial capacity as a scientist, he has given the Air Force an earful on the–as it appears to him–absurdity of it all…If a man tells me that two and two equal five–or that he has seen a flying saucer, I don't feel I have to prove he is wrong. I feel the burden is on him to prove that he is right…I asked Dr. Langmuir what he would advise the Air Force to do about flying saucers. He snapped his answer, "Forget it!

An Extraterrestrial Possibility

Up to this point, Shalett was probably quite satisfied with the tone of his article, and with what he was attempting to get across to readers. But suddenly, he brought up an incident that most readers were well aware of from the previous summer. Perhaps he felt obligated to bring it up, especially since the Air Force had no explanation for it, and it was mentioned in the SIGN Technical Report. He described the incident in detail, and it served as a reminder to readers that there really was something to those flying objects after all. With possible awareness that this would provide a pivotal point for the article, and for his readers, Shalett stated the following:

> *The Air Force, I suspect, would like to forget it. But then something new comes along like the strange adventure of two Eastern Air Lines pilots with what seemed to be a flame-shooting, double-decker space ship, and Wright Field has to send out another team of investigators.*
>
> *…Chiles and Whitted still say they do not know what to think; they say they are certain that they were not suffering from hallucinations and that what they saw was a manufactured object–not a meteor–and unlike any aircraft or missile known to them…Both men are married and fathers. They had nothing to gain by their story, for it was bad publicity. Neither has tried to profit by the incident, nor were they responsible, so far as I could learn, for the story being given out to the newspapers.*

Shalett ended his article on that note, and it was probably something he felt he needed to do. His discussion with Chiles and Whitted, and his review of their incident, was a bit of an eye-opener for him. He then mentioned that the Air Force wanted to seriously hear from such credible people who had information about other flying object sightings or incidents. And then Shalett provided a detailed listing of descriptive information the Air Force would like to have, which was possibly obtained from "AFOIR's Memorandum Number 4." He stated:

While the Air Force finds it difficult to believe that the heavens are populated with inexplicable skimming saucers, diving disks, bounding balls or spooky spaceships, even when the testimony comes from such excellent witnesses as pilots Chiles and Whitted, it does want to know about such things…sit down and write a letter containing all this information to Technical Intelligence Division, Air Material Center, Dayton, Ohio. At the same time, maybe you'd better buttress yourself with an affidavit from your clergyman, doctor, or banker. If you've really seen something and can prove it, you may scare the wits out of the United States Air Force, but it will be grateful to you.

೨৯২

It would appear that Shalett mostly toed the line in his article by expressing Project GRUDGE's view, except for a sudden turnabout at the end. He gave much detail on sighting incidents, while also going overboard in dismissing them as no cause for concern, and he called it the Great Flying Saucer Scare. He was highly negative about the Air Force investigating nothing of substance, although he ended up creating reader interest with an incident that many people recognized as a serious and credible encounter.

There is dichotomy in Shalett's article, and one wonders how he was maneuvered by officials at Air Force Headquarters, and those in Project GRUDGE at TID. He obviously ran into high officials who were in-the-know, and who were debunking the idea of flying objects. They suggested to Shalett that he "accept assurances" that the Air Force had no interest in the objects. And he must have been barraged by those telling him that flying objects were a fantasy, or were manufactured by mass hysteria. He talked with those whose job it was to investigate the objects. He studied the SIGN Technical Report. He reviewed many of the classic incidents, and he managed to talk with credible witnesses in some of the cases. Basically, however, Shalett thought he could reconcile the flying object incidents with "logical solutions."

On the other hand, Shallet came upon a "civilian technical consultant" at TID who was convinced that "saucers are visitations." He must have given Shalett an earful, although Shalett equated the person to a witness who "wants to believe," and "all the logic in the world will not convince" otherwise. He then ended his article with the amazing Chiles-Whitted incident, and he indicated that the Air Force actually "does want to know about such things." He said the Air Force "will be grateful to you" if you "sit down and write a letter" on what you witnessed. The contrast exhibited at the very end of Shalett's article, compared to the previous, was astounding.

It is curious how his article was put together. But it is quite obvious that Shalett was heavily indoctrinated at Air Force Headquarters, and at TID. He wrote his story from the perspective provided to him, while leaving an opening for an "extraterrestrial possibility" at the end. The possibility was made evident because of the Chiles-Whitted incident, which didn't have a logical or reasonable solution. Also, the Air Force made it quite clear to Shalett that it wanted to know more about such incidents. Perhaps this particular incident gave Shalett a tinge of doubt at the end, although he didn't think Air Force officials had been dishonest with him. He also knew of "quite sober citizens who insist that there is 'something funny' about the saucer business." With all the negative debunking in his article, perhaps he felt a need to leave the door ajar at the end. But the turnabout in tone of his article would have a major impact on readers, which was something the Air Force was not prepared for. And neither was the Control Group, who was now possibly flirting with the idea of suppressing Air Force investigation of flying objects, or stopping it totally. But would that really be possible, since the general military and public were concerned about continued sightings and encounters?

CHAPTER TWENTY

Damage Control

Project GRUDGE: May - August 1949

Missed Opportunity

One wonders how Sidney Shalett was directed and maneuvered by Air Force officials when working on his April and May 1949 two-part article for the *Saturday Evening Post*. One can envision that Secretary of Defense General Forrestal had reason to enlist Shalett in ferreting out the UFO question so that Truth might be revealed. Shalett was guided around Air Force Headquarters, taken to the Special Projects branch of TID at AMC, given access to documents such as the SIGN Technical Report, and provided access to a TID individual who was committed to the extraterrestrial view. In fact, the SIGN Technical Report was created about the same time that Shalett was at TID. That report was not the EOTS, but Shalett was presented with a grand opportunity to discover Truth, which Project SIGN was previously closing in on. Shalett, however, primarily chose the "debunking" direction, which high officials insisted on.

Press Memorandum Shakedown

Based on its message and how it was composed, it's curious why Generals Cabell and Vandenberg allowed release of the preemptive "Memorandum to the Press." It leads one to speculate that neither of them reviewed it, although Cabell probably did. If they did review it, one would think they would have preferred that it totally discount or debunk the extraterrestrial idea, which is what Vandenberg did when Shalett interviewed him. When considering that they would not want Shalett to have access, or a clue about the "top secret" EOTS, which was supposedly destroyed, it would be expected that the press memorandum would also echo Shalett's debunking. This mysterious inconsistency has been troublesome to researchers for years. But there is good reason for what occurred in this situation.

It seems quite evident that Shalett began his initial investigative research at Air Force Headquarters. This is where he was originally given permission for his investigation, and then granted access to secret materials by General Forrestal. Shalett interviewed several high-ranking Air Force officials there, and they severely debunked the UFO question. They, however, had no idea that Shalett would later encounter and obtain access to the "secret" SIGN Technical Report. This significantly changed the situation, and the Air Force was caught in the middle "between a rock and a hard place." Air Force mischief with the cover-up, and critical miscalculation, left nothing more the Air Force could do except echo most of the secret SIGN Technical Report in its press memorandum.

It must be assumed that there was careful but frustrating consideration by Air Force officials on how to handle this situation, but this consideration involved deeper overtones than one might want to contemplate, especially when looking at the scenario of events. Investigation by Shalett began in late January or early February with interviews of higher officials debunking the UFO question. Shalett then came upon the secret SIGN Technical Report, possibly at TID, and he discovered that serious and intensive Air Force investigation actually took place on the UFO question. Those in Project GRUDGE, however, were indicating that they were bored with the "utterly unfettered and mysterious" sighting reports. After this, on 1 March, General Forrestal was fired from his position as secretary of defense. Upon leaving the Defense Department on 28 March, Forrestal entered the government's Bethesda Hospital on 2 April, and that is when a draft of Shalett's article became available to the Air Force. This is also when the preemptive "Memorandum to the Press" began to be prepared and made ready. Its content chagrined and upset Air Force officials, but no good options were available for them. Cabell was probably bothered by it, but he knew a press memorandum was needed.

The Control Group, CIA, President Truman and others must have considered Forrestal's actions highly treasonable. He purposely provided Shalett the opportunity to reveal the monumental secret, or so it would appear. And there was no way that Forrestal could be put on trial for his treasonous actions without the "secret" being publicly exposed. Two weeks after Shalett's second and last installment was published, Forrestal died under curious circumstances, and this was after being secluded and isolated in the hospital for a total of about seven weeks.

Details of how the press memorandum was created are lost, but it is noted that it "originated" in AMC, with possible involvement by the Office of Public Relations at Air Force Headquarters. One might expect that someone in Project GRUDGE would have taken charge of producing a write-up for the press memorandum, but GRUDGE was in a state of stupor, where no one cared to be involved. Could it be that the task of producing the write-up was tossed in the air and someone with an interest at TID, but not involved with GRUDGE, caught hold of it? Did Colonel McCoy have something to do with it, with possible help from Albert Deyarmond?

With General McNarney returning to Air Force Headquarters in mid-February and working for Forrestal, and with General Chidlaw newly in charge of AMC, it is doubtful they were involved. But it's tempting to contemplate that those partial to Truth may have had a hand in creating the memorandum. Did someone from the terminated Project SIGN organization manage to step in and do the write-up? It's easy to imagine that a Truth coalition was involved with this, but there are no facts or specific details to confirm it. Of course, the intrigue regarding this could go wild with all kinds of conjecture and speculation, especially with knowledge that Forrestal met an unfortunate end to his life. If some sort of intrigue was involved, it's a good bet that others bent on revealing Truth would have chilling consequences to think about.

High Air Force officials were likely stymied on just what to do in regard to creating the press memorandum. The choice was to either continue with debunking the UFO question, or admit to the Air Force investigative activity that Shalett described in his draft. The main thing to note in all this is that General Cabell was the one who wanted the preemptive press release. In light of Shalett's draft article, Cabell wanted an accurate report to the public in order to inform on Air Force progress with investigating the UFO question. He did not like Shalett's intrusion and involvement, but this was primarily because of the need to keep Air Force's investigative activity quiet. Cabell was interested in subduing press and public anxiety over the UFO question, which would prevent reporting of unscrupulous and nebulous sightings. But now, with a draft of Shalett's article available, it was also necessary to confirm to the public and general military that serious investigation of sighting reports was continuing. Cabell had a stake in this, and he was definitely interested in continued Air Force investigation, even if Vandenberg and a few others chose to believe that there was nothing to the flying objects. Cabell knew they were real, but he didn't know what the objects were, or what the mysterious Green Fireballs were. In connection with this, higher officials could not afford to be two-faced in the memorandum in regard to debunking. It was better for the debunking to stand alone in Shalett's article.

Some researchers are of the opinion that a draft of only the first installment of Shalett's article was available for Air Force review. This required the Air Force to guess about Shalett's second installment, and then write the press memorandum accordingly. It's believed there was indecision and concern by the Air Force on what the total article would contain, but it's not reasonable that the Air Force would leave itself open without reviewing an entire draft of the article.

The Air Force likely reviewed Shalett's entire article in draft form, and then formulated its response by echoing the SIGN Technical Report. The actual problem the Air Force had in this situation is that high officials severely debunked the flying objects, and then Shalett reported on extensive Air Force investigative activity on the objects, and this is what created the dilemma for the Air Force.

While satisfied with Shalett's debunking, there was begrudging embarrassment about the extensive Air Force investigative activity he exposed, which contrasted with the official Air Force debunking. It was an uncomfortable position for high officials to be in, but the Air Force could not effectively deny the investigative activity that Shalett revealed. And high Air Force officials were left upset and angry, and without options.

The only question remaining here is what the Control Group was thinking in regard to this mess. Were they now contemplating abandonment of Air Force investigation of flying objects? If so, how would they then handle future serious sightings by the general military, or the public?

An Aroused Public

There can be no mistake about the severe impact this episode had on Air Force Headquarters. From the end of April through May, leadership was shaken to the core with what had transpired. The Air Force was in a quandary and could not respond to further public inquiry. The entire military was confused, the public was aroused, and blame for it all rested within Air Force Headquarters.

Shalett's article made it very clear that the Air Force did not relish being in the business of investigating flying objects. That is what Shalett gleaned from those he talked with, especially high officials. His article was mostly devoted to debunking flying objects, but the Air Force "Memorandum to the Press" expressed no debunking, and this stood out in sharp contrast to Shalett's article.

That contrast brought significant attention upon the Air Force, and the tide began to turn, but not like Air Force Headquarters, TID, and Project GRUDGE had anticipated or expected. The public was now excited by the UFO question. Flying object sightings increased greatly in number, quite possibly because the public sensed a "snow job" by Shalett in his debunking. The public was not buying the explanation of mass hysteria, sensory illusions, self-hypnosis, hallucinations, and illogical and absurd manifestations as reasons behind sightings by credible witnesses.

With no such excuse presented for the Chiles-Whitted incident, by either Shalett or the Air Force, the public took notice, especially with Shalett leaving open the fact that extraterrestrial flying objects were a possibility, and that the Air Force was interested in receiving credible reports. As a result, the Air Force "Memorandum to the Press" served to convince many that something must really be going on with flying objects. Tremendous suspicion was laid at the doorstep of the Air Force, and the Air Force was tongue-tied.

As public curiosity started to grow, questions were being asked, more reporters wanted access to Air Force records, and more articles were being written. Magazines such as *Life, True, Argosy*, and others were showing interest, and researchers started knocking on the door at Air Force Headquarters. There were many people now clamoring for answers to additional questions.

Specter of Disarray

Previously, Vandenberg had the EOTS destroyed and wanted no suggestion of an extraterrestrial answer without physical evidence, or undeniable proof. Now, with release of the Air Force "Memorandum to the Press," people were wondering what was going on with the Air Force, and also thinking about an extraterrestrial possibility.

There was also a specter of disarray in the Air Force. Some of the public did not find Shalett's article totally worthy, and many were perplexed, disturbed, and concerned regarding Air Force internal policy, especially as presented by Shalett. Many people who experienced previous encounters, or were involved with sighting incidents, were now upset to think they would be thought of as mental cases. Many more were thinking twice before reporting an incident, or they were just refusing to do so. This was especially the case within the military, who were becoming disillusioned with higher leadership in the Air Force.

One might consider that Cabell was perplexed about the situation he found himself in. On the other hand, it's also possible he was satisfied with information given separately to the public in the memorandum, and then to military and intelligence agencies with his JIC report. All were now updated on current Air Force status regarding the UFO question.

Cabell was diligently attempting to find answers, and his people were working very hard in that endeavor, except at TID. He had a consensus in both his divisions that flying objects were real, although AFOAI disputed the extraterrestrial idea. Project SIGN, however, previously labeled flying objects as extraterrestrial in origin, and Cabell was also leaning in that direction because of his close involvement with AFOIR investigators. For him, it must have been stressful knowing that Vandenberg quickly vanquished the extraterrestrial idea, along with some valuable and talented people in Project SIGN. Cabell knew his role was to find out the origin of flying objects, what they were up to, and what their intentions were, but it was necessary for it to be out of the public eye. The significant thing about this is that the Air Force "Memorandum to the Press" put the public on notice of something going on with flying objects, and they understood the Air Force was quite serious about learning more.

The Air Force was now presented with a messy situation on the UFO question. Some were in-the-know about Roswell. Some from Project SIGN believed the objects were extraterrestrial. Some in the Directorate of Intelligence knew the objects were real. And everyone in Project GRUDGE were instructed that flying objects did not exist. A majority in the public and military were confused and didn't trust the Air Force, primarily because of the contrast between Shalett's article and the Air Force "Memorandum to the Press."

Confusion in Project GRUDGE

In regard to people within Project GRUDGE, they were experiencing a serious dilemma. They were aware of the SIGN Technical Report and the Air Force "Memorandum to the Press," and neither of those debunked the UFO question. And they were aware of the *Saturday Evening Post* article, which severely debunked flying objects, while leaving the door slightly open. They were instructed to disregard the extraterrestrial possibility, and they understood that professional experts in Project SIGN lost jobs because of their extraterrestrial conclusion. They also knew that, as part of their job, they were to maintain vigilance to assess foreign involvement with flying objects.

For them, it also meant they must use pure simple science, and current scientific intellect, to diagnose "unknown" super technologies for some of the more puzzling sightings. But this presented a problem, and it relegated them to a "do nothing organization." Most of all, the SIGN Technical Report, and the Air Force "Memorandum to the Press," would be thorns to be reckoned with. There was no clear way they could do their investigative job, as instructed, with these documents around.

People in Project GRUDGE would now pay less attention to reported sightings, with no attention paid to Green Fireballs, which they considered to be natural phenomena. The Air Force OSI was now expected to handle sightings requiring special attention, with follow-up information sent in a report to GRUDGE's Colonel Hemstreet. Otherwise, time spent in Project GRUDGE would involve the filing of new cases, reviewing old Project SIGN cases, and revising summary conclusions to make them more prosaic. Little manpower existed to do much else.

Donald Keyhoe Begins Investigation

Turmoil and confusion were building at Air Force Headquarters, and it was not about to let up. Public curiosity continued to brew, and publishers of other magazines were wanting in on the action. Ken Purdy, the editor of *True* magazine, was getting nowhere in his inquiries to Air Force officials, and he called upon Major Donald Keyhoe, a retired Marine Navy pilot who previously wrote for the magazine. Purdy knew that Keyhoe had many contacts and associates in the military, especially at the Pentagon. Keyhoe previously worked at the Pentagon, and was once an escort for famous pilot Charles Lindbergh. Like Purdy, Keyhoe was perplexed with the contrast exhibited between the Air Force "Memorandum to the Press," and the *Saturday Evening Post* article.

After the second *Saturday Evening Post* issue on 7 May, Purdy discussed with Keyhoe the fact that General Forrestal managed to get the magazine to run two articles debunking flying objects, after keeping the lid on "Project Saucer" for more than a year. Purdy noted that Shalett debunked the flying saucer problem by relating it to mass hysteria, and he noted that Shalett was obviously guided by the Air Force throughout his investigation. Purdy was mystified as to why the Air Force would state, in its preemptive "Memorandum to the Press," that "existence of some yet unidentified flying objects necessitates a constant vigilance on the part of 'Project Saucer' personnel, and on the part of the civilian population." It seemed that the Air Force was concerned that flying objects were real, and Purdy was curious about why the Air Force felt compelled to contradict the debunking in Shalett's article prior to its publication. Was the Air Force attempting to conceal or hold back something, while privately debunking the subject to Shalett?

As Keyhoe proceeded to research and ask questions, he talked with civilian pilots, military pilots and others who were involved in incidents with flying objects, and it became clear to Keyhoe that something definitely was going on, especially after talking to his sources in the Pentagon. That is where he received his most convincing information, and Keyhoe would continue to be involved with this effort for several more months in a quest to find answers and leads on the UFO question. He kept meticulous notes regarding his inside information, and then he became convinced that flying objects were unearthly. He knew the Air Force was holding back information.

Navy Commander McLaughlin Sighting

On 12 May, less than three weeks after the Moore sighting at White Sands, which was reviewed in Chapter 18, Navy Commander Robert McLaughlin sent a two-page letter to Dr. James Van Allen of Johns Hopkins University describing his sighting at White Sands. McLaughlin was in charge of the Navy missile program there, and with high altitude balloon research, where flying objects were interfering with his work. He stated the following in his letter:

...managed to see one of these flying saucers along with three other officers...I was clearly able to see the white outline of the object with the naked eye. This object appeared almost directly overhead at the White Sands Proving Ground, gradually gaining velocity to the west and disappearing in a blinding burst of speed to the west...My first assumption led me to the conclusion that the object must have means of accelerating itself. The second assumption naturally is that no one on this planet is sufficiently far advanced to fly such an object.

This sighting made a huge impression on McLaughlin, but he was not finished getting his story out. The whole world would soon learn of his experience, which will be reviewed in Chapter 22.

Colonel Edward Porter Arrives

Also in mid-May, Colonel Edward Porter arrived to take over as chief of Air Estimates in AFOAI, replacing Colonel Brooke Allen. Porter would work closely with AFOAI Chief General Ernest Moore, and with Defensive Air Chief Colonel C.V. Hearn, who previously replaced Colonel Harris. Working for Colonel Hearn was Major Arron Boggs who joined Defensive Air about a year before. These four individuals (Moore, Porter, Hearn, and Boggs) would play a debunking role in close coordination with Project GRUDGE at TID. And they were not partial to the idea of extraterrestrial flying objects.

Because Colonel Porter replaced Colonel Allen at Air Estimates, a mystery unfolds, and there seems to be a strange lack of information on Allen until February 1951 when he is assigned as commander of Air Pictorial Service. His Air Force biography indicates he remained "for the most part" as chief of Air Targets Division until his new assignment in 1951. Actually, he only remained for about two months in Air Targets (the Strategic Vulnerability/Defensive Air Branch) before becoming chief of Air Estimates about the middle of July 1948 (see Chapter 14).

Previously, when Allen became chief of Strategic Vulnerability, there was a question about him taking the same job supposedly held by Harold Watson, and there was also a question on whether Watson actually held that position. Could it be coincidental that Watson now comes into the picture at the same time Allen departs? One is tempted to contemplate about where Allen was heading to next.

Colonel Harold Watson Arrives

When one reviews the chain of events and circumstances regarding Air Force handling of the UFO question, it is not surprising that Colonel Harold Watson now enters the picture at TID. Previously, in Chapter 13, a question was raised about Watson's whereabouts between October 1947 and July 1949. He now appears in full action at TID, beginning about the middle of July. Years later, he would comment that he came to TID at this time because General Cabell told him he "had a job for him to do." This statement has the suspicious sound of "damage control," but it might actually have been Watson's way of providing self-amusement regarding the circumstance of his subsequent removal in July 1951.

One might consider that Colonel McCoy, the previous chief of TID, was quite disturbed by the chain of events in his Special Projects branch, and termination of Project SIGN. Although TID was a very large organization, it's likely that McCoy was not comfortable "living a lie" with his small investigative unit, which was now decimated and more-or-less defunct. And it is more than likely he did not earn favor with higher Air Force officials who were in-the-know, especially if it was figured out that McCoy had a hand in the preemptive "Memorandum to the Press." On the other hand, McCoy was probably happy to be out of the flying object investigative business, and just manage the magnificent secret.

Could it be that Colonel Watson was suggested as someone who could effectively fill McCoy's TID position and Project GRUDGE? Watson was probably happy to oblige, but it might be expected that his management of the organization would not be in the same manner anticipated by Cabell. One can rightly suspect that Watson probably had other motivations and incentives, along with special directives, which were somewhat different than what Cabell would want or expect.

GRUDGE's First and "Final" Report

On 10 August, only a few short weeks after the appearance of Watson, Project GRUDGE released its first "periodic" report. Lieutenant Howard Smith, chief of Project GRUDGE, and Mr. George Towles were the authors. Colonel Hemstreet and Colonel Watson were the signers. The huge, voluminous report provided a very good perspective on how Watson would handle the UFO question. The abstract at the beginning of the report gave an immediate sense of the report's unprofessional and sloppy nature. It stated:

> *A Technical Intelligence Report covering the method of investigation of unidentified flying objects and results obtained to date is presented... Since the project is continuous in nature, this report comprehensively treats reports of sightings only up to January 1949.*

The contradictory and purposely obtuse statement, involving "results obtained to date," admitted that only reports of sightings up to "January 1949," which ended seven months previously, would be addressed. Those sightings were solely within the framework of the Project SIGN's investigation. Sightings for the following months, associated with Project GRUDGE, were not going to be considered. The report was simply be a rehash of the Project SIGN investigation during its final months, and this new GRUDGE Report was intended to serve as a replacement for the last Project SIGN Technical Report. It would present a totally different slant in interpretation by different individuals who had no involvement with the reported sightings or investigations. In consideration of this, one can infer that the Project SIGN Technical Report was a great thorn in the side of Project GRUDGE and the Control Group, and it needed to be replaced. Under new management and a new policy, it needed to be properly degraded and discarded. And the Control Group probably required this.

The Introduction to the GRUDGE Report was quite revealing in regard to the project's expected future. It stated:

> *This report may be considered as final for the period reported upon. It also indicates the probable future trends of reports of unidentified flying objects...In gathering and evaluating material for the report, it was found, and will still be seen, that the conclusions evolve without effort. No attempt has been made to force evidence into a pattern that was not clearly indicated.*

The obvious implication here is that the GRUDGE Report was intended as a replacement for the final SIGN Technical Report, and one's attention is immediately drawn to the report's purposeful lack of objectivity. It appears to exude Colonel Watson's confident rhetoric.

The final summary in the report is quite astounding. In reference to flying object sightings since the beginning of Project GRUDGE, it states:

> *...many have not yet been investigated, few have been completely tabulated, and none have been submitted to consulting agencies. It is certain that better over-all results will be obtained in the analysis of the later reports, as these incidents generally have been more completely investigated.*

The verbiage used here is almost unbelievable. It admits that little has been done since the first of the year in investigating sighting incidents under GRUDGE, and then it indicates that these particular incidents "have been more completely investigated." This probably alludes to the fact that it intends to produce "better over-all results" than Project SIGN, but the prospect for doing that would appear exceedingly slim, with only a skeleton workforce in GRUDGE consisting primarily of Lieutenant Smith, who was without technical investigative skills, or any ability in that area.

In regard to Green Fireballs, an important subject of interest to many in the military, Project GRUDGE was not about to consider this phenomenon for investigation, even though official directives were issued for those sightings to be reported to TID. The GRUDGE Report states:

> *Since 5 December 1948, a series of recurring phenomena described as "green fireballs" have been reported in the general vicinity of Albuquerque, New Mexico. Dr. Lincoln La Paz, noted meteoric expert, has been directly, though unofficially, associated with the investigation of these sightings and has himself observed the phenomena. Dr. La Paz states he is convinced the green fireballs are not ordinary meteors. This group of incidents has little or nothing in common with other incidents on file with Project "GRUDGE", therefore, these incidents are not considered in this report.*

It's truly astounding how this report was put together, primarily by Lieutenant Smith. If Colonel McCoy had been involved with its review prior to release, he would have required massive hasty revisions and corrections, although he would never have allowed creation of the report in the first place. It defies the senses on how lightly things were adjudged in producing the report. Not only that, but many serious flying object sightings and incidents had been reported since the beginning of Project GRUDGE, and more were still coming in, but none were being investigated. It's obvious that there was no intention to ever investigate reported sightings again. And this was probably an edict from the Control Group. The conclusion of the report stated:

> *Evaluation of reports of identified flying objects to date demonstrate that these flying objects constitute no direct threat to the national security of the United States.*
>
> *Reports of unidentified flying objects are the result of:*
> *a. Misinterpretation of various conventional objects.*
> *b. A mild form of mass hysteria or 'war nerves'.*
> *c. Individuals who fabricate such reports to perpetuate a hoax or to seek publicity.*
> *d. Psychopathological persons.*

The significant thing about the above statement is that it slammed many credible people. It included many from the military who reported very important sightings, which were placed in the "unknown" category. This was about 23 percent of all sighting cases documented.

The conclusion of the GRUDGE Report ended with the following statement:

> *Planned release of unusual aerial objects coupled with the release of related psychological propaganda could cause mass hysteria. Employment of these methods by or against an enemy would yield similar results.*

After reading this statement, one can be excused for finding it exceedingly weird, especially as a final conclusion to the report! But there is something quite unique about the statement. General Walter Bedell Smith, who was involved with psychological warfare and in-the-know on Roswell, looked upon flying objects as a mechanism that might be used for creating mass hysteria. He felt it important that GRUDGE include this warning in the report. Mass hysteria could be created by an enemy responsible for mysterious aerial objects, and in the spread of propaganda. But in this GRUDGE Report, it is very curious that "mass hysteria" is brought up in such a manner. It was obviously a concern, but it was never previously mentioned in previous reporting. It leads one to wonder if Green Fireballs were determined to be of Soviet origin at this particular time, or that this was the incentive for the warning. The psychological aspect of mass hysteria was so important, that the main recommendation stated:

> *...That psychological Warfare Division and other governmental agencies interested in psychological warfare be informed of the results of this study.*

A further recommendation stated that all other conclusions, "with sufficient supporting data, be declassified and made public in the form of an official press release." Of course, only a few months previous to this, an official Air Force "Memorandum to the Press" stated that flying objects were real, and advised the public and military keep vigilant. But this GRUDGE Report now gives notice that another Air Force "Memorandum to the Press" should be released, obviously to counter the previous one.

Because of different messages made public, the Air Force was in serious disarray with the UFO question, and the military was very confused with Air Force handling of the situation. Evidently, the voluminous GRUDGE Report was intended to resolve this. It indicated that all military bases should be advised of a new change in policy, and it stated:

> *It is apparent that further study along present lines would only confirm*
> *the findings presented herein.*

The intent of the GRUDGE Report was to bury the UFO question, and confirm a new Air Force policy (or Control Group policy). There would be no more confusion after this.

৵৶

The question remains how such a bungled and decrepit situation turned into such a monster. Was there really an element of intrigue involving the Control Group, which dictated that ultra-strong measures be taken? Was Colonel Watson a very shrewd individual who was in-the-know about Roswell and aware that it was his role to get the UFO question quickly removed and out of mind? Were those his orders? That would seem to be the case. The scenario documented to this point in time is so blatant, that such speculation cannot be put aside. It is so blatant that it becomes exceedingly difficult if not impossible to ignore.

When contemplating the GRUDGE Report, and putting it in context with what had occurred previously, it staggers the mind, and provides perspective to a full blown cover-up in progress. It was simply because Air Force investigators came too close to the extraterrestrial answer. And this was apparently compounded because of a critically serious internal conflict in regard to revealing the secret of Roswell, and Truth of the UFO question.

The GRUDGE Report was essentially a damage control effort designed to fit in with debunking by the *Saturday Evening Post* article, which had been published several months before. It was also necessary to produce an overriding document to negate everything done in Project SIGN, and to nullify the previous Air Force "Memorandum to the Press," which was contrary to interests of the Control Group and others.

In regard to General Cabell, one can either conclude he never knew the Truth of Roswell, and he felt that a full investigative effort was necessary to uncover the riddle of flying objects, or he learned the Truth and was bucking authority to continue with investigation, although that does not seem reasonable. Based on events to come, it is very likely that Cabell was not in-the-know, but he was highly frustrated with the chain of events that hit the Air Force. He knew that much work was now necessary to restore the Air Force's image. And he needed to continue with investigation of flying objects.

One can conclude that there was great internal turmoil within high Air Force and government circles regarding the UFO question, and with much discussion on whether to come clean with the Truth. It was brought to a head with Project SIGN and the EOTS, and then with actions by Defense Secretary Forrestal to enlist reporting services of Sidney Shalett.

After the death of Forrestal, which some sources deemed "necessary and regrettable," the UFO question needed to be extinguished, and there was a rush to produce the final GRUDGE Report. Then, for Project GRUDGE and Colonel Watson, it was time to button-up the matter.

One might wonder if Colonel Watson would find success with what he accomplished. Would the continuing excitement by the public regarding the UFO question be extinguished? Were flying object sightings and encounters by competent and credible people about to disappear? Were independent researchers and the press going to find other things to write about? Was the Control Group now going to be satisfied?

CHAPTER TWENTY-ONE

Project Saucer "Discontinued"

Project GRUDGE: August - December 1949

Nothing to Investigate

After the August 1949 release of the GRUDGE Report, there was a period of relative calm as far as the project was concerned. But it was not that way for an interested public involved with increased sightings, or for some in the military bothered by reoccurring mysterious encounters. The OSI, FBI, Army Counter Intelligence, and a few scientists at secret atomic facilities were still concerned and puzzled by the Green Fireballs still making frequent appearances. But very little investigation was taking place except for independent work by Dr. Lincoln LaPaz of the University of New Mexico.

At TID, there was a sense that Project GRUDGE was in the process of wrapping up and preparing to shut down operations. For all practical purposes, in the short time since Colonel Watson took charge, GRUDGE was now finished. No investigation was being accomplished on reported sightings, including Green Fireballs. Project GRUDGE, as well as Major Jere Boggs in the Directorate of Intelligence, was telling everyone who inquired about the UFO situation that there was nothing to investigate, since all sightings could easily be explained.

Sightings that Project GRUDGE was pressured to look at were quickly discarded for lack of reliable evidence, or they were explained as being a hoax, or misidentification of a balloon, plane, or natural phenomena. Other sightings were just simply filed away.

Under Colonel Watson, the professionalism once seen with Project SIGN was not evident. The attitude of those in the project was to put in the time, wait for the next duty assignment, wait to get out of the service, or wait for retirement. Allen Hynek characterized this particular time in Project GRUDGE by stating:

> *It was always considered far wiser not to 'rock the boat,' to please the superior officer rather than to make waves.*

National Security Bill

The intensifying Cold War with the Soviets, and expansion of communism, was a worrisome situation for government officials, and it became evident that better coordination and efficiency were required to manage an increasing and cumbersome military bureaucracy. As mentioned previously, in Chapters 14 and 17, this was the reason that Ferdinand Eberstadt's Study Group was attempting to find solutions, although his committee was not the only one giving this some review. Now, however, government officials became convinced that a better approach was needed to manage intelligence gathering in the Air Force, Army, Navy, and other intelligence agencies, and this would require an amendment to the *1947 National Security Act.*

On 10 August, the same day the final GRUDGE Report was released, President Truman signed the *National Security Bill*, which amended the *National Security Act* by instituting changes desired and deemed necessary. The biggest hurdle planners had to tackle was how to maintain the nation's democratic civilian rule and control, but also to entertain some sort of centralized decision-making in the military. This needed to be done without compromising a national democratic structure involving economic safeguards, security, and other aspects of a civilian controlled democratic society. It was not an easy thing to consider, but it was generally agreed that the secretary of defense should have more authority than originally envisioned. It would be with restricted "directional control" of military services. Secretaries of each of the military services would no longer be part of the NSC, and this meant that each would have to cooperate and coordinate under the secretary of defense in order to function effectively. Also, under authority vested in the secretary of defense, the RDB chairman was given ability to make decisions over all matters within its jurisdiction, but only where RDB board members were not unanimous.

Tombaugh Sights Object

At this particular time, Astronomer Clyde Tombaugh, who is famous for discovery of the minor planet Pluto, was living in Las Cruces, New Mexico, and had observed several Green Fireballs. But on 20 August, a different kind of flying object caught his attention. For an accomplished astronomer, this was something that created an indelible impression on him. He later wrote:

I saw the object about eleven o'clock one night…I happened to be looking at [the] zenith, admiring the beautiful transparent sky of stars, when suddenly I spied a geometrical group of faint bluish-green rectangles of light…The group moved south-southeasterly, the individual rectangles became foreshortened, their space of formation smaller…the intensity duller, fading from view…Total time of visibility was about three seconds. I was too flabbergasted to count the number of rectangles of light, or note some other features I wondered about later. There was no sound. I have done thousands of hours on night sky watching, but never saw a sight so strange as this.

Tombaugh's sighting made him a believer, and he confided this to Allen Hynek, and to astronomer Dr. James McDonald. But when famous Harvard astronomer Dr. Donald Menzel heard of Tombaugh's sighting, he labeled the sighting a trick of the atmosphere. This did not sit well with Tombaugh, but he continued to confirm his observation by calling it "so flabbergasting," and he stated:

I think that several reputable scientists are being unscientific in refusing to entertain the possibility of extra-terrestrial origin and nature.

Of course, Tombaugh was not aware of Menzel's background, which was explained in Chapter 17.

Soviets Announce Bomb

On 29 August, the Soviet Union announced successful development of an atomic bomb, with a monstrous blast code named "First Lightning." This was a shock to the United States, because it was not expected the Soviets could create a nuclear weapon quite so soon. It served to shake up the country and cause further examination of its defense posture, the number of nuclear weapons in its arsenal, and the spread of communism around the world. It was during this time that "containment of communism" became a high priority, as it would be for many years to come.

Request to Discontinue GRUDGE

In early September, Major Boggs of AFOAI received a copy of the GRUDGE Report from AMC with a request that project activity on sighting reports be discontinued. Colonel Clifford Macomber in AFOAI reviewed the report, and then prepared a memo to the JIC. This was the organization that General Cabell briefed at the end of April upon release of the press memorandum designed to blunt Shalett's article. Macomber's memo concurred with discontinuance of Project GRUDGE, and he then requested feedback from the Army and Navy regarding a "proposal to return reports of flying saucer incidents to normal intelligence interest."

It is quite possible that General Cabell was unaware of this memo by Colonel Macomber, and that Cabell probably never viewed the GRUDGE Report. It's likely he still considered his "Memorandum Number 4" of last February to be in force, which requested that sighting reports be sent directly to Project GRUDGE. As far as Cabell was concerned, Project GRUDGE was still in business. He gave no directives otherwise, and he expected that sighting reports would continue to be investigated, with results forwarded to AFOIN. It does appear, however, that Major Boggs, and others in AFOAI, were helping Colonel Watson put the UFO question and Project GRUDGE away for good.

Although there was no immediate reply to Colonel Watson regarding discontinuance of the project, he achieved his objective, and there was no longer a need for him to keep GRUDGE active. The project would continue in the person of Lieutenant Howard Smith, its chief single "investigator," but it was as if the project really didn't exist. The majority of sightings coming in were just filed away, ignored, or lost altogether.

The military was previously informed, within the GRUDGE Report, of a new policy regarding flying objects, but the matter of increased public interest still needed to be subdued, and the independent media still needed to be corralled. The specter of an extraterrestrial possibility was increasing in the consciousness of the public, and this was not under control or going away. But Colonel Watson was not concerned with that.

Additional Concern about Green Fireballs

Dr. Joseph Kaplan of the AFSAB, who had attended a meeting on 27 April at Sandia Base, and also at Los Alamos (see Chapter 18), then visited the Geophysical Sciences Branch of the Air Force Research and Development Department where he stressed the importance of investigating Green Fireballs. He next sent a letter to General Cabell about them, which then made its way to Vandenberg. All of this resulted in a 14 September letter from Vandenberg to General Benjamin Chidlaw of AMC on the subject of "Light Phenomena." It directed that AMC begin an investigation of Green Fireballs, and to also provide a representative at a 14 October meeting in regard to the matter.

Attendance at the October meeting, along with Dr. Kaplan and Dr. LaPaz, was Major Frederic Oder, who was director of the Geophysical Research Laboratory in Boston, which was under contract with AMC. Also attending were concerned representatives from Los Alamos, including Dr. Edward Teller, Stanislaw Ulam, George Gamow, and other physicists. Sandia Base was represented by base security personnel and intelligence people, along with Air Force OSI, Army intelligence, and the FBI. But representatives from Project GRUDGE did not attend. The result of this meeting was a recommendation that special instrumentation be placed in strategic locations to conduct observations of Green Fireballs on a continuous basis, which would provide more "quantitative information" so that the fireballs could be identified.

On 3 November, another meeting was convened at the Pentagon where the Green Fireball situation was introduced to members of the AFSAB by Dr. Kaplan. Forty-seven people attended this meeting, including Dr. Von Karman, James Doolittle, Dr. Howard Robertson, Dr. George Valley, Colonel Benjamin Holzman, and General Donald Putt. Kaplan stated the following at the meeting:

> *It has to do with the observations of some 46 incidents which have been picked out as real–the observation of what we are referring to as "green fireballs"...incidents that are laid essentially, or reportedly almost entirely, in the West Texas or New Mexico area...the presence there of our atomic energy activities...such people as Dr. Gamow, Dr. Teller, Dr. Reines, and other have been concerned with this...a report from a no less competent observer than Dr. Donald Menzel...has described them completely and is somewhat puzzled...Dr. LaPaz, who is one of the best meteoriticists in the country...is one of the people who has seen this...His thinking, I think, is that this is man-made...you can guess as to what nationality the men are who are supposed to have created these phenomena...proposed studies of these things have been delegated to the Geophysical Research Directorate.*

Results of this meeting made it apparent that action would be underway to conduct an official study of Green Fireballs.

RDB in Disarray

The RDB appeared to be foundering in November when Dr. Karl Compton resigned as its chairman. In his letter of resignation to President Truman, Compton mentioned an untenable situation in managing the RDB. Even with his enhanced power, as a result of amendments to the *National Security Act* in August, Compton became dissatisfied with the myriad of miscellaneous details forced upon scientists and contractors, which detracted from work at hand. Most of it involved budgeting and estimating processes, hassles between various committees, duplication of effort within committees, and lack of timely strategic guidance from the Joint Chiefs. There was no immediate replacement for Compton, and the vacuum in leadership added to RDB's weak and diffuse operation.

It is known, however, that discussion did take place in December about Green Fireballs, and about a plan previously approved by the AFSAB. The RDB affirmed the plan on 20 December. It involved the creation of Project Twinkle, which would undertake observation and study of the fireballs. It would be managed by the Geophysical Research Directorate of AMC's Cambridge Research Laboratories out of Boston. It is not known what the situation was in backwaters of the RDB where the UFO question was being handled, but there must have been some concern about this, at least from Vannevar Bush.

True Magazine - "Flying Saucers are Real"

Just before year-end, on 26 December 1949, radio networks and major newscasts suddenly announced an impending release of *True* magazine's January 1950 issue, which contained an article by Major Donald Keyhoe titled "The Flying Saucers Are Real." This began a period of very active commotion in the press, but nothing compared to what was being stirred up in the Pentagon.

When the issue hit newsstands, every copy was quickly snapped up by a clamoring public. At the time, the magazine was famous for its authoritative and accurate factual reporting, as well as its truthful analysis. The article stated very forcibly at its beginning that:

> *This is the most interesting and the most important true story we have ever published. It is utterly true...It is our sober, considered conviction that the conclusion arrived at in this story is a fact, that...THE FLYING SAUCERS ARE REAL.*

The article continued by stating that its conclusions were formed after an intensive eight-month investigation. It reviewed contradictory statements by the Air Force, which dismissed flying objects "as of no real significance," while also providing warning to the public to maintain "constant vigilance." The article also reviewed various puzzling classical cases the Air Force previously labeled as unidentified, and it went into detail demonstrating Air Force lack of competence in its investigations.

One amusing statement in the article is as follows:

> *True's conclusions are logical and reasonable in the light of the full facts. They have long since been fully accepted by informed authorities.*

That particular statement had a certain "ring" to it. It was a ring that pointedly exploited the same rhetoric contained in the GRUDGE Report, but it contained a more sensible and opposing perspective. Keyhoe obviously obtained a copy of the GRUDGE Report, and he fully capitalized on its nonsense. Additional statements in the article contained the following:

> *...the planet Earth has been under systematic close-range examination by living, intelligent observers from another planet...The only other possible explanation is that the "saucers" are extremely high-speed, long range devices built here on Earth. Such an advance (which the Air Force has convincingly denied) would require an almost incredible leap in technical progress even for American scientists and designers.*

> *True learned that a rocket authority stationed at Wright Field has told Project Saucer personnel flatly that the saucers are interplanetary and that no other conclusion is possible...It is the opinion of True that the flying saucers are real and that they come from no enemy of earth... There has been no sign of belligerence in any of the saucer cases...*

> *It would seem wiser, if space visitors are suspected, to tell Americans the truth. Having survived the Atomic Age, we should be able to take the Interplanetary Age, when it comes, without hysteria.*

Keyhoe's article concluded with a very appropriate comment, which emphasized an Air Force statement from its previous "Memorandum to the Press" on 29 April (see Chapter 18). It stated:

> *...no matter what you suspect is behind the secret curtain of Project Saucer, you can believe the laconic Air Force warning: "The saucers are not a joke."*

Air Force Hiding Something

When the issue of *True* magazine hit the newsstands, the article caused a public sensation, and it became one of the most widely read articles in publishing history. It was a reaction quite opposite to public panic created by Orson Welles' radio episode of 1938. Of course, the article by Keyhoe made a lot of money for the magazine, but it also brought attention to the fact that the Air Force was not presenting an honest face to the public. The implication and reaction to it was that the Air Force was hiding something, and everyone knew it. But as long as the Air

Force was not about to admit duplicitous behavior, or clear up the mountain of duplicity it was accumulating, suspicion would remain and continue to build. This would mark the beginning of widespread public awareness of the UFO question, with great suspicion that the Air Force was instituting a massive cover-up. And this suspicion continues to this present day.

After highly-professional people from Project SIGN left the scene around the end of 1948 to mid-February 1949, there was pressure placed upon people in the new Project GRUDGE to dismiss the UFO question, which was accomplished with the GRUDGE Report. But at the same time, General Cabell in the Directorate of Intelligence needed to keep investigation alive in order to determine origin and purpose of flying objects, since they still appeared to be real and also threatening. The objects were continuing to generate questions and inquiries from many in the military.

Based on Colonel Watson's GRUDGE Report, it would appear that Watson had strong incentive to dismiss flying objects, although General Cabell was probably not aware of this. It would also appear that General Cabell must have been frustrated by the fact that many in the military continued to exhibit concern about the UFO situation, while the technical intelligence "people" at TID were evidently not investigating, analyzing, or forwarding reports to the Directorate of Intelligence at Air Force Headquarters.

GRUDGE "Discontinued"

When *True* magazine hit the streets, the Air Force immediately came out with a countering press release, which stated that there was "no basis to reality of flying saucers," and then it gave reasons why. But along with this, and quite amazingly, it was also announced that Project Saucer was "discontinued." And this was also the case for Project GRUDGE. The next day, on 27 December, Major Boggs at AFOAI advised TID's project officer Lieutenant Howard Smith of the news by telephone. Supposedly, this call constituted "official approval" to stop the project, which was in answer to TID's previous request in early September. There is no indication, however, that General Cabell was ever advised or aware of this. And General Donald Putt of Air Force Research and Development, who originally authorized the project, never gave authorization for its discontinuance.

Air Force Taken by Surprise

The Air Force was obviously taken by surprise with the *True* magazine article, and with the commotion it caused, but some people at Air Force Headquarters knew it was coming. Keyhoe's contacts knew of his investigation and that he was working on the article, but there were other officials in-the-know who were not aware and overlooked what Keyhoe was doing. They then became highly distressed.

As far as Vandenberg and the Control Group was concerned, they had previously wanted to better understand the purpose and intentions of flying objects, but they could no longer afford to let public hype run rampant. They were hiding a magnificent secret, and they were not about to unleash Air Force investigators.

This placed a huge burden on Cabell, who was left in the middle with his frustration. One would think that Cabell would have discussed the turbulent situation with Vandenberg, and he probably did, but Vandenberg probably protected the secret. Vandenberg and the Control Group now needed to keep Air Force investigators away from flying object investigations, and they wanted the public quieted, the media subdued, and Project GRUDGE closed down.

GRUDGE Report Available to Public

Amazingly, on 30 December, a few days after the popular article from *True* magazine came into the hands of many millions of readers, and three days after Project Saucer was discontinued, TID formally announced that Project GRUDGE's "Saucer Report" was available to the press and public at the Pentagon, although it was not for general distribution. Of course, release of this report was intended to counter the sensational claim by *True* magazine, thereby putting the entire matter of the UFO question to rest – Forever!

It was assumed that the six hundred-page report would put most everyone to sleep who looked at it. It explained most sightings in detail, with application of the usual explanations. Nearly one third of them were attributed to astronomical misidentification, thanks to Hynek, and many were chalked up to mistaken identity of familiar objects. Others were very detailed but mystifying, and those were the ones composing 23 percent of cases labeled as unidentified. They were lumped into a "psychological category" relating to fantasies or illusions of the eye and mind, and because of this nebulous and indistinct category, the unknowns took on a unique "flavor." They were expected to be ignored and cast aside by readers of the report. This was obviously the intention by creators of the GRUDGE Report, but Hynek would later comment on it as follows:

> *Twenty-three percent unknowns! That could have been considered extremely newsworthy. But hardly a peep out of the press.*

Amazingly, nearly one-quarter of sightings documented in the GRUDGE Report were unexplained, yet no one in mainstream media questioned it. Those unknowns, however, did attract interest of independent investigators such as Donald Keyhoe, who then capitalized by pointing out the reality and unearthly nature of the flying objects in these incidents, which were observed and encountered by highly credible witnesses. It then became extremely obvious that the Air Force was truly hiding something, especially when it had the audacity to state in the Project Saucer Report that:

> *It is readily apparent that further study along the present lines would only confirm the findings herein.*

Nevertheless, the Air Force and Watson were finished with the UFO question, at least for now. Air Force officials and the Control Group were highly confident that release of the Saucer Report (GRUDGE Report) would quiet the public, and also "hobbyist" investigators such as Keyhoe. But that was no guarantee that antics of flying objects, or sightings by competent witnesses would be quieted. Information on flying objects would continue to be released by the media and independent investigators, which would continue to attract additional public attention.

ം൭൮

Again, a pertinent question arises. How does such a conflicting and contorted scenario within the Air Force come about?

Previously, there were Air Force technicians, engineers, scientists, and professional experts at TID involved with Project SIGN. And they determined that flying objects were real, and also extraterrestrial. Their "Estimate" of this was immediately rejected by General Vandenberg, who was motivated to do so. Project SIGN was then renamed Project GRUDGE, and the Special Projects branch was purged of its professional investigators.

Next, a preemptive Air Force press memorandum was released indicating that flying objects were real, but not necessarily extraterrestrial. Its release was timed to publication of a *The Saturday Evening Post* article by Sidney Shalett, who debunked reality of flying objects due to direct Air Force influence on him.

Then, when Colonel Watson took over Project GRUDGE, he immediately produced the GRUDGE Report, but without investigating sighting cases, and he proclaimed nonexistence of flying objects. He evidently had great motivation for doing so.

Then an article by Donald Keyhoe in *True* magazine was issued stating that flying objects were definitely real, and undoubtedly extraterrestrial. It effectively pointed out Air Force duplicity in handling the UFO question.

Then another immediate Air Force press release indicated that Project Saucer (Project GRUDGE) was discontinued because of a determination that flying objects did not exist.

Then the Air Force released the Project Saucer Report (GRUDGE Report) to the public to prove that flying objects didn't exist, but it only verified that Keyhoe was right about Air Force duplicity, because Keyhoe took much of his information from the GRUDGE Report.

The whole aspect of this scenario is terribly incredulous, and it suggests that certain elements in the Air Force, or the Control Group, were extremely overwhelmed with protecting the secret of Roswell, and also Truth of the UFO question.

CHAPTER TWENTY-TWO

Air Force Investigation Crippled

Project GRUDGE: January - February 1950

Intelligence Requirements Cancelled

A confirming letter for discontinuance of Project GRUDGE was written by Major Aaron Boggs, and sent to AMC on 4 January 1950 with the subject heading "Project Grudge." It stated:

> ...*In view of the findings and subsequent concurrence by all Services through the JIC, it is recommended that special project action by the Intelligence Department, Hq. AMC, on "flying saucer" reports be discontinued. Any future reports on this subject should be accorded the same consideration as that given to intelligence on other subjects...This headquarters is taking action to cancel all its outstanding intelligence requirements issued for collection of information on "flying saucers" including Air Intelligence Requirements Memorandum Number 4, Unconventional Aircraft", Department of the Air Force, Hq. USAF, dated 15 February 1949...*

The letter went on to mention that the GRUDGE Report would be "reviewed to determine potentialities for psychological warfare application as recommended by AMC," and would be kept as "the official record" regarding flying object investigations.

It is assumed that General Cabell was aware of the letter since it was stamped and authorized by his executive officer, Colonel John Schweizer. The letter was "acknowledged" by branches within AFOIN, and it included Colonel Garrett's signature representing AFOIR. This did not mean, however, that everyone in AFOIN was in concurrence with the "recommendation" to discontinue "special project action" on "flying saucer reports," or to discontinue Project Grudge.

The previous public announcement that Project Saucer was discontinued was intended specifically for public consumption, but the general military did not totally understand this, which provided a confusion factor for them. Directives still existed regarding flying object reporting requirements, and the military would still need to be specifically advised on cancellation of "intelligence requirements issued for collection of information on 'flying saucers'," which was mentioned in the second half of Major Boggs' statement above. A "pending" notification to the military regarding reporting directives for flying object sightings would clear up the confusion.

This "official" letter confirmed to AMC and TID in writing, at least in their interpretation, that "special project action" was officially discontinued for Project GRUDGE. But there was still no actual "official" directive in that regard. With Cabell away from the office, the letter was simply a "recommendation" by Major Boggs to stop the project. It was a very clever piece of wording, but GRUDGE would still be an existing functioning project until officially "discontinued" or "cancelled" by appropriate authority, and that authority was either General Putt or General Cabell. Even if Cabell was aware of Boggs' letter, he was not about to discontinue the project, although he would later see validation in temporarily changing previous directives for reporting on flying objects. General Cabell was depending on TID to continue with monitoring and analysis of critical flying object incidents for national security reasons. This would not change.

When one contemplates this situation further, it's mind-boggling to realize that the Air Force decided it could finesse its problem by implying that its fictitious investigative project (Project Saucer) was finished, and then also make the previously secret GRUDGE Report available for public viewing. The motivation for doing this is somewhat understandable, but there is a contemptuous side to it. In reality, the Air Force had much reason for continuing investigation of flying objects, but the *True* magazine article tied Air Force hands on current investigation going forward. All investigation needed to be out of the public eye and confined.

The sloppy GRUDGE Report, directed and produced by Colonel Watson in August 1949, was obviously intended to rectify the fiasco of the previous April concerning the preemptive press memorandum, and also Shalett's *Saturday Evening Post* article. The contrived conclusion in the report confirmed that flying objects were not only non-threatening, but that they did not exist. This, however, conflicted with many unexplained sightings by highly credible military personnel, and also with the *True* magazine article by Donald Keyhoe, who strongly and convincingly pointed out Air Force incompetence. Contrary to naïve Air Force judgement, the released GRUDGE Report was now available to confirm those allegations of incompetence, and this was not the intention Air Force officials had when releasing it to the public. Some sources claim that General Cabell believed that AMC and TID were taking Project GRUDGE underground with announced cancellation of Project Saucer, and release of the GRUDGE Report. That does not track, however, with Boggs' interaction with Colonel Watson to completely close down Project GRUDGE investigative activity.

The Air Force previously bungled its preemptive response to Shalett's article, and now there was a determination to put Donald Keyhoe to shame. That was the naïve and deceptive thinking of Air Force officials, although they were lacking a better way to safeguard and maintain their unearthly secret. Again, the Air Force appears to have "shot itself in the foot" with its actions. The GRUDGE Report was an obvious exercise in debunking, which attempted to suggest that investigation of flying objects was a useless endeavor. But it would take more than "discontinuance" of Project Saucer, and release of the GRUDGE Report, to make a difference in dissuading a now very skeptical and re-awakened public.

Twenty-three percent of sighting cases in the GRUDGE Report, which were labeled unknown, were now exposed for everyone to see, and many of those sightings were of objects displaying flying characteristics not possible with earthly engineering technology. The sloppy GRUDGE Report did nothing to contradict this, except to enhance public skepticism of the Air Force. With good reason, the public and independent researchers became totally suspicious.

Colonel Watson did not want to be bothered with the UFO question any longer. He felt he had effectively challenged the previous TID conclusion that flying objects were real, including the Air Force press memorandum from last April stating that the public "must remain vigilant." He was determined to hold fast to contrived conclusions presented in the GRUDGE Report, even with the 23 percent considered to be "unknown." To him, Project GRUDGE was finished, especially with the announcement and letter from Major Boggs that Air Force headquarters was supposedly taking action to discontinue "special project action," and cancel previous reporting directives for flying objects.

One would think that General Cabell was beginning to realize that the situation he found himself in was extremely confounding, and he was probably having much difficulty wrapping his mind around it. He knew flying objects were real, but there was also a decision to not distress the public and general military with an extraterrestrial possibility, especially without real evidence. It was still necessary, however, to investigate and determine "tactics" and "probability" of the flying objects. It was a curious situation for him, but his confusion was compounded in not having knowledge of the magnificent secret, or of the Control Group's desire to close down Project GRUDGE. He would press on and manage the best he could.

Suppression of Sighting Reports

On 12 January, General Cabell found it necessary to send a letter to various commands, including the FBI and CIA, to rescind memoranda and letters previously sent early in 1948, which instructed commands to send sighting information on flying objects directly to TID. The new directive indicated that those reports should now go through the same channels as "other intelligence information." Cabell was bound to this action because of announced discontinuance of Project Saucer, and that this would reduce existing confusion in the military.

Cabell, however, still wanted to keep a handle on what was going on with the objects, even if the public and general military thought that investigation by the Air Force was finished. He was particularly aware that the Air Force OSI was involved with investigation of mysterious sightings, especially those around military installations in New Mexico and other places, and he needed current sighting updates and analysis on those. Commanded by General Joseph Carroll, the OSI was a special component of the Air Force under Air Force Secretary Symington. Its activity was controlled by the Air Force inspector general, and not by Air Force Chief of Staff Vandenberg. But the OSI did interface with Air Force organizations, including the Directorate of Intelligence.

On 30 January, AFOSI headquarters sent out a message to its field offices stating:

> *In the future any information received on this subject will be reported to Hq OSI through the media of spot intelligence Reports (See AFCSI Letter No. 106, 18 Oct. 49). Active investigation of incidents concerning 'Unconventional Aircraft' will not be conducted unless a specific request is made by competent authority.*

This message may have been confusing to some commands, but it was generally understood that only reports of legitimate sightings of "Unconventional Aircraft" would be made, rather than the obviously less than credible ones. The directive was then followed on 8 February with AFSCI Letter #85, which stated the following:

> *Spot Intelligence Reports concerning sightings of unidentified aerial objects need not be forwarded by TWX unless considered to be of priority Counterintelligence interest to this [AFOSI] headquarters.*

It's obvious that those at higher levels at Air Force Headquarters were now assisting with total suppression of sighting reports. It was now up to individual commands to decide if reports would be sent in, but the message was now clear that flying objects didn't exist. The Air Force would not waste time on them. Accordingly, Air Force OSI would play dead, supposedly, and this fitted with comments previously made in Shalett's *Saturday Evening Post* article many months previous, where high Air Force officials individually articulated that excitement about flying objects was a fantasy gone wild.

Cabell, however, would proceed with investigation of flying objects in his Technical Capabilities Branch (TCB) headed by Lieutenant Colonel Milton Willis, but it would be a token investigation effort. Cabell's plan was that TCB would forward sighting reports it received from normal intelligence channels to TID. This is where real investigation was "supposed" to be taking place. He was of the opinion that TID's Project GRUDGE was still in operation and processing "priority interest" sightings related to national security. Major Boggs and Colonel Hearn, however, would provide interference with that.

Unfortunately, Cabell's assumption that TID would be looking at sighting reports was a false impression. Project GRUDGE was essentially dead. TID had no interest in sighting reports, and would accomplish no investigation except to log sightings, bundle them for storage, or lose them. Nothing would change in that regard, and Cabell lacked direct control with that organization. It is quite possible that Watson was very pleased with this chain of events.

True Magazine - "How Scientists Tracked A Flying Saucer"

On 22 February, *True* magazine again caught the public's attention with its March 1950 issue, and it was another sensational story. It contained an article titled "How Scientists Tracked A Flying Saucer," and it was written by Navy Commander Robert McLaughlin. He was chief of the Navy's guided missile program at White Sands Proving Grounds in New Mexico, and he was the rocket and balloon specialist connected with high flying objects mentioned in Chapter 20. The magazine introduced the article by stating:

> In its January issue TRUE said that the flying saucers are real and interplanetary. Its story was widely supported by the nation's press and radio. TRUE's findings are here confirmed by Commander McLaughlin, a rocket expert at White Sands Proving Ground, who worked independently of this magazine's investigation. He reveals how a troop of Navy men and scientists tracked a flying disk with a precision instrument and tells of flights he and others witnessed.

Magazine editors then ended their statement with the following:

> *The Editors of TRUE believe that the admitted existence of 34 "unexplained incidents" indicates that the announced discontinuance of Project Saucer was premature. TRUE further believes that the official announcement was ill-advised, made in haste, and that the United States Project Saucer has in fact been continued without interruption under another code name. The 34 "unexplained incidents" which the Air Force concedes—there are no doubt others—are the best documented and the hardest to explain in any fashion other than by the interplanetary thesis first set forth by Donald E. Keyhoe in TRUE for January and here repeated by Commander McLaughlin.*

True magazine editors wasted no words, and they certainly hit the "mark" in their comments. They were clued in to "unexplained incidents" in the GRUDGE "Saucer Report," to the "hasty" announcement of "Project Saucer" discontinuance, and to the continuation of "Project Saucer" with "another code name." They must have discovered the secret name of "Project GRUDGE," without revealing it, which used the name "Project Saucer" for public disclosure purposes.

McLaughlin stated the following:

> *I am convinced that it was a flying saucer and, further, that these disks are space ships from another planet, operated by animate, intelligent beings.*

> *I think it is safe to say that it wasn't any type of aircraft known on Earth today. Even if, as is likely, there are top secret models which you and I know nothing about, there is no human being in this world who could take a force of 20 G's and live to tell about it.*

> *I think that the saucers are piloted space ships, first, because of their flight performance. The White Sands Saucers were most definitely capable of changing their direction while above our atmosphere. This extreme maneuverability—plus their large size—eliminates for me the likelihood of their being operated by remote control.*

> *My own experience with rockets leads me to feel that a Saucer with such characteristics is far beyond the technical powers of anyone on earth.*

The article went on to describe how their scientific endeavors, using high altitude research balloons, were interrupted on several occasions by high flying objects. In an incident observed by several technicians and scientists, an elliptical disk about 105 feet in diameter was seen flying at an incredible elevation of about 296,000 feet, and at a speed of five miles per second. It then abruptly swerved upward, which calculated out to about twenty times the force of gravity within the sharp turn. The object was tracked by their special telescope, and it was moving in an opposite direction, and much higher than the balloon they were tracking.

The interesting thing about the reported sightings is that they took place months previously, but the flying objects were still being reported from White Sands at the current time, and also afterward. Sightings of flying objects were relatively common in the area, with many technical and scientific professional people observing them at the same time.

McLaughlin's story, which was cleared by the Navy, was an obvious contradiction to the Air Force position on flying objects, and the debunking by Project GRUDGE. There was an obvious "iron barrier" in regard to communications between military services. The Navy was not necessarily on the same page as the Air Force, which became a frustration General Cabell needed to deal with. It served to remind him again of the reality of flying objects, a subject that was always on his mind.

As for the public and media, this particular *True* magazine article provided a confirming answer to the UFO question, especially with it authored by a high-ranking active duty Naval officer. There could not be a more respected or credible witness. In addition, he was backed up by many scientists and technicians who confirmed the objects.

The two articles in *True* magazine, within two months of each other, proved that the Air Force was avoiding Truth and honesty. But there still remained a nagging question about the objects, which was due to a lack of tangible evidence. A major mystery still remained that needed further resolution, and there were many in the civilian community who were eager to get on with finding additional answers. Accordingly, flying objects seemed happy to oblige. It was just the beginning of many more credible sightings to be encountered, which Keyhoe and others would later latch on to.

Iron Barrier

The "iron barrier" between military services was specifically noted by government officials and intelligence agencies. Collection of necessary scientific intelligence was very important, but if service branches could not cooperate and coordinate together on this, the nation could find itself in deep trouble. There was a danger that other nations might accelerate far ahead with new technology and armaments. And this situation was particularly noted by the RDB and CIA-OSI.

The CIA was giving this situation serious review and consideration, since it was already deeply involved in monitoring the situation at TID, RDB, and elsewhere. It's possible that some in the Control Group understood that the great secret they were protecting might become compromised due to continued military involvement with flying object investigation, and particularly in management and protection of recovered materials. Military members of the Control Group, as expected, would take exception to this, but it was also realized that only the military could handle and sort out sighting reports, especially with so many occurring within the military's sphere of operations.

Rivalry between military services was not yet resolved, and neither was competition for intelligence, but the Air Force was now putting itself in a mode of backing off from investigation of the UFO question. In fact, it closed itself down to the point where it was now impossible to take investigative action, although Cabell was hoping TID was involved to some extent for national security reasons. Flying objects, however, were not going to stop making appearances, and the number of very important sightings by credible witnesses was about to dramatically increase.

❧❧

In regard to the UFO question, the position that the Air Force found itself in at the beginning of January 1950 is truly remarkable. The Air Force had tied itself into a great knot with no way to struggle loose. General Cabell desperately wanted to investigate flying objects and determine what they were, but now the Air Force was crippled by its own mishandling and weak attempts to safeguard the monumental secret, and the Truth of Roswell. It was a grand distorted effort to cover up Truth of the UFO question.

One might attempt to conjure up excuses for the situation the Air Force found itself in, but no excuses can be conceived or make sense. All excuses circle around the fact that the Air Force was protecting or hiding something. But what was it? Was it an effort to prevent public panic if Truth were revealed? Was it an effort to prevent Air Force investigators, who were not in-the-know, from discovering Truth? Was it an effort to prevent revelation of the secret of Roswell and the UFO question? Was it an effort to protect the secret keepers?

Those excuses only confirm Truth. The Air Force was simply involved in a desperate effort to deceive the public and general military with their cover-up. But the Control Group was also heavily involved with this, and it was becoming entrenched in a self-perpetuating, compartmentalized, ultra-secret monster that could not be tamed, manipulated, or compromised.

In his *True* magazine article, Navy Commander McLaughlin essentially confirmed and verified that the Air Force was holding back Truth. His article was granted official clearance by the Navy, and this offered a strong contribution toward Truth, but this "contribution" would become a serious internal concern in high government circles.

Truth would continue to be a difficult thing to suppress, especially since sightings and encounters of flying objects by competent witnesses would continue to challenge the monumental secret being protected and managed by the Air Force and the Control Group.

CHAPTER TWENTY-THREE

Sightings Proliferate

Project GRUDGE: March - June 1950

Significant Changes in AFOIN

In March 1950, significant changes were in the works at AFOIN. This included leadership changes, and also the changing or adding of division names and branches. The Defensive Air Branch became the Air Targets Division, and the Technical Capabilities Branch became part of the Evaluation Division. General Ernest Moore was elevated to the position of assistant for production, and he would work directly for Cabell as before. People previously mentioned would generally remain in their jobs, but leadership changes took place in a number of positions in the revamped organization.

Sighting by TID

On 8 March, personnel at TID took the opportunity to view a flying object just by stepping outside their building at Wright-Patterson AFB. The object was first reported by TWA pilot Captain Bill Kerr when and he and fellow pilots, Miller and Rabeneck, were landing at Vandalia airport, which was about eleven miles northwest of Wright-Patterson AFB. Control tower personnel were already watching the brightly-lit object, which was observed for several hours high in the morning sky between heavily scattered clouds. It was the control tower that contacted TID at Wright-Patterson.

Radar contact was soon established with the object, which indicated a strong solid return. Four F-51 pilots were quickly sent up to intercept, and two of the pilots spotted a circular object. They described it as "huge and metallic," but they returned to base when it disappeared high above and behind clouds.

This incident was investigated by TID's surrogate investigators, James Rodgers, Roy James, and Albert Deyarmond. But they concluded that the object was planet Venus, and they also determined that radar indications were simply reflections from ice-laden clouds. The experienced radar operator, an Air Force master sergeant, vehemently disagreed with the assessment because he knew without doubt his radar detected a single solid object, which tracked upward when he picked it up. One of the F-51 pilots complained bitterly that the object "was no planet," because it became larger and more distinct when he approached it, and the object was not visible or apparent in clear skies immediately afterward, or on subsequent clear days following.

William Webster Takes Over RDB

On 15 March, William Webster officially took control of RDB, which was more than four months after Karl Compton resigned. At this particular time, the organization was overwhelmed with budget problems. Its oversight committees were challenged in accomplishing certain agendas, primarily because defense-related programs were increasing in number. There was no money to pay for them all. There was debate raging about giving more focus to the more important scientific and technical programs, but the military was not giving appropriate attention to this, or participating in technological strategic policy. The RDB was also relying more on the CIA, rather than the military, to provide scientific intelligence. One can only speculate on how this situation was being discussed in the Control Group, which was embedded in the RDB.

Interestingly, RDB leadership was becoming an unchallenged dynasty. This was evident because Webster was well-connected with Vannevar Bush, Lloyd Berkner, and Karl Compton. They were all connected to MIT and to Vannevar Bush, who was their previous leader when working at the wartime OSRD, then the JRDB and RDB. Berkner was Bush's executive secretary in all three organizations, and all of these people were in-the-know about Roswell.

Sighting at Farmington

On 16 March, and for the following two days, a huge sighting occurred over the town of Farmington, New Mexico. On the 18th, the *Farmington Daily Times* displayed a front page headline stating, "Huge 'Saucer' Armada Jolts Farmington," and the related article mentioned the following:

> *For the third consecutive day, flying saucers have been reported over Farmington, and on each of the three days their arrival here has been*

*reported between eleven and noon. Three persons called the Daily Times
office to report seeing strange objects in the air just before noon. Persons
along Main Street once again could be seen looking skyward and
pointing...According to residents shortly after the incident, government
agents from an unspecified organization arrived and encouraged them
not to discuss what happened for national security reasons.*

The *Seattle Daily Times* also reported on that date that:

*...All of the 250 persons interviewed by Walt Rogal, editor of the
Farmington Times, admitted seeing what generally was described as a
mass flight of flying objects...Regal estimated 85 percent of this towns
population of 6,100 saw the objects...Most folks who would talk about
it thought the things were flying at about 15,000 feet. They told of
watching them hover over the town for about 20 minutes then speed
away to the northeast...each object was "at least the size of a B-36,
maybe larger."*

People in Farmington who witnessed the objects reported they appeared
silver in color, flew in formation in an easterly direction at about 20,000 feet, and
were moving 1,000 mph or more. One, moving faster and lower than the others,
appeared to be the leader and was red in color.

Saucers Found in New Mexico

On 22 March, FBI agent Guy Hottel wrote a memorandum to the attention
of FBI Director J. Edgar Hoover stating the following:

*An investigator for the Air Force stated that three so-called flying
saucers had been recovered in New Mexico. They were described as
being circular in shape with raised centers, approximately 50 feet
in diameter. Each one was occupied by three bodies of human shape
but only 3 feet tall, dressed in metallic cloth of very fine texture...the
saucers were found in New Mexico due to the fact that the Government
has very high-powered radar set-up in that area and it is believed the
radar interferes with the controlling mechanism of the saucers.*

Of the many case files and memos archived over the years by the FBI dealing
with flying objects, the FBI indicates that this memo is the most popular of those
made available to the public for online Internet access. There is also something

about this memo that has a very familiar ring to it, and it pertains to three bodies dressed in metallic cloth (or silver suits) that were recovered. This not only harkens back to the Roswell incident, but also to additional incidents, which prompts one to take a second look at reality and Truth of this memo. Another curiosity pertaining to this is that it brings forth a recall of the Aztec incident described in Chapter 14. Further discussion on this will be found in Chapter 25.

Also of note in this FBI memo is mention of a government high-powered radar that was believed to interfere with the flying objects. This was mentioned in regard to the Roswell crash, and also other flying object crashes. Missile-tracking radars were known to exist at White Sands, but there were also other high-powered radar stations, built in 1946, which were located at El Vado (AFS-P8), Continental Divide (AFS-P51), and Moriarty (AFS-P7), in New Mexico. These formed a wide triangular area with Los Alamos near its center.

This memorandum must have been very interesting to FBI Director Hoover, not just for the fact that some saucers were recovered, which he very much wanted to be aware of, but he now had a question about current Air Force status regarding investigation of flying objects. Like most people, he was under the impression that Air Force investigation was discontinued, and he now needed an update. That update was provided on 28 March when D. Milton Ladd, assistant director and head of FBI's Intelligence Division, provided another memorandum to Hoover that stated:

> ...*Special Agent Reynolds obtained the following information today (3/28/50) from Major Boggs and Lieutenant Colonel C. V. Hearn of Air Force Intelligence. The Air Force discontinued their intelligence project to determine what flying saucers are the latter part of last year... The reason for the discontinuance was that after two years of investigation over three-fourths of the incidents regarding flying saucers proved to be misidentifications of a wide variety of conventional items such as lighted weather balloons and other air-borne objects... They reiterated that the Air Force is conducting no active investigation to determine whether flying saucers exist or what they might happen to be...*

The interesting part about this memo is that it was confirmed that "the Air Force discontinued their intelligence project [Project GRUDGE]." Major Boggs previously wrote a letter with a "recommendation" that "special project activity" be discontinued, but there was no directive from official authority to discontinue or cancel the project. And Cabell was still under the impression that Project GRUDGE was investigating cases of priority concern.

Project Twinkle Begins Operations

The study of Green Fireballs, which was named Project Twinkle, was initiated by Dr. Kaplan of AFSAB in coordination with Major Oder of the Geophysical Research Division (GRD), and it was given approval by General Donald Putt of the Air Force's Research and Development Department. A letter from AMC requesting approval for the project was previously sent in December 1949 to General Putt. He then affirmed the project on 20 December, with a request to AMC that project plans and a budget be presented.

On 5 March, an AMC meeting at Wright-Patterson, between personnel from Holloman AFB and the GRD, resulted in a 16 March AMC letter with the subject heading "Light Phenomena." The letter indicated that a $20,000 contract was granted to Land-Air Inc., which currently operated phototheodolites (movie cameras) at White Sands by experienced, professional technicians. They would now operate two stations for a continuous Green Fireballs watch during a six-month period under the general control of GRD, and with sponsorship by the Cambridge Research Laboratory (AFCRL) in Boston, which was attached to AMC. With minimal funds provided, it was not the kind of project initially envisioned and proposed, but it would be a limited operation using available personnel and equipment from Holloman AFB.

Prior to this, on 21 February, an observation post was set up at Holloman, manned by two people during daylight hours, using a theodolite, telescope, and camera. Evidently, this was done because of frequent reports of "unexplained aerial phenomena" observed in February around Holloman, and also at Vaughn, New Mexico. The commanding officer at Holloman directed setup of the post on base. During a month of observation, however, nothing out of the ordinary was observed.

On 24 March, upon opening an observation post installed near Vaughn, Project Twinkle became active in the search for Green Fireballs. This site was chosen due to an abnormal number of sightings received from there, although no reports were ever previously "documented" from that location. Establishment of the site, located about 85 miles east-southeast from Sandia Base, 110 miles southeast from Los Alamos, and 135 miles north-northeast from Holloman AFB/White Sands, did not appear to be an appropriate place to establish an observation post, especially when compared to other places where a majority of more recent sightings had been reported. This place, however, attracted attention of Holloman's base commander after receiving notice of sightings from Vaughn, and also after sending several investigators to question witnesses. Also, the choice of Vaughn may have been influenced by Colonel William Hayes, an Army Affairs officer based at Kirtland AFB, who observed bright balls of white light near Vaughn on three occasions. Continuous observation, on a contractual basis, was officially scheduled to begin on 1 April for Project Twinkle.

The plan was to have Land-Air Inc. manage Askania phototheodolites at Vaughn, and also at White Sands where they were already set up. Holloman AFB personnel would manage spectrographic cameras and radio frequency instruments at the post set up on base. The spectrographic cameras, however, became unused because personnel assigned to operate them were transferred, and the radio frequency instruments obtained from the Army Signal Corp were deemed too expensive to operate.

Dr. LaPaz indicated he could not participate in the project because of other pressing duties, but he also expressed his new idea that the fireballs were possibly U.S. guided missiles under test, which were supposedly designed to defend sensitive installations in the area. He admitted he might be wrong about that, but one could also speculate he was preparing to receive credit for his idea, since Project Twinkle was now in the perfect location (White Sands) to prove him right. It was evident, however, that LaPaz considered the fireballs to be of man-made origin, even when others considered that they might be something else from the heavens.

Captain Edward Ruppelt (of Project BLUE BOOK fame) would later state that he once had lunch with a group of scientists at AEC's Los Alamos Laboratory where the subject of Green Fireballs came up, and he spent several hours with them in an interesting discussion. He stated:

> *All of them had seen a green fireball, some of them had seen several... When the possibility of the green fireballs' being associated with interplanetary vehicles came up, the whole group got serious. They had been doing a lot of thinking about this... The green fireballs, they theorized, could be some type of unmanned test vehicle that was being projected into our atmosphere from a "spaceship" hovering several hundred miles above the earth.*

In light of this, it is quite interesting to learn that top U.S. scientists would offer the interplanetary explanation. There is, however, some sense to this because many professional scientists and technicians at White Sands had previously seen the mysterious disk-shaped objects, although rarely a Green Fireball, and they were quite convinced the flying disks were not of this earth.

Air Force Secretary Symington Resigns
Stuart Symington was continually requesting additional funds to sustain the Air Force, but his budget and allocation for increased forces continued to be reduced, and he considered this a threat to national security. This concerned him

greatly and he complained that the proposed budget offerings "in the Cold War with Russia, was not buying military superiority." He felt the Air Force needed to be modernized and strengthened because Soviet forces were quickly becoming stronger. He previously won his fight to get the Convair B-36 put into production, but that was not enough. Potential disaster loomed, and Symington was not about to take responsibility.

Because of the Soviet nuclear bomb threat, Symington also suggested that funds be allocated for research, development, and future buildup of new intercontinental ballistic missile (ICBM) defense systems for nuclear deterrence. He felt this was necessary because of the Berlin crises, continued Soviet belligerence, advancement of communism, and buildup of Soviet forces.

His advice and pleas were ignored by both President Truman and Secretary of Defense Johnson. Symington felt he had no other choice than to resign, and on 30 March he submitted his resignation. He performed his last duty for the Air Force on 24 April, and then assumed his new appointment as chairman of the National Security Resources Board (NSRB). In his new job, he was hopeful of advocating for necessary resources to fight the Cold War.

Symington was replaced as secretary of the Air Force by Thomas Finletter, who equally believed in a stronger Air Force, although he was helped along with that belief three months later due to subsequent onset of the Korean War. It would appear that Truman should have heeded sound advice from various people attempting to catch his attention.

Sighting by TWA Flight 117

On 27 April, at about 8:25 P.M., Trans World Airways Flight 117 was heading for Chicago, just south of South Bend, Indiana, after passing Goshen, when a bright reddish-orange disk shaped object (the color of "glowing hot metal," a "blood red moon," or the "glow of a refinery stack") was spotted by pilots Captain Robert Manning and Captain Robert Adickes. They were cruising at about 2,500 feet altitude when the object approached the plane from the right. It then paced the plane for several minutes at a slightly lower altitude, with the horizon positioned above the object. This provided a good view from about half a mile away. The pilots, who wanted more witnesses to the event, called stewardess Gloria Henshaw to come forward. She saw the object, and then Adickes instructed her to go tell passengers so they might also see it. He wanted to make sure everyone on board was aware and witness to what was happening, and he also went back to the cabin for a moment to be with passengers.

The pilots attempted to maneuver close to the object several times, but it still kept a distance away. In one attempt to bank toward it, the object made a "sharp-angle turn as if it was a non-inertial object." In an attempt to turn the plane directly toward it, the object dived lower and then sped quickly ahead at an estimated speed of about 400 miles per hour. It then disappeared out of sight.

According to Captain Adickes, an Air Force general visited him at his home about three months later and interviewed him for several hours. It is not known who the general was, but it was curious that an Air Force general would take it upon himself to conduct such an interview, rather than an intelligence officer of lower rank. It suggests that someone at Air Force Headquarters thought the incident to be of very high priority. Could it have been General Cabell, or some other general from Air Force Headquarters who had incentive to conduct the interview? This sighting incident would soon be brought to public attention by Major Keyhoe, and it is reviewed in the next chapter.

Sighting at McMinnville, Oregon

On 11 May, Paul and Evelyn Trent experienced a sighting about nine miles southwest of McMinnville, Oregon, in the small community of Bellevue. The sighting is now a classic.

Evelyn was walking back to their farmhouse after feeding her rabbits, and she noted a "slow-moving, metallic disk" in the distance, which was heading in her direction from the northeast. Upon seeing it, she called to her husband who then came out of the house. When he saw the object, he quickly grabbed his camera and managed to take two pictures of the object before it sped off, disappearing toward the west. After the pictures were developed and stored away for several weeks, the local newspaper found out about them, and so did the nation. They were published in the 26 June issue of *Life* magazine, and also in newspapers across the country. The pictures have been extensively analyzed by scientists using photometric analysis, and they are considered to be some of the best photographic evidence ever taken of a flying object. Even with all the publicity generated by this incident, Project GRUDGE paid no attention, but it's quite likely that General Cabell became aware of it.

Project Twinkle Sights Objects

Project Twinkle began its official operation on 1 April. The Geophysical Research Directorate's managing agent in charge of the project was Dr. Anthony Mirarchi, and he made a visit to Holloman AFB at the end of May. He wanted to look at the operation, but he also heard of two sightings. When he arrived, he requested specific information of object sightings made on 27 April and 24 May. His request was possibly made to Administrative Officer Lieutenant John Albert. Later, his request was answered in a letter, which stated the following:

...Per request of Dr. A.O. Mirarchi, during a recent visit to this base, the following information is submitted...Sightings were made on 27 April and 24 May 1950 of aerial phenomena during morning daylight hours at this station. The sightings were made by Land-Air, Inc., personnel while engaged in tracking regular projects with Askania Phototheodolites. It has been reported that objects are sighted in some number; as many as eight have been visible at one time. The individuals making these sightings are professional observers therefore I would rate their reliability superior. In both cases photos were taken with Askanias...

Attached to the letter was a data sheet from Wilbur Mitchell of the Data Reduction Unit at White Sands that stated:

Objects observed following...test of 27 April 1950...the following information is submitted directly to Lt Albert...the following conclusions were drawn:
a. The objects were at an altitude of approximately 150,000 feet.
b. The objects were over the Holloman range between the base and Tularosa Peak.
c. The objects were approximately 30 feet in diameter.
d. The objects were traveling at an undeterminable, yet high speed.

Although these Project Twinkle sightings were not Green Fireballs, the objects sighted at the Holloman/White Sands station on these two days could be considered much more significant. On 27 April, four objects were tracked and filmed with a theodolite instrument at one location, and with a cinetheodolite at another. This allowed triangulation of the objects in the Data Reduction Unit where the altitude and size of the objects was confirmed. It must be kept in mind that no technology on Earth at the time could position a plane more than 50,000 feet or a manned balloon more than 76,000 feet (in 1956). These objects were at 150,000 feet and moving at high speed, and they were filmed! On 24 May, eight more objects were observed. Two of them were tracked and filmed by two stations. Unfortunately, it was determined that the two stations were each tracking different objects on this date and time, so no triangulation was accomplished.

Project Twinkle sightings at Holloman AFB and White Sands bring to mind previous sightings by Navy Commander Robert McLaughlin, and also the sighting by Charles Moore in April 1949 (mentioned in Chapter 20). The sightings recorded by Project Twinkle serve to confirm the sightings of McLaughlin and Moore, but no one in official capacity was apparently astute enough or cared enough at the time to make a connection with the recent 22 February issue of *True* magazine (see previous chapter). If the Project Twinkle objects had been properly reported through Air Force channels, they probably would have been ignored at TID. Of course, these sightings were never made public, but it's a very good possibility that Project Twinkle crews at theodolite stations at White Sands were making a connection with the *True* magazine article, while also recalling many other previous incidents.

Eddie Rickenbacker - "They Are Ours"

Eastern Airlines President Eddie Rickenbacker, who was a former flying ace in World War I, issued the following statement about flying objects on 16 May in Savannah, Georgia:

> *There must be something to them, far too many reliable persons have made reports on them. I am duty-bound not to say what I know about them or what I don't know about them. However, if they do exist, you can rest assured that they are ours.*

Not surprisingly, Rickenbacker knew of many pilots who experienced sightings and encounters with flying objects, including his famous Chilles-Whitted pilots. And he may have been aware of the TWA pilots from the 8 March incident near Wright-Patterson AFB, and TWA pilots from Flight 117. Rickenbacker was under public pressure to comment on these incidents, but government pressure was also being applied to him, and probably to officials of other airlines. He said he was "duty bound" to keep silent about reported sightings of flying objects, and then he was cynical and contemptuous in his statement that "they are ours." Later, he would comment that, "Flying saucers are real. Too many good men have seen them who don't have hallucinations." According to Edward Ruppelt, more than "thirty-five good reports" were submitted by airline crews during this time, from April to June. It's quite likely that Rickenbacker was encouraged to speak out about them, possibly with backing from Senator Richard Russell, also of Georgia.

More than a month earlier, on 6 April, an article in the press indicated that Senator Russell stated the following:

> *I am completely baffled by Flying Saucer stories. It seems inconceivable that so many pilots would have hallucinations or be fooled by cloud or atmospheric formations. From their testimony, it seems they do exist. But our Air Force says they do not. I just can't understand it.*

It's quite likely that Senator Russell received an earful from Rickenbacker, who may have urged him to initiate a Senate investigation. Later, however, Senator Russell would have significant reason of his own for organizing a Senate investigation.

Book by Donald Keyhoe - "The Flying Saucers Are Real"

On about 21 May, a pre-release of Keyhoe's new book, *The Flying Saucers Are Real,* came out in a few Midwestern states. Five days later, the book was suddenly withdrawn from newsstands when the Department of Defense found out about the book and applied pressure. It was also demanded that official release of the book, scheduled for 5 June, be cancelled. According to the book's editor, Jim Bishop of Fawcett Gold Medal Books, the book would be released anyway unless a restraining order was issued proving that the "publication will do the country harm," or "involve national security." In a subsequent response, the Department of Defense stated it had "no interest" in the book. One might wonder how the Department of Defense initially found out about the book, and then decided to apply pressure. Did the CIA have something to do with it?

The book was a further expansion of Keyhoe's *True* magazine article, which was published at the beginning of the year. In the book, Keyhoe put forth a very convincing and factual account on reality of the "saucers," and he provided much information gleaned from his Air Force contacts, accumulated records, and interviews with witnesses on various reported sightings. He made good use of recent sightings, and he emphatically stated that the American people "would not panic" upon release of the Truth. This book served to further awaken and influence an interested public, and also many in the general military, which then added many more people to increased ranks of believers. Keyhoe was not finished with his quest, however, and continued to gather more evidence to discredit and embarrass the Air Force, which will be outlined in the next chapter.

Summary of New Mexico Green Fireballs

On 25 May, Lieutenant Colonel Doyle Rees, the AFOSI district commander in the New Mexico area, sent a confidential letter to AFOSI director Brigadier General Joseph Carroll at Air Force Headquarters with a "Summary of Observations of Aerial Phenomena in the New Mexico Area." Rees, like many other officials at sensitive sites in New Mexico, was involved with the mysterious Green Fireballs and critically aware of many credible sightings, especially quite recently. He wanted to make sure that OSI Headquarters was fully informed and up to date, particularly in light of restricted reporting procedures previously issued. His letter stated, in part, the following:

...This compilation of sightings is not a complete record of all reported observations, but includes only those in which sufficient information was available to justify their inclusion. The observers of those phenomena include scientists, Special Agents of the Office of Special Investigations USAF, airline pilots, military personnel, and many other persons of various occupations whose reliability is not questioned...conferences were held at Los Alamos, New Mexico, for the purpose of discussing the green fireball phenomena...continued occurrence of unexplained phenomena of this nature in the vicinity of sensitive installations is cause for concern...This summary of observations of aerial phenomena has been prepared for the purpose of re-emphasizing and reiterating the fact that phenomena have continuously occurred in the New Mexico skies during the past 18 months and are continuing to occur, and, secondly, that these phenomena are occurring in the vicinity of sensitive military and government installations.

Korean War Begins

In a surprise attack, on 25 June, North Korean tanks and troops crossed the 38th parallel, which separated them from South Korea. Unknown to the United States at the time, the Soviets were responsible for planning the invasion by North Korea, which included a guarantee that Chinese troops would enter the war if necessary. The United States was not just totally unprepared, despite tough talk from Stalin with his new atomic bomb, and Symington's dire warnings, but the political factor within America was taking the path of looking inward toward home and family life. There was a drastic reduction in armed forces and defense spending, and this was fostered by the current Truman administration. During the last few months, however, the United States knew about buildup of North Korean forces at the 38th parallel, and it was known that something big was imminent. But nobody wanted to face reality.

This was the situation involving unconcerned leadership, and a nation based on democracy, where the focus was not on gaining an upper hand on a country outside its borders, but for providing a good life for people within its own borders. A future conflict with communism was a concern, especially with Soviet expansionism demonstrated in Berlin, and with Chinese Communists involved in an abusive revolution to subdue and subjugate their people. The focus of the United States, however, was directed more toward European countries and Japan, and helping those countries rebuild so they could become good neighbors within the world community.

Lessons learned from Korea, however, can be easily forgotten. Tyrants still exist in this world whose goal is to increase dominion over neighbors, as well as their own people. It is still happening today and made evident by massive migrations of refugees, bloodshed inflicted by terrorists, and vindictive rhetoric by despotic leaders who figure they are above the will of their people, or of other countries. In the Korean War situation, the United States and other peace loving countries in the United Nations became totally engaged in stopping North Korean aggression, and the United States took a leadership role.

In TID, Colonel Watson did not want to be bothered with pesky flying objects, even though sightings continued to be forwarded by AFOIN. He and others at TID were involved with a much higher priority, with emphasis on the enemy in Korea. Besides, Watson's project to investigate flying objects had been previously "discontinued," and there was no way he was going to get involved with that again.

General Cabell Perplexed

For Watson and others at TID, the investigative effort on flying objects was finished, but Project GRUDGE still lingered in the background. This, however, was a situation that could not last. The UFO question was not disappearing, and a great many credible sightings were being observed and reported across the nation. These included reports from air controllers, military pilots, civilian pilots, commercial pilots, radar operators, and many civilians in both rural and urban areas. Nothing was being accomplished by the Air Force to make the situation go away. Flying objects still existed, talk of them was increasing, and they continued to fly around for the public and others in the military to observe and encounter.

One can envision that Cabell must have been very concerned, disappointed, and upset about the circumstance of events, and with the fiasco that had developed over the past year. One wonders if he again sensed a loss of control with what was going on at TID/AMC. A year previously, it was a difficult situation when Colonel Watson came to the "rescue" to take over TID, but now the situation was untenable and unmanageable.

Cabell was at a loss on how to deal with the Air Force's lost public image regarding the UFO question. Some researchers suggest he was probably content to just maintain a small underground investigation, but that is nonsense. Cabell was probably stewing about those Air Force elements, even in his own command that seemed to have overt disregard for flying objects. He knew the objects were real, and he was not about to waver. It was Keyhoe's *True* Magazine article that triggered Air Force release of the GRUDGE Report, along with a high level Air Force response stating that flying objects did not exist. Cabell was not actually involved with circumstances that led to this situation, but he was left "holding the bag." Because of that, one might consider that Cabell had good reason to think about possible manipulation of the UFO question behind his back. Or maybe not.

One can also wonder if Cabell was possibly in-the-know regarding the UFO question. It is difficult to conceive he wasn't at this point in time, especially with events transpiring over the last year and a half. One would think he must have been privy to the secret, but future events suggest that he still wasn't in-the-know. It is very likely, however, that Cabell was not at all happy with the chain of events, and that Colonel Watson was probably on his mind. Cabell knew that serious investigation was necessary to find out what the flying objects were, but absolutely nothing was progressing toward that goal.

Cabell's situation resulted from the fact that the Air Force lied to the public about flying objects, which was necessary in order to stifle public clamor and protect the secret regarding the UFO question, but flying objects were not going to disappear. To compound the situation, the general military was thoroughly confused and discouraged by strong language from high Air Force officials stating that flying objects were nonsense, but the flying objects continued to play games with pilots and others. They knew that the harassing objects were real and needed investigation, but follow-on directives by Cabell and the OSI regarding the rescinding of reporting procedures were still confusing, and it did not serve to clear up matters. This presented a road block to the previously efficient method of obtaining sighting information and conducting investigations. And many could see that the Air Force was now deficient with this.

General Cabell remained quiet on the situation, except for his previous directive to the military to include sighting reports to Air Force Headquarters with other general intelligence. He understood that TCB was forwarding reports to TID, and he was assuming that proper investigation was taking place. Cabell knew that something was definitely going on with flying objects, as confirmed in *True* magazine articles by Donald Keyhoe and Navy Commander McLaughlin. But how to proceed from here was perplexing, and he was stuck in a situation that was nearly impossible to deal with.

Perhaps Cabell was thinking his only option was to investigate sightings more closely at Air Force Headquarters, while keeping watch on what was going on at TID. The problem with this, however, is that he didn't have necessary technical expertise under his command to investigate flying objects effectively. For him, the situation at the end of June 1950 must have seemed both chaotic and ironic. There was a constant parade of flying objects observed by the public and military, with the public taking flying objects seriously, but with investigative hands of the Air Force tied behind its back. Things needed to change, and Cabell must have been thinking of some way to get his investigation restarted. Would he be able to do it? Would the Control Group allow him to make a fresh start?

ॐॐ

From the beginning of March to the end of June 1950, flying object sightings increased dramatically and attracted much attention. Flying objects were continuing to flirt with many in the public and military, and a great number of sightings were observed by very competent and reliable people.

Project GRUDGE was "stagnate," but General Cabell recognized that many new sightings were attracting attention, and he took notice. He was hoping that GRUDGE was still conducting some sort of investigation for the purpose of national security, but the previous chain of events involving Project GRUDGE was bothersome. With so many significant sightings and encounters taking place, something needed to be done, but it was a difficult and perplexing situation to consider.

There were many in the public and general military who did not know the Truth, but they knew that flying objects were real because of many sightings and encounters reported in the press and reviewed in books and magazines. There were also those in-the-know who were privy to Truth, and they were duty bound to protect the secret. But some of them believed it was a mistake to live the lie. It was a lie that was too difficult to manage and maintain, and not morally justified to keep from humankind. This was proving to be the case in the existing situation.

CHAPTER TWENTY-FOUR

Sightings Produce Anxiety

Project GRUDGE: July - August 1950

Cabell's Strong Demand to Watson

Early in July 1950, Cabell knew he needed to take action to correct the state of affairs regarding the UFO question. He now understood that no investigative activity was taking place on flying objects at TID, and that Colonel Watson was the crux of the problem. Cabell was not pleased, and he decided to provide Watson with detailed instructions on what was expected of him by the Directorate of Intelligence. He gave this task to Colonel Edward Barber for relay to Watson.

Colonel Barber was an intelligence officer attached to the Supplemental Research Branch in the Office of Intelligence Collections (AFOIC), and he proceeded to formulate a letter with instructions provided by Cabell. He then bypassed official channels to hand-deliver the letter directly to Watson. As intended, the secret letter was friendly and slightly conciliatory, although it was straight to the point. In part, it stated the following:

> *Gen Cabell's views regarding the "Flying Saucer" project are in substance as follows...He feels that it probably was a mistake to abandon the project and to publicly announce that we are no longer interested...he feels that it was incumbent upon him not to overrule, a least for the time being...Our instructions...dated 12 January 1950...requested all recipients to continue to treat information and observations received as intelligence information and to continue processing in a normal manner...Gen Cabell's views are that we should reinstitute,*

if it has been abandoned, a continuing analysis of reports received
and he expects AMC to do this as part of their obligation to produce
air technical intelligence...He specifically desires that the project, as
it existed before, be not fully re-implemented with special technical
teams traveling around the country interviewing observers...and he
is particularly desirous that there be no fanfare or publicity over the
fact the USAF is still interested in "flying saucers"...General Cabell
desires that if circumstances require an all-out effort in this regard at
some future time, we will be able to announce that we have continued
quietly our analysis of reports without interruption...we will continue
to receive from USAF sources reports of "flying saucers" and we will
immediately transmit these reports to AMC.

Cabell knew that a restart of active investigation into flying objects was imperative. Many sightings and encounters were taking place in the civilian community, and also in the general military. The FBI, OSI and AEC were expressing concern, and action was needed to find out what the objects were.

For Cabell, this new communication to Watson was a "very strong demand," and he made it perfectly clear that investigation of incidents would be "reinstituted" immediately on a continuing basis, but it would be done in a way not to arouse public suspicion of Air Force interest in the subject. Cabell left no doubt on this and he made reference to his previous 12 January letter where he requested that sightings and encounters be treated as intelligence information, which would be sent through normal intelligence channels and then processed by TID. He mentioned that if investigation of received information on sighting reports was previously abandoned, he now expected AMC to fulfill its "obligation to produce technical intelligence."

Because of previous necessity to keep Air Force interest in flying objects undercover, one can understand Cabell's desire for that to remain. It was necessary to do this because high officials, including Vandenberg, publicly ridiculed the idea of flying objects, and then the Air Force announced that flying objects did not exist. It's also likely that Vandenberg instructed Cabell to keep flying object investigation quiet. But, since much active concern now existed in the general military, and other agencies, it now gave Cabell an opportunity to open up much needed covert investigative activity.

For Cabell, the thing that stood in the way of full restoration of investigative activity was the previous staunch denial by the Air Force regarding existence of flying objects. By opening up covert investigative activity, this would still not diminish sightings and encounters of flying objects, or settle the anxiety of a curious public, or appease concern of certain government agencies. But it would at least open the door to some investigation, although it would only work if there was full cooperation from Colonel Watson.

Cabell had always suspected flying objects were real, from early on, and one would think he would have challenged Vandenberg by this time. It's likely, however, that he still didn't suspect the Truth of what Vandenberg and the Control Group were hiding. He understood he was required to honor Air Force denial of flying objects, while also wanting to seriously investigate the objects harassing them and attracting public attention. Air Force high officials, and the Control Group, were seriously concerned about the objects, but they continued to prefer that the Air Force stay out of the investigation business. They were stymied on this, however, because of strong public and military concern.

One is tempted to speculate that certain Air Force officials thought that Truth of flying objects was too frightening for the public to absorb, or that captured technology was too important to acknowledge, or that they didn't want to be implicated in a serious lie. That may be true, but the scenario of previous events, and the crude Air Force reaction to it all, does cause one to wonder why the Air Force would absorb such turmoil. Many in the Air Force, who were involved with sightings or encounters, knew that flying objects were real. Those in-the-know knew they were extraterrestrial. They also knew that a great many competent and credible people were experiencing sightings and encounters, and there was also realization that some pilots and others were paying the price, but it was still critically important to maintain the secret.

As might be expected, Watson took Cabell's instructions and demands in stride, but without commitment. He had more important things to consider within TID than to discreetly chase down what was strongly proclaimed to be "non-existent" flying objects. It was not his fault that flying objects were catching public attention. It was the fault of the press, media, and everyone else. Besides, he was doing his job to protect the secret of the UFO question. A much more urgent priority was presented by the Korean War, which was an argument that could not be denied. To appease Cabell, Watson assigned Lieutenant James Rodgers as project monitor, and he put Captain Roy James on his official investigative staff for flying objects, but those two individuals were totally influenced by Watson in regard to non-existence of flying objects.

General Twining Returns

In mid-July, General Nathan Twining finished his stint as chief of the Alaskan Air Command, and he returned to Air Force Headquarters where he replaced Deputy Chief of Staff of Personnel General Idwal Edwards. Twining was now working for Vandenberg, and the Korean War was a top priority for both of them, but they both had a vested interest in the events that occurred at Roswell three years previous. Because the public and military was aroused at this time by many flying object sightings, one would suspect that a change of policy was in the works, and that Twining might be assisting with that. In fact, he may have been called back from Alaska for that very reason, because Cabell certainly needed help. But Twining would also be coordinating with the Control Group, and attempting to maintain the secret for them.

It is possible that Twining was handed control and maintenance of the Roswell secret for the Air Force, and this would definitely provide challenges, especially if he maintained an internal desire for revelation of Truth. With this responsibility, he definitely needed a good working relationship with General Cabell, who was determined to find answers to the UFO question. There were also many others that Twining needed to pay close attention to in order to insure that the right people were positioned to best advantage in maintaining the monumental secret. But there was also the matter of continued harassment by flying objects that needed monitoring, and he would rely on Cabell to insure that this was taken care of. It also meant that Cabell needed total cooperation from Colonel Watson. But would he receive that cooperation?

Sightings at Hanford, Washington

At the end of July, and into August, reports were coming in about flying object sightings near the Hanford Nuclear Facility in Washington State. This was not necessarily a new thing at this location, or at other AEC facilities. Hanford was producing plutonium for atomic bombs, and security for this top secret facility was a high priority.

On 4 August, a memorandum received by Ulysses Carmen, who worked for the General Survey Corp (GSC) at the Pentagon, described flying objects that were discovered by radar, and then visually observed as they hung motionless over the Hanford facility. Fighter interceptors from Larson AFB (previously Moses Lake AFB) were then scrambled to intercept them. The memo stated the following:

> *Since 30 July 1950 objects, round in form, have been sighted over the Hanford AEC Plant. These objects reportedly were above 15,000 feet in altitude. Air Force jets attempted interception with negative results. All units including the anti-aircraft battalion, radar units, Air Force fighter squadrons, and the Federal Bureau of Investigation have been alerted for further observation. The Atomic Energy Commission states that the investigation is continuing...*

Prior to the end of WWII, in both January and July 1945, Navy pilots based 30 air miles away from Hanford, at the Naval Air Station in Pasco, Washington, chased down elusive flying objects in their Hellcat F6F fighters. They described them as "saucer-like," "reddish-orange," "very bright," "extremely fast," and "very high." Secret documents that describe their encounters confirm this. And this documentation is quite significant, since both events happened before the more famous incidents of Kenneth Arnold near Mount Rainier in June and at Roswell in July 1947.

Quite obviously, the flying objects now being reported at Hanford were a big deal to everyone, especially to certain federal agencies. The objects were real, as confirmed by radar, by visual observation from the ground, and also by pilots who chased them. General Cabell would now come under additional pressure to deal with the matter.

True Magazine - "Flight 117 and the Flying Saucer"

In its August issue, *True* magazine again came out with a flying object story written by Donald Keyhoe. This was the third article on flying objects in eight months, and the magazine was applying additional pressure on the Air Force. The article was titled "Flight 117 and the Flying Saucer," and it confirmed to the public that the Air Force was still not doing its job.

The article provided a complete factual and detailed account of TWA's Flight 117 incident that occurred on 27 April, which was reviewed in the last chapter. Keyhoe interviewed people from the airline and people who were on the plane, which included pilots, stewardesses, and passengers, including two Boeing employees. Nine other airline encounters were also described, including the Chiles-Whitted incident. And Keyhoe reviewed other recent incidents occurring on 8 March (observed by TID), 12 March (observed by forty passengers), 21 March, 18 April, and 29 May. In regard to these sightings, Keyhoe also mentioned the following:

> *...there are incomplete reports of others...The Air Force either denies knowledge of C.A.A. reports or refuses permission to see them. Concerning its own data, it announces: "There is no investigation going on. Flying saucers simply don't exist."...Any thinking person who examines the mass of evidence can reach but one conclusion: the Saucers are real.*

At TID, Colonel Watson commented on the airline sightings by stating, "It's all a bunch of dammed nonsense. There's no such thing as a flying saucer." Regarding the pilots, he stated, "They were just fatigued. What they thought were spaceships were windshield reflections."

Keyhoe ended his *True* magazine article with a very blunt and up-front statement, which even today (more than sixty-five years later), could equally be stated:

> *Was the release of this Project Saucer report* [GRUDGE Report] *a blundering slip-up? Or was it part of a slow, halting program to prepare us for a dramatic disclosure? After a year's investigation, I believe that Air Force denials and contradictions have been due to fear of public reaction. I believe that Project Saucer was created to cover up the facts until the American people could be prepared. Apparently, there are some men in the Air Force who still think we are not ready. For almost three years, the answer to the Flying Saucer mystery has been a cautiously hidden secret. If we are not ready now, we never shall be. It is high time that the American people were trusted with the truth.*

This amazing and very appropriate statement from Major Donald Keyhoe is actually a very simple summary and restatement of what this particular book is all about—the Air Force needs to come forth with the Truth. The Air Force needs to confess to elected high authority why they continue to contrive ignorance. It cannot be stated more clearly! The public is tired of being lied to, and this situation reflects general mistrust that the American public has with elected leadership today. In this case, however, it all derives from the fact that career government officials, not necessarily elected officials, still believe that a naïve public can be manipulated as desired. But elected officials need to recognize this situation, and take action.

The Nick Mariana Incident

On 15 August, a movie film of flying objects was taken by Nick Mariana, who was a local owner of a minor league baseball team in Great Falls, Montana. He was conducting business that day at the clubhouse of the town's Legion Ball Park. Upon entering the grandstand, and looking up toward a towering flag, he spotted two silvery objects moving from the northwest toward the south. When the objects stopped, he realized they were unique, and he quickly rushed a short distance to his car and grabbed his sixteen-millimeter movie camera. He proceeded to film the objects that were continuing to pass overhead, and he could see that the objects were disk-shaped, with the appearance of bright shiny metal. Both objects were the same size and traveling at the same rate of speed, but they were slower and distinctly different than two jet planes he saw a few minutes later. Each object appeared to be spinning like a top, were about fifty feet across, and about fifty yards apart. As they continued to move toward the southeast, they were viewed passing behind supports of a black water tank directly south of the ballpark. His secretary, Virginia Raunig, also saw them and corroborated the sighting.

After Mariana showed his movie several times at community events, and also to friends, the Air Force OSI became aware of the film and picked it up. It was then forwarded to Project GRUDGE at TID for review and analysis. Evidently, the film gathered enough publicity that Project GRUDGE was pressured to make an analysis of it. It was first announced that "they couldn't make out any flying saucers" in the film, and then that it was "too dark to distinguish any recognizable objects," but Project GRUDGE officially labeled the objects as sun reflections from the two F-94 planes that were in the area. A couple years later, upon request from the Pentagon, Captain Edward Ruppelt was able to obtain the film again from Mariana for Project BLUE BOOK analysis. This time, it was determined that the planes were in a completely different area of the sky, and photometric analysis of the film revealed that the objects were not birds, balloons, meteors, F-94 planes, or sun reflections. The sighting was then placed in the unknown category.

After the Air Force returned the film the first time, it was discovered that several feet from the beginning of the film were missing. This footage was the best part showing the flying objects most distinctly, and revealing their spinning motion. It was the kind of motion sometimes observed in other object sightings.

The rights to Mariana's film were later bought by a movie production company, and then the film was sent to Douglas Aircraft Company for analysis by scientist/engineer Bob Baker. After spending weeks analyzing the film, which was now half as long as the original, Baker's final report simply indicated that the "natural phenomena explanation (sun reflections) did not have merit." He later confided that he was personally convinced the objects "were vehicles from outer space."

There is also much more to this story, including lawsuits, which may have been the reason for the Pentagon's request that Project BLUE BOOK get re-involved in looking at the film. Mariana was also later enticed to move away from his baseball team and the town of Great Falls, and not much was heard from him again. The most significant part of this story, and the answer to what happened to the missing part of the film, is reviewed in the next chapter.

FBI Memorandum on Project Twinkle

On 23 August, an internal three-page FBI Memorandum on Project Twinkle, from A. H. Belmont to D. M. Ladd, provided a "Summary of Aerial Phenomena in New Mexico." It mentioned that Dr. LaPaz considered the Green Fireballs and discs to be U.S. guided missiles under test in the New Mexico area. If he was wrong about that, LaPaz suggested that an immediate systematic investigation must be done. The memo also stated that LaPaz suggested the objects might also be guided missiles launched from USSR bases in the Ural region, which would give the missiles an arrival time of "less than fifteen minutes."

The memo further stated that information provided by AFOSI, on 19 July, indicated that Green Fireballs and discs were still being observed in the "vicinity of sensitive military and government installations." It stated that eight to ten objects were sighted by Land-Air Incorporated on 24 May, and a twenty-four-hour daily watch was being maintained by Project Twinkle. A review of those sightings, which were not fireballs, was provided in the last chapter. There can be no doubt that information on both the Green Fireballs, and observed discs, was creating much interest in high places, and that a Soviet connection was not being ignored for Green Fireballs.

Project Twinkle Sights Objects Again

On 30 and 31 August, flying objects were again sighted by Project Twinkle observers, which would later be mentioned in a final Project Twinkle report in November 1951. That report stated, in part, the following:

> On 30 August 1950, during a Bell aircraft missile launching, aerial phenomena were observed over Holloman Air Force Base by several individuals…no results were acquired…On 31 August, the phenomena were again observed after a V-2 launching…much film was expended, proper triangulation was not effected, so again no information was acquired…Generally, the results of the six-month contractual period may be described as negative.

These sighting incidents at the end of August were very significant, even though results for Project Twinkle were described as "negative." In the 30 August sighting, eight Bell Aircraft employees observed two circular objects high in the sky near a B-50 aircraft, which was the launch platform for a Bell MX-776 missile. The objects exhibited a bright non-reflective glare as they alternately sped away rapidly for short distances, hovered, and made sharp square turns. The next day, Askania theodolite crews filmed and tracked many high flying objects traveling at high speed from several directions during a three-hour period. Although no effective triangulation was made, thirty-five millimeter color film recorded the antics of the three-dimensional objects. Also, an audio recording of observer comments was made in the process. Four F-86 Sabre jets were sent to intercept the "aerial phenomena" to no avail, but the objects then returned when the jets were gone.

The interesting thing about these sightings is that they were responsible for extending the contract of Project Twinkle for a follow-on six-month period. Although project results were considered "negative" for the first six months, the activity of the "aerial phenomena" was considered "sufficiently significant." This double-talk seems inconsistent, but the semantics involved would suffice for those in charge. And the one primarily in charge at this time was Dr. Anthony Mirarchi of the Geophysical Research Division, who had previously inquired about sightings occurring in April and May. He became convinced the objects were real, including Green Fireballs, and he would later become quite outspoken on the matter.

It must be understood that Project Twinkle was authorized to look primarily for "natural phenomena," which is what Green Fireballs were "agreed" to be. The disk-shaped objects sighted by Project Twinkle observers, however, were considered real objects, or "aerial Phenomena" under apparent intelligent control, and this didn't qualify them for Project Twinkle "reporting." Only "natural phenomena" qualified. This would let Project GRUDGE off the hook so that TID personnel did not have to get involved. The following is what was stated in regard to further continuance of Project Twinkle:

> *...the phenomena activity over Holloman AFB 150 miles south of Vaughn, N. Mexico during the latter part of August 1950 was considered sufficiently significant so that the contract with Land-Air (Askania cameras only) was extended for six months ending 31 March 1951...*

As an aside to this, and with further consideration, one is motivated to contemplate the possibility that Green Fireballs were of Soviet origin, as Dr. LaPaz had suggested. Did TID miss the opportunity to conduct real technical intelligence investigation in this regard, or did the government already know something? Walter Winchell, the famous national news reporter, previously reported (twice, within about a year) that the objects were Soviet, but no one ever determined where he received his information, and the FBI was ordered not to investigate him. One might also be reminded that the Soviets managed to launch the world's first satellite seven years later, and also launch the first man into space four years after that.

The final statement in the GRUDGE Report also comes to mind regarding "psychological warfare" and "mass hysteria." It gave the report an "obtuse" ending, but it would serve as a reminder to those in the military involved with psychological warfare that military attention could be distracted with flying objects (or Green Fireballs). One can imagine that serious U.S. government concern could be generated with Green Fireballs targeted at New Mexico's secret facilities, and occasionally at other sensitive locations, especially with the Korean War in progress. Fireball construction, designed to be relatively lightweight, creating very noticeable mysterious effects, and leaving no trace, would constitute real psychological warfare against the U.S. government! On second thought, perhaps the government already figured out the answer to those Green Fireballs, which were very bothersome to those managing sensitive military areas in New Mexico. Now, however, the sightings were becoming less frequent, but Dr. Mirarchi was certain that this was because the Soviets were aware of Project Twinkle, and were avoiding potential U.S. examination of their Green Fireballs.

৩৯৫

By the end of August 1950, many important sightings had been reported by competent and reliable witnesses. Questions were being asked by the press and public, and there was also much concern expressed by government agencies. Also, Colonel Watson personally received a strong demand from General Cabell to immediately reinstitute Project GRUDGE with active investigation of flying objects, but it would be performed discretely so that the public would not become aware. The objective was to obtain answers regarding the objects. But Watson was not about to do this without specific orders regarding a specific sighting, such as the Nick Mariana incident.

Much anxiety regarding the UFO question would continue, with increasing concern by the public and general military.

CHAPTER TWENTY-FIVE

GRUDGE Revival Attempted

Project GRUDGE: September - October 1950

Cabell's Mandate to Watson

Colonel Frank Dunn, who was mentioned in Chapter 12, was part of the Collections Branch in AFOIR, which was commanded by Colonel Robert Taylor. In September 1950, Colonel Dunn was transferred to TID and assigned as Colonel Watson's deputy commander. The exact time of Dunn's arrival is not certain, but it is significant that he now appears at TID where little investigative activity was happening on the UFO question. One might recall that the Collections Branch is where Lieutenant Colonel Garrett and others worked who were partial to the extraterrestrial answer.

General Cabell understood that current handling of sighting reports was inefficient. This was due to the fact that reports were first reviewed at Air Force Headquarters, and then forwarded to TID at AMC for investigation. This inefficiency needed to be changed, but it was also apparent that Colonel Watson's TID was not performing the job required per Cabell's strong demands in July. In his previous July letter to Watson, Cabell specifically demanded that Project GRUDGE be "reinstituted." Since that time, major incidents involving flying objects were reported by credible people, but with no feedback from Watson on investigation results. Cabell was determined to end this with application of further pressure on Watson, including an announcement to all commands that TID was back in the business of investigating flying objects.

This occurred on 8 September when Cabell, once again, sent notice to various commands in a letter with a subject heading titled "Reporting of Information on

Unconventional Aircraft." This was the same heading as his previous 12 January letter, but this time it was an instruction to resume sending flying object sightings directly to TID at AMC. Cabell's previous demand to Watson was now a "strong mandate," but Cabell would need to continue monitoring TID to see if any action was taking place. Also, Cabell must have had strong thoughts on possible replacement of Watson, and he must have communicated with General Chidlaw at AMC about it, or at least considered doing so.

One can imagine, from Watson's perspective, that he was being unnecessarily harassed. He had put the flying object situation to bed, which was his original priority and mandate. And he was already cooperating, to some extent, with very minimal review of certain special cases sent by TCB, but now he was being forced into full-fledged investigative efforts, and to "effectively" respond to sighting reports. This mandate was contradictory to his senses, and contrary with his previous task. He did not want to risk the huge secret he was attempting to maintain, and he did not want to forever masquerade investigative report results, which eventually would be placed under a microscope. It was too difficult a proposition to return to the old ways, and the past provided an example of the fallacy of going backward. He knew the day would come again when TID investigators, or others, would not be able to dispose of credible sightings without huge problems developing. And it was best to just ignore the UFO question. Far more important, however, was the fact that the Control Group desired that the Air Force, or at least TID, remain out of the business of investigating flying objects.

Watson possessed a strong conviction that the only option remaining for the Air Force was to remove itself from the investigation process. With the new mandate, however, Watson needed to devise a plan to forever delay becoming involved with such investigation, but it would be a tough road to tread.

His initial objective was to first obtain specifics on exactly what could be said to the press and public regarding sighting cases, and about TID's investigative analysis of them. Otherwise, he wanted authority to release all details concerning incidents brought to his attention, including investigative results, but he was certain such authority would not be granted. He needed protection from releasing the secret, but he also needed safeguards to protect him when delivering false or vague information. He needed recognition from Air Force Headquarters that inadequate analysis and conclusions would not appease a press and public that was becoming increasingly savvy. He knew that the Air Force, and Control Group, wanted less public awareness and attention on the UFO question, and he hoped his plan would get him off the hook and prevent a "round robin" of investigation and denial. It was much easier to just provide an answer to the public indicating that the Air Force had previously investigated flying objects, and concluded that they did not exist. There would be no more flimsy excuses and denial, or continuing public reexamination, if sighting cases were just simply ignored.

When one ponders this situation, it seems quite absurd that flying objects, which were considered "figments of the imagination" or objects that "didn't exist," would cause such turmoil within the Air Force. But this was the case, due to active Air Force participation in a cover-up of Truth. On the other side of this were credible witnesses of flying objects, strong concern in the general military about the objects, and a public clamoring for direct answers to the UFO question.

Quite typically and unfortunately, Watson was now in the kind of situation that many officials at high levels are often faced with, which requires they be on guard and astute enough to protect themselves and their careers. Watson needed to protect himself, and also preserve the secret he and the Control Group owned. He was finding himself in a very difficult position, and he needed to press on with his own specific demands, which needed to be initiated sooner rather than later.

Book by Frank Scully - "Behind the Flying Saucers"

Frank Scully was a known and respected journalist, who was a writer for *Variety* magazine. He dealt with stories tinged with light humor, entertainment, gossip, and stories catching public interest. He definitely was not a science writer or news reporter, and he usually did not deal with serious subjects. But he was quite distrustful of government intervention in peoples lives, and he did not shy away from controversy. In that light and quite coincidental to Cabell's mandate on 8 September, Scully came out with a book on the same day titled "*Behind the Flying Saucers.*" To an interested public, the book was viewed as very profound, entertaining, and mysterious. And it became an instant national sensation for Scully and his publisher.

The book appeared to be absolute fact, with a serious account regarding the crash of several flying objects, which resulted in deceased aliens. It was referred to as the Aztec incident. It contained many references to magnetism (magnetic lines of force, magnetic frequency, magnetic disturbance, magnetic flux, magnetic zones, and magnetic waves), and it suggested that everything was impacted by magnetism. This was the energy source that gave speedy movement to flying objects. The unfortunate thing about this book, and for Scully, is that two of Scully's close friends were integrated into the story, and they had provided Scully with his material. They were later proven to be conmen, and subsequently convicted of fraud. Supposedly, they managed to indoctrinate the incredibly naïve and gullible Scully with fiction and fantasy, and they pulled off a great hoax.

Scully wrote his story based on what his friends told him, but some people have indicated that it is hard to believe Scully didn't realize his material was mostly fiction, or at least unreliable hearsay. Some say there are so many facts in his book filled with double talk, and scientific gibberish, that one would think an editorial review might have verified it as fiction. But they believe it is quite likely that many in the uninformed public (and press), even today, would not bother to notice because of the skillful writer Scully seemed to be.

Public acceptance between Keyhoe's new book, previously released in June, and Scully's is interesting, especially when considering that Scully's book was subsequently interpreted to be fiction. The public was attracted to Scully's supposed fantasy and mystery, which contrasted to Keyhoe's attempt to reveal Truth. But was Scully's book totally fiction?

It is important in this case to be aware that there are two sides to consider in this, especially when recognizing that the CIA was becoming very adept at passing out disinformation and discrediting individuals and events, and also for steering things in a particular direction. There is another side to Scully's story that provides a basis for additional consideration involving hidden Truth. When one refers back to the FBI memo of 22 March, in Chapter 23, regarding the crash of three objects in New Mexico, it could be presumed that the memo was derived from Frank Scully's fraudulent sources, but the information may not necessarily have been totally false. The crashed object or objects found near Aztec, New Mexico, on 25 March 1948, along with recovery of many bodies by the Air Force (see Chapter 14), could possibly lend credibility to the matter.

Dr. Wilbert Smith

Release of the book by Scully, along with the book by Keyhoe in June, caught much public interest. It also attracted the attention of Dr. Wilbert Smith, a Canadian who was a senior radio engineer for Canada's Department of Transport. While attending a conference in Washington, D.C., where he purchased both books, he was struck by the incredible things he read. Because of his scientific work, he was able to relate to information regarding "magnetism" in Sculley's book. For several years, he was conducting an investigation into radio wave propagation, cosmic radiation, aurora effects, and geomagnetics in his role as director of a small unofficial study named Project Magnet. The idea that flying objects might be using technology related to magnetism, as described by Scully, was exciting for Smith to learn, because it related directly to the type of preliminary research his project was involved with, which his government latter classified. Only to Smith would one's attention be attracted in this way, and it gave him great incentive to make specific inquiries while he was in Washington, D.C.

Dr. Smith was well-respected, and he was well-educated in electrical engineering, physics, and mathematics. He held a high security clearance in the Canadian government, and was connected with a number of well-placed individuals in high positions. He was responsible for regulating national radio communications (similar to the Federal Communications Commission), and was in charge of Canada's top secret radio station, Radio Ottawa, that gathered and relayed intelligence for its secret agents.

On 15 September, Smith visited the Canadian Embassy in Washington, D.C., where he made "discreet enquiries" through staff there to see if any additional information could be found about claims made in Scully's book. He then met with Arnauld Wright, a Canadian Defense Research Board (DRB) liaison officer, and he also met with Lieutenant Commander James Bremner, a Canadian military attaché. Bremner proceeded to arrange a meeting for Smith with Dr. Robert Sarbacher, who was a scientific consultant to the RDB with two offices in Washington, D.C., including one at the Pentagon. It was Sarbacher who provided Smith with startling information, which Smith subsequently documented in a Canadian top secret memo. This was mentioned in Chapter 15, and will be further reviewed in the next chapter.

In later interviews, Sarbacher confirmed discussions he had with Smith. He also indicated he never took part in major meetings of the RDB, although he did have access to all official RDB reports in his office, and he was aware of flying object materials in U.S. possession. Sarbacher was closely associated with many individuals in the RDB, including scientist Dr. Eric Walker, who was now the executive secretary of the RDB. As previously mentioned, Vannavar Bush was on its Oversight Committee and associated with the embedded Control Group. It will be recalled that Lloyd Berkner was also a previous executive secretary of the RDB, as well as the JRDB. There is little doubt that Sarbacher knew basic details of what he related to Wilbert Smith, and what Smith subsequently stated in his Canadian top secret memo, which is quite significant to the UFO question.

Project Twinkle Contract Extended

On 15 September, the contract for Project Twinkle with Land-Air Inc. was officially extended for another six months. As previously mentioned in the last chapter, sightings of disk-shaped objects at the end of August were "sufficiently significant" enough to extended a 24-hour sky watch for "natural phenomena." This was in spite of the fact that general project results were considered "negative" for Green Fireballs during the previous six months. It seems that "non-existent" flying discs provided substantial reason for extending the contract, which places great suspicion upon the Air Force. The extension of the contract also involved consideration of several questions brought forward by Land-Air, including whether observed objects were allowed to be fired upon and shot down.

George Marshall - New Secretary of Defense

On 19 September, General George Marshal took over from Louis Johnson as secretary of defense. Johnson was a supporter of cutbacks in the defense budget at the time when President Truman asked for resignation of James Forrestal, who staunchly believed in maintaining a strong military posture. Johnson, like Truman, believed that the Navy, and especially the Marines, were no longer needed as a military force. When North Korea invaded South Korea in June 1950, with sponsorship by the Soviet Union, Johnson stated:

> In South Korea, there has been developing a nucleus for a force, which in the absence of substantial external aggression, should assure the security of that country...Although we have given military aid to South Korea for a comparatively only a short time, those in a position to know feel it has been quite effective, and that they have made substantial progress in providing for a proper defense.

Of course, Johnson was the one "in a position to know," and he failed miserably in his job because North Korea quickly advanced to the southern tip of the Korean peninsula before U.S. and United Nations Troops were able to force a North Korean retreat. The importance of maintaining military strength, in the face of Soviet belligerence, became a hard lesson for the Truman administration to learn. It was now time for George Marshall, the top U.S. general who led the overall effort in WWII as "organizer of victory," to step in and fix the situation. By 30 September, he would give General Douglas MacArthur the go-ahead to "feel unhampered...to proceed north of the 38th parallel" with U.S. forces.

Air Intelligence Report Destroyed

On 25 September, Air Intelligence Report 100-203-79 ("Study No. 203"), produced by AFOAI on 10 December 1948, was ordered destroyed. The order stated:

> It is requested that action be taken to destroy all copies of Top Secret Air Intelligence Report Number 100-203-79, subject, "Analysis of Flying Object Incidents in the U.S., dtd 10 Dec 1948.

Later, the reason mentioned for this action was that conclusions reached in "Study No. 203" were "superseded by findings of Project GRUDGE," and the document was "no longer pertinent." Mention was also made that the destruction was not intended to "indicate reduced interest in the intelligence aspects of flying object reports," but that "such reports will continue to be treated as items of normal intelligence interest," which was outlined in Cabell's memo to all commands on 12 January, but it was without mention of Cabell's further revision on 8 September.

Like the SIGN Report, "Study No. 203" confirmed that flying objects were quite "real," although the report considered them to be of domestic origin, but this still provided great incentive for its destruction. The other excuse used was that "the true nature of flying saucers was not apparent" when the report was first produced, therefore a "forced consideration" was originally made in creating the report. Of course, this was not true, because the report was originally created to replace the EOTS, and to remove the extraterrestrial idea. In this report, as well as the previous EOTS and SIGN Technical Report, they all indicated that flying objects were real, which ultimately forced the GRUDGE Report to conduct damage control efforts, and take attention away from the UFO question.

All flying objects were now considered "figments of imagination," and the Air Force was now announcing to everyone that flying objects were not real. They were also indirectly indicating to the military that flying objects were going to be "considered real," because full investigation of sightings would take place anyway. It was to be kept quiet, however, without letting the public know. Again, it was a strange language of "double-speak." Although copies of "Study No. 203" were destroyed, an original report remained at Air Force Headquarters, which was discovered years later via the FOIA.

JANAP 146 Revisions

Previously, the Joint Army Navy Air Force Publication (JANAP) was originally created by a joint committee of the military services in the summer of 1948. The committee's intent was to set up, in the form of a publication, procedures for coordinating and effectively communicating vital intelligence information involving sightings of an emergency nature by pilots and others. In December 1948, with approval of General Cabell, it was distributed to Air Force locations worldwide, and sometimes used to report flying object sightings and encounters. This reporting action was evidently inspired by the Chilles-Whitted incident, although such reporting was not specifically directed in the document. Toward the end of 1949, and into 1950, especially with Colonel Watson in control of Project GRUDGE, this reporting on flying objects was thrown into a state of confusion, because "flying objects were not supposed to exist."

Now, on 25 September, a revision was made to JANAP 146, and it was renamed JANAP 146A CIRVIS (Communications Instructions for Reporting Vital Intelligence Sightings). This revised document, also approved by General Cabell, was significant in the fact that it became an official document for reporting flying object incidents, which specifically and significantly used the term "UFO Reporting." It provided instructions for both military and commercial pilots.

Although Cabell approved the document, there is a question on how well he understood its reporting directions, or how the document would mesh with his overall plan regarding Colonel Watson and Project GRUDGE. His 8 September letter to all Air Force commands was issued with instructions to send all sighting reports to TID, but JANAP 146A directed that all reports be routed to Air Defense Command (ADC). Adding confusion to this was Air Force letter AFSCI #85, previously released on 8 February, which was not yet retracted. It discouraged sending sighting reports, except in special cases, and it was originally distributed by AFOSI Headquarters to its field offices, which were acting as first responders to flying object sightings.

Because JANAP 146A directed that sighting reports be sent to ADC, it provided the potential for taking TID out of the flying object investigative business, and this would later be recognized and exploited by Colonel Watson in an effort to make that happen.

Direct Orders Issued to Watson

On 3 October, Cabell's executive officer, Colonel H. J. Kieling, sent a letter to Commander of AMC General Chidlaw, and it was directed to the attention of Colonel Watson. It referenced Cabell's letter of 8 September directing all commands to send sighting reports to AMC, and this new letter specified how AMC would handle the reports. It stated:

> *The following summarizes Air Material Command responsibilities in processing these reports, and supersedes all previous instructions on this subject...In investigating incidents or sightings of this character, Air Material Command will be responsible for utilizing such facilities as the importance of the incident may appear to deserve, including OSI personnel, AMC depot personnel, or special teams...it is important that AMC be especially aware of its responsibility for keeping Headquarters USAF currently informed in this field of technical intelligence...non-governmental sources may be acknowledged and follow-up information obtained by direct correspondence...Queries from the press as to USAF interest in "flying saucers" may be answered generally to the effect that the USAF has a continuing interest in all air technical and scientific information and therefor has an interest in any such reports... modification is further suggested to remove restrictions on investigation of incidents...*

This letter was countersigned by seven other branch heads in the Directorate of Intelligence, including Cabell. The significance of this letter is that it gave direct and specific "orders" to Colonel Watson, with confirmation to his boss General Chidlaw, that "full investigation" by Project GRUDGE would be restarted, including the use of "special teams" to pursue investigations. It also provided a general answer for responding to press inquiries regarding Air Force interests in flying objects.

The following day, a letter was sent from the Directorate of Intelligence to Air Force Director of Public Relations General Sory Smith (who previously replaced Stephen Leo in the month of May), and it relayed similar information that included how queries from the press would be handled regarding the UFO question. It requested that this information be forwarded to all public information officers, and to air attaches.

Watson again received strict "orders" to get back to work on investigation of flying objects full time, and in-depth. Watson, however, would now attempt to obtain a specific definition on what could be said to the press and public regarding sighting cases, and on investigative results. He knew that the "general answer" specified in Cabell's letter was vague, and it would not suffice for inquiries by the press and public in regard to specific sightings.

He would now implement his plan, and send exceptional quality photographs of a recent flying object incident to the Directorate of Intelligence for evaluation. To proceed in the investigation of incidents, he wanted concise and clear information regarding what could be released for any investigative case that TID might be involved with going forward. He knew the press would insist on specific information regarding the analysis of incidents, especially in regard to the incident he was now involved with. He knew that the press, and others, would not be satisfied with a vague response to an incident that stated, "the USAF has a continuing interest in all technical and scientific information."

CIA's Hillenkoetter Replaced by Bedell Smith

On 7 October, President Truman replaced CIA director Admiral Roscoe Hillenkoetter with Army General Walter Bedell Smith. Hillenkoetter was involved with Roswell, and like former Defense Secretary Forrestal, he was of the opinion that Truth of the UFO question should be publicly revealed. His departure from the CIA, however, was probably associated with the fact that he failed to provide early notice of invasion plans by North Korea, which would have allowed for better defense preparation. Hillenkoetter received orders to return to the Navy, and to a special command involving a cruiser division operating off the Korean coast. Interestingly, he would later retire and be elected to serve on the civilian board of the National Investigations Committee on Aerial Phenomena (NICAP).

General Smith was previously ambassador to the Soviet Union for three years, leaving the post in March 1949. On his return to the United States, he made the following statement in regard to the Soviet Union:

> …We dare not allow ourselves any false sense of security. We must anticipate that the Soviet tactic will be to wear us down, to exasperate us, and to keep probing for weak spots…we are forced into a struggle for a free way of life…

More than anyone, General Smith was very leery of the Soviets, and he was not surprised by North Korea's invasion of South Korea, but he also was not able to provide a more immediate warning of this either.

It will be recalled that Smith was mentioned in Chapter 20 as being concerned with possible public "mass hysteria" involving "aerial objects," and this concern was reflected in the conclusion of the final GRUDGE Report, which seemed like an isolated and strange interjection to the report. It was strange because the GRUDGE Report essentially indicated that flying objects did not exist, but then the conclusion indicated that "release of unusual aerial objects" against an enemy, along with "psychological propaganda," would have "similar results." In regard to this, one might question whether Smith was possibly referring to Green Fireballs. The strange twist at the end of the GRUDGE Report leads one to speculate on what prompted this peculiar ending to the document.

On further examination, one could conceive that Smith may have known about the origin of Green Fireballs in early 1949 before the GRUDGE Report was released, just like newsman Walter Winchell knew. In all of the sightings, most of which were investigated by Air Force OSI and Dr. LaPaz, there was great effort to keep Green Fireballs from public knowledge. There was also another side to Smith's GRUDGE recommendation, and that was his desire to not draw public and general military attention to the UFO question. Overt government involvement in sightings, encounters, and investigative actions would risk revealing Truth of the UFO question, or Green Fireballs.

As a close confidant to President Truman, and in-the-know on Roswell, one would expect that Smith would now work to insure a safe refuge for the Control Group, and that he would help with effective management of the monumental secret, which currently resided within the RDB. With a possibility of General Smith learning that Dr. Sarbacher had "spilled the beans" to Canadian engineer Dr. Wilbert Smith, and with concern about other potential happenings within the RDB, it could be speculated that a move of the Control Group to a new home would be hastened, sooner or later. The CIA's Office of Scientific Investigation would offer a possible safe environment for monitoring and controlling the UFO question, but no indication of that was currently evident in the minds of officials.

"Nothing of Value" in Mariana Film

On 9 October, Chief Assistant for Production General Ernest Moore directed Colonel Lester Harris to telephone Colonel Watson and inform him that photographs he sent to the Directorate of Intelligence for evaluation must be processed by AMC, and not forwarded to Air Force Headquarters. Watson was away for the day, so Harris talked directly with Colonel Frank Dunn and relayed Moore's request. He also told Dunn that it was "unnecessary to release full details resulting from analysis of incidents." It was stated that in cases specifically queried by the press, "the following may be released:"

> We have investigated and evaluated _____ incident and have found nothing of value and nothing which would change our previous estimates on this subject.

Colonel Dunn was further advised that if investigation results contained information of intelligence value, no information should be released and should immediately be sent to the Directorate of Intelligence.

After Colonel Dunn relayed the news, Colonel Watson returned the call to Harris, probably on 10 or 11 October. There are no details regarding the call, but it would have been interesting to learn what was discussed. Watson, however, received news he did not want to hear, which then raises additional curiosity about this situation. One might speculate that Moore, Porter, and possibly Harris were now accommodating Cabell to some extent, or perhaps they were interceding to keep the highly incriminating photographs away from Cabell and Air Force Headquarters. After the call, a confirmation letter dated 11 October was sent to AMC Commander General Chidlaw, with attention to Watson, and it was signed by five branches of AFOAI, including General Moore. The letter stated:

> This headquarters is cognizant of press interests in the so-called "flying saucers" reports, referred to by this headquarters as "unidentified aerial objects." Your headquarters has previously been advised as to the release of information concerning Air Force interest in this subject... Colonel Watson requested guidance in the matter of releasing results of investigations, analysis, and evaluation of incidents brought to his attention. This headquarters believes that release of details of analysis

and evaluation of incidents is inadvisable, and desires that, in lieu thereof, releases conform to the policy and spirit of the following: [same as indicated above from phone call]…incidents possessing any intelligence value will be forwarded to this headquarters for information and for any action relative to possible press releases.

Very coincidently, a news article released by the Associated Press from Dayton, Ohio (Wright-Patterson AFB) on 11 October stated the following in regard to the Nick Mariana incident, which was reviewed in the last chapter:

Air Force officers said today they couldn't make out any flying saucers in a film taken by a Great Falls, Mont., baseball manager. Officers said the film will be returned to Nick Mariana, the manager, if he wants it. Mr. Mariana took pictures last Aug. 15 of two objects he said buzzed over the Great Falls ball park. Later, he said, an Air Force intelligence officer confiscated the film and told him not to give out any more details about it.

It is obvious, as previously indicated by Nick Mariana, that Watson likely removed incriminating evidence from the film that was "confiscated," and this included several feet from the beginning of the film. He used pictures from the film to "test" Air Force Headquarters, and to obtain a response on how they would explain the pictures to the public. Watson was told to indicate "nothing of value" when reporting on flying object incidents, but rather than using those specific words, Watson stated that "they couldn't make out any flying saucers." He then proceeded to keep the "incriminating" part of the film, and return the rest to Mariana, since Mariana wanted the film back. There was no way that Watson would allow return of the most detailed and incriminating parts of the film.

The words Watson chose to use in communication with the AP must have been calculated to irritate the Directorate of Intelligence, because the film that was returned to Mariana still showed the objects clearly, although without detail. Watson was considerably disturbed with Air Force Headquarters, and he retaliated by immediately notifying the press of TID's analysis, which gained national coverage. His response, however, was found to be deficient by the press and public, and he was later forced to come up with the more plausible response that the objects were reflections from two jet planes in the area. This became the official conclusion of Watson's Project GRUDGE. And Watson was probably satisfied that he finally made his point to the Directorate of Intelligence.

The sequence of events in this Mariana episode is too coincidental to ignore, and it points out the considerable effort Watson was willing to make in order to get out of the business of investigating unearthly flying objects.

Sympathy for Watson?

For all intents and purposes, the last letter from the Directorate of Intelligence sealed the requirement put upon Colonel Watson regarding investigation of unconventional aircraft, unidentified aerial objects, and "flying saucers." This must have been devastating to Watson, and must have troubled him greatly. He wanted no part of the new policy, since he previously proved non-existence of flying objects, and he had been given concurrence by high level officials on that. His position was that the new policy from the Directorate of Intelligence put the UFO question at risk, and back into the spotlight. The secret needed to be maintained, especially with all the hype regarding credible sightings, and with all the books and articles being written about reality of the objects, but he was left with no alternative. The quality pictures he sent to Air Force Headquarters of an extraterrestrial vehicle were intended to force cancellation of his investigative activity on flying objects, but he didn't receive the response he wanted or expected. He was at an impasse, but there was no way he could concede on this matter, because the Control Group would not allow it.

One now wonders if Watson was receiving any feedback from Vandenberg, Twining, General Smith of the CIA, or others in-the-know on how to deal with the situation, but it is likely they were too busy with Korea and other problems to help resolve his specific situation. This may have terrified Watson to a certain degree, but General Chidlaw, or others in the Control Group probably knew of the situation, and may have sympathized with him, and backed him.

General Twining Named Air Force Vice Chief

On 10 October, General Twining was named vice chief of staff under Vandenberg. Within three months after returning from Alaska, he was now on track to eventually become Air Force's top man in the Pentagon, and he would be sharing duties with Vandenberg to control the UFO question. It was a huge task, not to mention all the other things involved with his new position, including the Korean War. But he was now in the driver's seat to guide the direction of his future career.

Oak Ridge Nuclear Facility Sightings

Beginning on 12 October, and in days following, including November and December, numerous sightings of flying objects took place at Oak Ridge Nuclear Facility, and the surrounding vicinity of Knoxville, Tennessee. But these sightings did not have the signature of Green Fireballs. They were of strange objects that hovered, sped away quickly, and maneuvered intelligently. Many of the sightings were reported by guards, scientists, top officials, and credible people with "Q" clearances. Many of the objects were tracked on radar, and some were chased by Air Force F-82 fighters, but no interceptions were made. This caused much attention, with additional involvement by the Air Force OSI, CIC, FBI, and others.

Project GRUDGE also became involved when James Rodgers at TID replied to a letter from Headquarters Third Army regarding sightings reported on 12 and 13 October at Oak Ridge. The letter from Rodgers stated that radar indications were probably spurious radar echoes from ice laden-clouds, similar to that diagnosed in TID's 8 March sighting. This finding, however, was disputed in a letter by Colonel Willis, which stated that weather was clear on those dates, and he indicated that AMC would be directed to send electronics personnel to Oak Ridge to investigate the radar.

Some of the local investigators at Oak Ridge also attempted to determine if object sightings coincided with unique increases in radiation levels. After instrumentation was devised to detect the possibility, which required close coordination on the sightings, positive correlation was obtained, and this encouraged additional study.

Credible people at Oak Ridge, and the Hanford Nuclear Reservation, probably suspected the Truth and the reality of what they often observed, primarily because they were witnesses to it. But they were also befuddled by an unresponsive investigative system that was haphazardly controlling the UFO question, courtesy of the Air Force and Colonel Watson.

IAC Convenes Under CIA's Bedall Smith

On 20 October, a meeting of the Intelligence Advisory Committee (IAC) was convened by new CIA Director Bedall Smith. He had previously attended a meeting of the NSC on 12 October to acquaint himself with his parent organization. Now he needed to meet with his advisory organization, which included members from the State Department, AEC, FBI, Joint Staff, and intelligence representatives from each military service. In attendance for the Air Force were General Cabell and General Ernest Moore. In subsequent meetings of the organization, both General Moore and Colonel Porter would attend.

సౌ✿ఌ

In September and October of 1950, there was no lack of activity regarding the UFO question. Sightings of flying objects were prevalent, and General Cabell was determined to get back into the business of investigating flying objects, although Colonel Watson of TID was putting up a fight to keep the Air Force, or at least

TID, out of the business. And this was despite direct orders and a mandate to fully reinstitute Project GRUDGE.

Also during this time, the public was entranced with the books by Scully and Keyhoe, while Dr. Wilbert Smith, a Canadian, discovered some startling information associated with the RDB.

Then, the film taken of flying objects by Nick Mariana proved to be a stumbling block for the Air Force. The film was used by Colonel Watson to get the Air Force to abandon flying object investigation, but mishandling of this situation provided further indication of the Air Force's struggle in cover-up of the UFO question.

CHAPTER TWENTY-SIX

Control Group Identified

Project GRUDGE: November - December 1950

Watson Goes on Offense

At TID, Watson was not happy after receiving a phone call, and also a follow-up letter regarding photos he sent to the Directorate of Intelligence. He did not agree with any of it. But he was determined to have his way by continuing with a plan to remove TID from the business of flying object investigation, although he needed help.

With the letter from Air Force Headquarters in hand, which provided a standard response to the press on queries regarding flying object incidents, Watson showed it to Wright-Patterson's Public Information Officer Colonel Clare Welch for his review. This was done with realization that Welch was mildly upset with routine Air Force handling of the UFO question, and with the poor state of affairs existing toward the public and press on the subject.

Colonel Welch then wrote a four-page letter of his own to Brigadier General Moore at the Directorate of Intelligence. His letter essentially defended Watson and criticized the idiocy of Air Force policy. The letter, dated 1 November, stated in part:

> *Your department carries these investigations in a classified status, especially until you have had an opportunity to get at least the basic facts of the reported phenomena. All the time you are doing this, however, the press in some part of the country is pursuing their story. The result is that they often feed to the public a lot of misinformation with just*

enough fact to make it embarrassing to us when we can not [sic] discuss it…Neither Skully nor Keyhoe have ever contacted this headquarters… If they had, we could have made available to them information which would have killed their story…In my opinion, security classifications should only be used when the material under classification would be of far greater danger to the public welfare in the hands of the enemy than it would be beneficial by free knowledge of our public. Certainly flying saucer stories do not come in this category…The statement which you suggested in your letter to Watson which you thought might cover the situation would not be acceptable to the press…let's be realistic about this situation and bring it out into the open where it now belongs.

It would have been interesting to listen to Watson's discussion with Colonel Welch prior to the letter to General Moore. Welch did not have a clue about Watson's real dilemma, and he was totally naïve as to the real reason for the security clamp-down on the UFO question. In his mind, why would nonexistent flying objects pose a threat to anyone, and why would it be "inadvisable" to "release details and evaluation of incidents" to the public? He felt that "flying saucer stories" do not come in the "category" of danger to "public welfare," and he felt that the UFO question should be brought "out into the open where it now belongs." Welch was confident enough on the subject to essentially talk down to a general at Air Force Headquarters by stating, "let's be realistic about this."

Welch played into Watson's hands, and it's obvious that Watson was no dummy. Watson was playing a very serious game, where everything depended on his ability to effectively prevent further investigation of flying objects at TID. He was hoping to manipulate Air Force Headquarters, and have them see things his way. In the meantime, over the next two months, it would take several rewrites by Air Force Headquarters to finally compose a response to Welch, which resulted in a reply stating:

The points covered in your letter are receiving full consideration. If this examination of the current situation indicates that policy changes are desirable, a new policy will be established.

In effect, Air Force Headquarters was stumped on how to properly answer Welch's letter. A reply was then returned to him indicating essentially nothing. This was a further example of how difficult it was for the Air Force to manage the UFO question. Watson, however, was not finished pursuing his offensive, and he would implement additional plans. He must have sensed that if he could not maintain a roadblock at TID, his future at TID was probably in jeopardy, and so was the Truth he was protecting. It was a critical time for Watson, but he had no choice in the matter.

Watson "Snows" Bob Considine

In continuing with his plan and his offensive, Colonel Watson made contact with a number of writers and members of the press to express his views on flying objects. One of them was Bob Considine, who was a staff correspondent for the International News Service, and a writer for *Cosmopolitan* magazine. On 12 November, Considine wrote the first of four articles for the news service, which stated, in part, the following:

> *The U.S. Air Force is still investigating reports of flying saucers, despite an announcement that its efforts to hunt down the mysterious objects were abandoned. To find the true story…Bob Considine went first to the men who know the story best—the Air Force operators of "project saucer." "I've seen a lot of flying saucers," calmly remarked Col. Harold E. Watson… "Plenty of them…And every single saucer turned out to be the sun shining off the wing or body of a distant DC-4, or a jet, or a weather balloon, or it was a reflection off a water tank, or something else that is readily explainable."…Col. Watson stated with emphasis that at the end of nearly every flying saucer report that can be tracked down stands a crackpot, a religious crank, a publicity hound, or a malicious practical joker… "We're still in the business," the Colonel added. "The Air Force naturally will always have a lively interest in whatever is reported in the sky. That's our job."…We asked him to explain the accounts of mysterious flying object reports by responsible pilots for TWA, United, Eastern Chicago and Southern, and other lines. "Fatigue, I'd say," he replied. "And the power of suggestion. And the optical tricks that windshields can play on an airman – especially, at night."… "The Air Force has in its possession no flying saucers, or parts of flying saucers. It has no bodies of 'little men,' nor any samples of the so-called clothes these imaginary creatures wore"…*

In subsequent reference to Colonel Watson, Captain Edward Ruppelt later stated that "I've overheard him tell how he completely snowed Bob Considine." Compared to previous negative talk by Watson regarding the UFO question, his rhetoric was becoming very abusive to a great many responsible, reputable, and credible people. This prompted much contempt toward Watson, and indirectly toward the Air Force and "government." This had the effect of shutting the mouths of many involved with sightings and encounters, which continues to this day. There is no joy in being labeled a crackpot, crank, malicious joker, or other highly unsavory characterization.

"Group Headed by Dr. Vannevar Bush"

On 21 November, Senior Radio Engineer Dr. Wilbert Smith of the Canadian Department of Transport, who was mentioned in the last chapter, sent a top secret memorandum to his Deputy Minister of Transport Charles Edwards to inform him of current geo-magnetic studies being undertaken, and particularly the study "of a means whereby the potential energy of earth's magnetic field may be abstracted and used." Smith indicated that the Chairman of Canadian Defense Research Board (DRB) Dr. Omond Solandt agreed that work should proceed as "rapidly as possible" on this, and Smith now recommended that his unofficial Project Magnet become an official project in their Radio Standards Lab.

Smith also mentioned in his memorandum that while attending a conference in Washington, D.C. in September, he obtained two books, one by Scully and the other by Keyhoe. He commented that the one by Keyhoe claimed that "flying objects were of extraterrestrial origin, and might well be spaceships from another planet." And he said that Scully's book stated that "preliminary studies of one saucer, which fell into the hands of the United States government, indicated that they operated on some hitherto unknown magnetic principles." Then Smith revealed the following in his memorandum:

> I made discreet enquiries through the Canadian Embassy staff in Washington who were able to obtain for me the following information:
>
> a. The matter is **the most highly classified subject in the United States Government**, rating higher even than the H-bomb.
>
> b. **Flying saucers exist.**
>
> c. Their modus operandi is unknown but concentrated effort is being made by **a small group headed by Doctor Vannevar Bush**.
>
> d. The entire matter is considered by the United States authorities to be of tremendous significance. [Bold print supplied by author.]

Smith's top secret memo, if truly authentic, could easily be considered one of the most significant and valuable pieces of evidence revealing cover-up of the UFO question by the U.S. government, which was actively protected by the U.S. Air Force. The document confirmed that a group headed by Vannevar Bush was involved in protecting Truth of the UFO question, and this secret was classified higher than anything in the nation. The Control Group was now identified as connected with the RDB. One must note, however, that Bush was not chairman of the RDB at this time, but he was still on its Oversight Committee and in charge of a "small group" imbedded within the RDB that was managing and controlling the UFO question.

The memo, obtained from Canadian government files, has created doubt among a few researchers, which is to be expected. But other released Canadian documents back it up with continuing dialog and correspondence. Also available are handwritten notes Dr. Smith took in the meeting he attended with Dr. Sarbacher, who was working as a consultant for the RDB in the U.S. Department of Defense. Both Smith and Sarbacher are on record with their verbal verification of meeting together, and passing information between each other.

The top secret Canadian memorandum by Dr. Wilbert Smith provides evidence to the fact that there was a Control Group operating with Vannevar Bush in charge, and it was involved with Unearthly Flying Objects. Many sources have confirmed the veracity, integrity, and authenticity of this highly significant document, which was evidently declassified by the Canadian government in 1979.

These documents, along with other secret and top secret documents from U.S. government files, all serve to indicate a cover-up of the UFO question, and that a Control Group was involved in managing the magnificent secret.

Canadian Government Honors Secret

A secret letter from Gordon Cox of the Canadian Embassy in Washington, D.C. to Dr. Smith stated the following:

> *On the Ambassador's instructions no one in the Embassy, apart from Wright and myself, is to discuss the matter with anyone…I have not yet had an opportunity to meet Keyhoe but I can easily do so through Bremner, and I will make an exploratory contact in the near future. I did not see his article after it was referred to you…The Ambassador and I would be particularly interested in any indication you may have heard when you were here on the possibility of an official U.S. Government statement. It is this political angle with which I will be principally concerned. Anything you can do to help in this respect will be welcome…*

Besides the mention of Keyhoe, the thing that is very interesting in this letter is the fact that the Canadian Embassy was holding the secret of the United States in strict confidence. There was also great interest and anticipation of a possible statement that might be released by U.S. officials relating to reality and existence of flying objects. The Canadian government wanted to be kept aware of what the United States intended in that regard, and possible release of information.

Keyhoe Learns of Canadian Involvement

In regard to Keyhoe, he was contacted by Dr. Wilbert Smith, possibly with help from attaché Colonel Bremner when Smith was in Washington, D.C. in September. Keyhoe and Smith periodically met together for about two weeks discussing the UFO question. After being informed of Smith's geo-magnetic studies, Keyhoe decided to write another *True* magazine article regarding a possible propulsion method used by flying objects. When his draft was ready, he forwarded it to Dr. Solandt of the Canadian DRB for review, who then sent it on to Dr. Smith. After Smith revised the document, in consideration of things of "extreme embarrassment" to the Canadian government, he then returned it to Solandt with a note stating:

> *I am quite sure that Major Keyhoe will accept the revised material in the same spirit in which it is offered and furthermore that the publication of this material, if permitted by the United States Research and Development Board* [RDB], *would be in the public interest.*

The revised draft was then passed on to the Canadian Embassy in Washington, D.C., with an understanding that Vannevar Bush would be contacted to also review the article and provide clearance for it. According to Keyhoe, the revised article was returned "after a check at the Pentagon" by Canadian Embassy liaison Arnold Wright, who was also a member of the Canadian DRB. The revised draft included a note from Smith to Keyhoe stating the following:

> *If you publish any of this, I want you to make clear. We're government engineers and scientists, but we are working on our own time. We've gone back to the fundamentals of electromagnetism and examined all the old laws. We know now it is possible to create current by collapse of the earth's magnetic field. Eventually, I think, we can achieve enough current to power a flying disk.*

Canadian officials understood it was appropriate to obtain approval from

Vannevar Bush in light of the information they acquired. Assuming that Bush, the RDB, or others were contacted regarding clearance, it may have terrified U.S. officials to contemplate that the Canadian government, and also Keyhoe, had associated Bush and his group with the UFO question. Due to certain delays, according to Keyhoe, his article was never published in *True* magazine, although it did make it into a later book he wrote in 1954, titled *Flying Saucers from Outer Space.*

In regard to Dr. Smith's Project Magnet, it supposedly withered away after a few years, although he was often called upon for information, and for analysis of materials by the CIA. Edward Ruppelt would later indicate that the U.S. Air Force took over Smith's project, which is further referred to in Chapter 30.

For Keyhoe, the feedback he received on his current book, and Scully's, must have been eye-opening. He was now aware of Canadian involvement with the UFO question, and he must have gained awareness that the RDB and Vannevar Bush were involved. Discussions with Dr. Smith made a lasting impression on Keyhoe, primarily because Smith's theory perfectly explained the nature of flying objects. According to Keyhoe, Smith stated:

> *I'm convinced they're real–that they're machines of some kind. We've weighed three possibilities. One, they're interplanetary. Second, they're a United States secret device. Third, they're Russian. The last two don't stand up. From the weight of evidence I believe the saucers come from outer space…our experiments indicate that the true discs, which are probably launched from large parent ships, utilize magnetic fields of force. And it's possible that the parent ships also use this same source of power…certainly the theory's been ridiculed…I'd have doubted it myself before our experiments…Most of the effects are caused by the disc's rotation, though sometimes a corona discharge is the cause…If they're not heating up from rotation, and there's not corona discharge, you wouldn't see one…probably many discs aren't seen at all, especially at night…assume a rotating ring begins to speed up, so that it overheats from its movement through the magnetic field. At first, out of the darkness, you'd see a pale pink –if the speed-up was not too rapid. Then the color would brighten to red, orange-red, through yellow to the glow of white-hot metal. If you slowly heat any metal you'll see the same changes…Somewhat higher, it would be green, or bluish green. Higher still, you might see all the normal corona colors–red, yellow, blue, and green…In the majority of cases, however, you could expect just the red-orange-white range, and the reports bear that out…It's fairly clear,*

*from the reports, that the discs are made of some silvery-colored metal.
In sunshine they gleam like conventional aircraft. But there are color
changes in daytime, when the saucers maneuver or suddenly speed up.
Many of them have been described as turning red or getting white-
hot–also the reverse. However, in bright sunlight it's harder to detect
the changes–and to recognize the disk shape, too.*

National Invasion Alert

On 6 December, a message from the Joint Chiefs of Staff was sent to commands
worldwide regarding increased possibility of general war, and that commanders
should "take such action as is feasible to increase readiness without creating
atmosphere of alarm." Also that day, the Air Force Directorate of Plans sent a
message to the secretary of defense titled "Air Alert," which stated in part:

*The ConAC Air Defense Controller notified the Headquarters USAF
Command Post that at 1030 hours a number of unidentified aircraft
were approaching the northeast area of the United States and that there
was no reason to believe the aircraft were friendly...By radar contact
it was determined that approximately 40 aircraft were in the flight, at
32,000 feet, on a course of 200 degrees, in the vicinity of Limestone,
Maine...The emergency alert procedure went into effect immediately...
The Office of the President was notified...At 1104 hours the ConAC
Air Defense Controller stated that the original track had faded out...*

It seems that much hasty action was underway, with absolute certain belief
the United States was about to come under attack, and this was quite reasonable
considering the status of U.S. involvement in the Korean War, and the existing Cold
War with the Soviets. The president and worldwide commands were notified, but
the concern "faded out" in less than an hour, and the "unmistakable" formations of
objects on radar simply disappeared. It would not be expected that the Continental
Air Command (ConAC) would make such a critical error regarding a high alert
notification unless they were totally certain of what they were observing, and they
were very certain that the United States was in the process of being invaded, and
that notification to highest levels was required.

On 8 December, a FBI "urgent" message regarding "flying saucers" was released
stating:

*This office very confidently advised by Army Intelligence, Richmond,
that they have been put on immediate high alert for any data whatsoever*

concerning flying saucers. CIC (Counter Intelligence Corps) here states background of instructions not available from Air Force Intelligence, who are not aware of reason for alert locally, but any information whatsoever must be telephoned by them immediately to Air Force Intelligence: CIC advises data strictly confidential and should not be disseminated.

Two days after a national high emergency alert for threatening incoming aircraft, there was now a high alert for "flying saucers." One is led to consider that perhaps Air Force Intelligence may have determined that the previous flight of "approximately 40 aircraft," which "faded out" on radar, may have been flying objects instead. Now there was strong concern about the potential for invading flying objects, and there was a desire to find out what they were up to. This idea was enhanced when researchers discovered reports of a flying object crash on 6 December, just across the Mexican border south of El Indio or Del Rio, Texas, where it was rumored that the Air Force recovered all remnants of a disk, and also body "remains."

Pilots Sight Green Fireball

On 16 December, pilots of a Frontier Airlines flight from Gallup to Albuquerque, New Mexico, spotted a Green Fireball racing through the sky. It occurred about 11:30 P.M. and lasted for about ten minutes. Captain Harvey, and copilot Merrick Marshall, spotted the object in the northeast near Las Vegas, New Mexico, and it was traveling toward Los Alamos at an estimated speed of more than 700 miles per hour. It made a turn over Los Alamos and then headed back toward Albuquerque while passing over and behind the Frontier plane. The fireball changed from bright green to bright white in color as it passed by, and then it disappeared.

This fireball seemed to be quite typical of the many previously reported, except for being in sight for a much longer duration. The pilots indicated it could not possibly have been another plane, and it certainly did not resemble a typical meteor, which normally lasts for a few seconds. Curious facts about this incident agree well with others observed in the past. It appeared to maintain a relatively constant speed, was faster than an aircraft, but slower than a meteor, and it was very bright with the unusual and unnatural green color. The other thing worth noting is that this object may have been under some kind of programed control due to its flat trajectory, constant speed, and a turning maneuver of more than ninety degrees. Was it man-made? If it was U.S. made, it's strange that would it pinpoint Los Alamos during part of its track, and also fly at altitudes frequented by commercial airlines.

۞

The year 1950 was an active time in regard to sightings and encounters with flying objects, but the last two months were very enlightening in regard to Air Force dealings with the UFO question. For Colonel Watson, the investigation of flying objects was resolved as far as he was concerned. Rather than cowing to orders from the Directorate of Intelligence to investigate them, he was now beginning a strong personal debunking offensive to demolish attempts by the Air Force Directorate of Intelligence to restart investigation.

Amidst this backdrop, the books by Keyhoe and Scully were attracting wide public attention, and they were given extra special attention by Canadian engineer Dr. Wilbert Smith. Because of his interest and eventual inquiries, he was able to obtain crucial information regarding a U.S. Control Group associated with flying objects, which was connected with Dr. Vannevar Bush and the RDB. He documented his findings in a top secret Canadian document, which is considered a major revelation involving Truth of the UFO question.

Over and over again, the scenario of an Air Force and government cover-up of the UFO question is revealed because of attempts to keep the press and public in the dark about intense Air Force interest in flying objects. The "non-existent" flying objects, or "figments of imagination," were raising havoc with Air Force attempts to identify what they were, and what they were up to. And they were also frightening the Air Force with mock invasion attempts. Or so it seemed.

CHAPTER TWENTY-SEVEN

Flying Objects Debunked

Project GRUDGE: January - April 1951

Watson's Motivation

At the beginning of 1951, the UFO question continued to be downplayed by TID, with little or nothing done about investigation of reported sightings. Colonel Watson's GRUDGE Report, created a year and a half previously, in August 1949, essentially stated that all sightings of flying objects could be explained, and those that did not have an explanation could be resolved if given enough time or more complete data. Watson's basic premise was that there were no sightings worth investigating, and this was despite the ones that some would consider significant sightings by credible witnesses.

This was the staunch attitude preached by Watson, which was the driving force for lack of investigative activity within the Special Projects branch of TID. Although sighting reports continued to be received, because of General Cabell's request to various commands to send reports directly to TID, it did not mean that these reports were being investigated. Most sighting reports receiving attention were treated as a joke, which created "belly laughs" among cohorts of Lieutenant James Rodgers, who was Watson's primary investigator. Even if some in Watson's organization were inclined to differ with his views, there was no way Watson would permit a return to the past. For him, the potential of exposing Truth of the UFO question with further investigation was just too risky.

Many in the military, including Director of Intelligence General Cabell at Air Force Headquarters, were very concerned about continued sightings by credible observers. Although Colonel Watson's Special Projects branch in TID was a field unit dedicated to serve General Cabell, and responsible for investigation of sightings, Watson's organization continued to remain in a dormant state. Watson's incentive and attitude did not change with continued issue of specific directives, mandates, or orders from Cabell. But he did increase his rhetoric against flying objects, and about people making sightings.

Watson must have been aware that a groundswell of controversy was brewing in the military about the UFO question. Many were high officials not in-the-know and wanting answers, but not being satisfied. Watson knew it was a matter of high importance, but he was not about to impart that importance into his organization. He knew that the Directorate of Intelligence was not happy, yet he was not about to accommodate General Cabell in any way. In fact he concealed TID's inactivity even more. His audacity to essentially ignore a general at Air Force Headquarters was very strange indeed, and it prompts one to contemplate how Watson thought he could continue with this tactic and get away with it.

One might consider that Watson's extraordinary hubris, based on his fame from WWII efforts to recover German high technology, was the reason for persisting in what seemed to be reckless abandon, but he must have also felt a measure of confidence and protection from somewhere. Was it from the Control Group? Normally, a Colonel would fear consequences in disregarding a commanding general, especially one at Air Force Headquarters. One cannot forget, however, that Watson's immediate boss, General Benjamin Chidlaw, was chief of AMC and a man previously in charge of receiving Roswell debris at Wright Field. Chidlaw had reason for protecting the secret.

Watson had many friends and connections in TID/AMC whom he had worked with during WWII. Some were highly qualified scientists and engineers who worked on Project SIGN, and had proclaimed the extraterrestrial answer. They were removed, with most reassigned to other duties. Some of the civilians who attempted to stay neutral were still in his command, such as Al Deyarmond, George Towles, and his executive assistant John Honaker, but others were gone. Based on this, one might consider that Watson was not about to sacrifice additional people to the UFO question.

Aside from that, one can surmise that Watson had serious motivation for protecting Truth. If there was nothing to protect, his obvious inaction and disregard of directives from higher headquarters were self-destructive, which no reasonable officer of his rank would attempt. There was a hidden reason for doing what he was doing, and he was not about to let up on debunking of flying objects. He would continue to take the offensive, and one of his first moves was to talk with Bob Considine, who proceeded to write the four part-article for the International News Service, which was discussed in the last chapter. Now, the debunking would continue with another article by Considine.

"The Disgraceful Flying Saucer Hoax"

The January 1951 issue of *Cosmopolitan* magazine featured an article by Bob Considine titled, "The Disgraceful Flying Saucer Hoax." This was an extension of articles he had previously written, but now his rhetoric had increased substantially, and many people took notice. Taking credit for getting the article into the magazine was Wright-Patterson's Public Information Officer Colonel Clare Welch, and it must have been something that Watson soaked up with great pride. Considine did not waste words in his amplification of what Watson related to him. In part, the article stated the following:

> *It has cost millions of dollars and some lives. Our dreams have been haunted by little men from nowhere. Here is the truth about the most wild-eyed fake of our time…Pranksters, half-wits, cranks, publicity hounds, and fanatics in general are having the time of their lives playing on the gullibility and Cold War jitters of the average citizen… And every time a newspaper or radio news bureau falls for their gag, or dementia, another legion of screwballs is mobilized…Airmen (and airwomen) employed and trusted by such commercial air lines as TWA, Eastern, United, and Chicago and Southern, speak of unidentifiable winged things blazing by their ships…cynical saucer-hoaxers will continue to go scot free, with a cackle of delight, until a penal act is created to check such offenses…Nick Mariana…looked up from the grandstand of the ball park and saw what he later described as two bright saucers, streaking across the clear Montana sky. He raced outside the park, unlocked his car, took out his home-movie camera, ran back to the stands, adjusted the camera and exposed about fifteen feet of film… The Air Force came into the case, received the film…was able to tell Mariana that the bright disks on his film were sun reflections from the ball park's water tower. And when he insisted that he had seen two bright things blazing across the sky, the Air Force agreed…and found that two F-84's (Air Force jets with a top speed of 600 mph) had landed at the nearby field…*

For many, the article published in *Cosmopolitan* was taken as a direct assault on witnesses of flying objects. It was very caustic and malicious, and it was difficult to swallow that all the very credible and responsible people involved in many incidents over the past year were half-wits, cranks, fanatics, or screwballs. It didn't add up.

And it sure didn't add up for Nick Mariana who was hit very hard by the magazine article, which resulted in the loss of several supporters of his baseball team. It was enough for Mariana to consider filing a lawsuit.

Lieutenant Edward Ruppelt Enters Picture

In early January, TID received a new investigative analyst by the name of Lieutenant Edward Ruppelt, who has been mentioned in previous chapters. Ruppelt, however, was not directly assigned to investigation of flying objects, but was situated at a desk very close to where Lieutenant James Rodgers was working. Ruppelt was curious about sightings being talked about, and he occasionally witnessed amusement generated by others gathered around Rodgers. In a short time, Ruppelt became familiar with the prevailing attitude regarding the UFO question, but he found that some people were still definitely, but quietly, believing in the extraterrestrial idea, especially those few who were previously around and observed Project SIGN activity. There were also a few who once participated in Project SIGN, but were now careful to toe the party line. People who were open to the extraterrestrial idea, however, were also very careful not to voice opinion around those scoffing the idea, or those directly involved with Project GRUDGE.

Ruppelt was receiving a good education on the status of the UFO question within the organization, but it was a curious situation for him to realize that there was an actual difference of opinion within TID. Little known to him at the time was that he would later become heavily involved with the UFO question, and also gain notoriety in what would become a more famous investigation called Project BLUE BOOK. In his current job, he was now deeply involved with looking at and analyzing performance of Soviet MiG-15s, which were beginning to challenge the Air Force in Korea, but he was also without tangible evidence in that regard.

Watson and Chidlaw Visit Korea

Because TID was in desperate need of a MiG-15 to analyze and study, Colonel Watson and General Benjamin Chidlaw traveled to the Korean War zone in an effort to obtain an enemy plane. In an amazing successful effort to recover pieces of one shot down, Watson and Chidlaw prepared the wreckage for shipment, and then quickly returned back to AMC.

Watson and Chidlaw worked well together, and it's noteworthy that both were in-the-know about Roswell, with one involved in a desperate endeavor to cover-up the UFO question. Chidlaw was promoted to acting commander in chief of AMC in February 1949, which was when Project GRUDGE replaced Project SIGN, and General McNarney returned to Air Force Headquarters to work for Defense Secretary James Forrestal. With Chidlaw in charge of AMC, Watson soon entered the picture and quickly debunked the UFO question. And he produced the first and final GRUDGE Report within a month of his arriving.

Sighting at Artesia, New Mexico

On 16 January, a sighting of flying objects took place at Artesia Municipal Airport in New Mexico by six individuals. Two of them were members of the General Mills Aeronautical Research Laboratory, which included pilot Ray Stiles and photographer Ray Dungan. They were connected with the launch of a Skyhook balloon earlier in the morning at Alamogordo, New Mexico, and they had flown to Artesia to track the balloon. At the airport, they were joined by four others, including the airport manager and the editor of the local newspaper. They all sighted the flying objects near the balloon.

A report of this sighting was documented a year later by a "revamped" Project GRUDGE, which specified the year of observation as 1952, although the pilot's log confirmed the date as January 1951. The incident was described in an investigation report as follows:

> ...two members of a balloon project from the General Mills Aeronautical Research Laboratory and four other civilians observed two unidentified aerial objects in the vicinity of the balloon they were observing. The balloon was at an altitude of 112,000 ft. and was 110 ft. in diameter at the time of the observation...Two objects at apparently extremely high altitude were noticed coming toward the balloon from the northwest. They circled the balloon, or apparently so, and flew off to the northeast...no further investigation is contemplated. The observers are known to be very reliable and experienced.

Flying objects in this incident were described as disk-shaped, and they were two or three times the size of the balloon. One witness described the objects as having a convex surface that was dull grey in color with a hazy rose tinge. When the objects departed, they took off at extreme speed.

Flying Object Controversy Hits Press

Previously, during the month of November 1950, Bob Considine authored news articles debunking the UFO question, and then he followed up with an article in the January *Cosmopolitan* magazine. It all began as a debunking campaign started by Colonel Watson, and aided by Colonel Clare Welch. Now, in the middle of February, additional debunking suddenly expanded in the press, and one cannot help suspect that Colonel Welch may have had a hand in it also, or maybe it was the Control Group or CIA that found an opportunity to help out. But it was also with cooperation from Dr. Urner Liddell.

On 13 February, the monthly issue of *Look* magazine hit newsstands with an article quoting Liddell, who was a scientist with the Office of Naval Research (ONR). He quite "convincingly" debunked flying objects in a ten-page ONR report. The Associated Press (AP) and United Press (UP) then came out the same day with articles also quoting Liddell, and the AP article stated the following:

> *…Flying saucers are real—but they're only huge balloons used in cosmic ray studies, Look magazine says in today's issue, quoting Dr. Urner Liddel [sic]…The balloons, called "skyhooks," were first sent aloft in 1947 and it was then that flying saucer reports began. There were more balloons in the next two years and more "saucers" seen…When this project began, it was kept secret, the magazine quotes Dr. Liddel… The Liddel report is considered to be the most authoritative scientific explanation of the flying-saucer phenomena. As far as Dr. Liddel is concerned personally, he considers his answer incontrovertibly right… He added that secrecy was "no longer" necessary…*

The UP article stated:

> *… The Air Force two years ago told the world that the flying saucer was a myth compounded of false identification of conventional objects—such as balloons, of mild mass hysteria, or hoax. Today Dr. Urner Liddell, chief of the navy's [sic] nuclear research program, added another footnote to the story of the whirling disc. He said flying saucers are not flying saucers but huge plastic balloons used in cosmic ray research. William Webster, chairman of the research and development board, then added his bit to the debunking. He agreed with the Air Force and Liddell that the flying saucer, as such, does not exist… "Accounts of flying saucers," Dr. Liddell said, "were either tall tales or reports from reliable observers who were either looking at the balloons or misinterpreting what they saw." Dr. Liddell said the balloons are bags 100 feet in diameter which travel at windswept speeds up to 200 miles an hour and sour [sic] as high as 19 miles. The balloons, known as "skyhooks," carry delicate instruments to learn about conditions at such heights.*

Balloons used in Project Skyhook were nothing new, because ONR had been launching them over the past few years, and they were becoming synonymous for balloons used in high altitude research. They were not generally known publicly, although they were previously mentioned in published articles. Also, many projects utilized different balloons of one type or another, including weather balloons, and their prevalence in the sky leaves no doubt that sightings of balloons were often mistaken for something else, although there can be little doubt that reliable trained researchers recognized the difference between their balloons and other objects cavorting near them.

Dr. Liddell of ONR indicated that all sightings, and future sightings, of flying objects could be laid to misinterpretation of Skyhook balloons. He stated that, "There is not a single reliable report of an observation which is not attributable to the cosmic balloons." One wonders what Project Twinkle observers might have thought upon learning this, or what the observers on 16 January at Artesia, New Mexico, thought, or Navy Commander Robert McLaughlin, or Charles Moore before that. The wildly maneuvering objects that scientific researchers observed, which were cavorting around balloons at extremely high altitude, could not have been other balloons, birds, or planes.

When one considers that ONR had received many reports in the past from its own project managers and researchers at White Sands and Holloman AFB regarding the appearance of flying objects, and that ONR was aware of them, suspicion arises about possible outside assistance for Liddell's report. It is quite apparent that claims by Dr. Liddell were very much flawed, especially when contemplating the many sightings closer to ground level by pilots who chased them, and others who observed them in detail. It leads one believe that Liddell's presentation must have been manufactured for the purpose of debunking. But who put him up to this? Was it ONR, or did he have a connection with the CIA?

Liddell's statement in the AP article, which mentioned that secrecy was "no longer" necessary, caught the attention of certain people. Those involved with sightings and encounters recognized that the statement was designed to debunk, which enhanced the suspicion of a cover-up. But those in-the-know regretted the statement, because an element of secrecy was necessary on all flying object investigations in order to protect Truth of the ultimate secret.

The mention of RDB chief William Webster in the UP article is a bit surprising. He gave a press briefing at the Pentagon on the day the article appeared, and he berated ideas that flying objects were a secret U.S. weapon, or an interplanetary Martian invasion. He backed up high Air Force officials, and also Liddell, by confirming that "there is nothing to the flying saucers." Ironically, as one who was in-the-know on Roswell, and providing a home for the Control Group, Webster was using his position of authority to add his weight to the matter. But it really does cause one to wonder who put him up to this.

On 16 February, an article from the *International News Service* quoted a response to Liddell from Donald Keyhoe, and it stated:

> ...*Retired Marine Corps Major Donald E. Keyhoe, author of the book "Flying Saucers Are Real," sent Liddel [sic], chief of the Nuclear Physics Branch, office of Naval Research, a four page telegram contradicting his theory. Keyhoe told Liddel that his story "deceives the American people." The Liddel article was published in a nationally circulated magazine, it stated that all reports of saucers that could be authenticated could be traced to sighting of the Navy's 100-foot plastic balloons used to study cosmic rays. Liddel said that the suns rays reflected on the lower side of the balloons when the sun was low in the sky, illuminated them in such a way that they appeared to be discs. Keyhoe pointed out that "veteran pilots of Eastern, Trans-World, American and other airlines" had sighted glowing or illuminated unidentifiable objects moving at high speed across the night sky. The retired Marine said that the appearance of the objects at night refutes the reflected sunlight theory of illumination... "If this constant smokescreen were designed to cover some super secret American weapon, I would be the first to say 'Thank God.'"*

Keyhoe quickly picked up on Liddell's intentions, based on obvious omissions and inaccuracies, which weren't likely to be recognized by the public, or a naïve press. Liddell's unabashed claim that all sightings were the result of mistaken Skyhook balloons was enough to clue Keyhoe to the fact that Liddell was a fraud, and he intimated that Liddell was purposely involved in a "smokescreen" effort. There was no way that a conscientious scientist could legitimately expound in such a careless manner, unless there was ulterior motivation. Keyhoe wasted no words in his criticism of Liddell.

On 26 February, Dr. Anthony Mirarchi waded into the conversation, and his mind was centered on Green Fireballs. He was the Air Force scientist sponsoring Project Twinkle for the GRD. On that date, the UP reported the following:

> *Dr. Anthony Marrachi, former Air Force scientists [sic], urged today that radar and spotter observation posts be set up to track down flying saucers "that may lead to another Pearl Harbor... The results of my own investigation," he said, "indicate that we cannot exclude the possibility that the so-called flying saucer is the result of experiments by a potential enemy of the United States."...*

On the next day, the UP also stated:

> *Former Air Force scientist Dr. Anthony Mirarchi says flying saucers may have been missiles launched by a foreign power to carry out photographic missions over the nation's atom testing grounds...Dr. Mirarchi's theory conflicts with that of Dr. Urner Liddell, a Navy scientist. Dr. Liddell said that flying saucers were plastic balloons sent into the upper atmosphere for radiation research...The former Air Force scientist said, however, that the Navy opinion "does not tell the whole story...there is the high speed of the reported saucers which is not consistent with the balloon theory...there is the color seen with them - yellow, green or orange...I'm dead certain they weren't balloons," he said. "There is a strong possibility that they might be missiles from some foreign land."*

Clearly, Dr. Mirarchi objected to Dr. Liddell's assessment of flying objects, and this was due to Mirarchi's experience with Project Twinkle. He additionally stated that "his theory was substantiated by his experience in New Mexico when he set up phototheodolites...and nothing happened in three months." This was during Project Twinkle's attempt to observe Green Fireballs, and this comment was prompted by the fact he was convinced an enemy spy effort knew of his project, which enabled fireballs to purposely avoid detection during that time. He was certain the objects (fireballs) were foreign reconnaissance missions, and he implied they were Soviet missiles.

Mirarchi was knowledgeable of disc sightings made by theodolite crews in April and May 1950, and also aware of disk sightings made at the end of August, which were "sufficiently significant" to extend his project. Mirarchi's motivation now, however, was the fact that he recognized Liddell's effort was intended to deflect attention from flying objects. Mirarchi definitely knew that disc sightings were real, and not balloons, but he primarily wanted to bring attention to a potential Soviet menace that he considered more important. As a result of Mirarchi's comments, the Air Force was not happy with his revelation regarding Green Fireballs, since the subject was secret and had been successfully kept away from the public. In fact, Mirarchi was seriously considered for prosecution due to his statements, but that idea was later discarded, which causes one to speculate that the Soviet "menace" was possibly real after all.

Termination of Project Twinkle

Project Twinkle's first contract with Land-Air covered a period from 1 April to 15 September 1950. Its second began on 1 October 1950 and was finished on 31 March 1951. With termination of the second contract, the 24-hour watch for Green Fireballs was completed. Green Fireballs, however, would continue to be periodically and independently observed. A final Project Twinkle report stated the following in regard to the second contract:

> *During this period, occasional reports were received of individuals seeing strange aerial phenomena, but these reports were sketchy, inconclusive, and were considered to be of no scientific value. No sightings were made by the Askania cameras. Nothing whatsoever was reported by the Northrup pilots…In summary, the results during this period were negative.*

The mention of "strange aerial phenomena" was in reference to disk-shaped objects often seen cavorting with Skyhook research balloons, but these objects did not qualify for Project Twinkle's reporting of Green Fireballs, and so the detailed reporting of this "phenomena" was ignored. Results for Project Twinkle were considered "negative" since no Green Fireballs were observed. Ironically, on 14 March, just prior to termination of the project, sightings of strange disk objects over Holloman AFB were observed by several employees of Bell Aircraft Corporation during a Bell aircraft test, and by Project Twinkle tracking crews on the ground.

Watson's Letter to Cabell

On 23 April, Colonel Watson wrote a letter to General Cabell to offer a few suggestions. This was his last big effort to try and convince Cabell to remove TID from the business of investigating flying objects. He wanted to make a strong appeal with a better idea. It was an appeal that made sense to Watson, and probably to others at high levels connected with him. It was his "ace in the hole." In part, Watson stated the following:

> *…Extensive investigations of many incidents were made and conclusions were drawn on each incident and insofar as the facts available would permit…Many of the incidents cannot be fully explained because of the lack of facts upon which to base a technical investigation. However, a great number of the incidents were found to be the result of cloud formations, balloons, meteors, sunlight reflecting from aircraft, etc… In August 1949 a report was prepared entitled "Unidentified Flying Objects Project Grudge," and the project was cancelled. In October 1950 the project was reinitiated at the request of your headquarters. Since*

that time hundreds of reports have been received and investigated. The conclusions which have been drawn since the re-initiation of the project are for all intents and purposes identical to those drawn in the earlier investigations…little if any results being obtained which are significant from the standpoint of technical intelligence…Notwithstanding this conclusion, it is considered that it would be impracticable in connection with Air Force responsibilities to say that we are no longer interested in any incidents of the aforementioned nature…Accordingly, it is felt that the project requirements should be revised to assure that all unidentified aircraft are reported without delay and by expeditious means to Air Defense Command…it is obvious that if an unidentified flying object turned out to be an enemy aircraft, current procedures would be entirely inadequate…it is recommended that the directives governing responsibility related to investigation of unidentified flying objects be reviewed and consideration be given to shifting the emphasis to Air Defense Command responsibilities, using AMC as a technical intelligence service organization as required.

Cabell, prior to considering the primary subject of Watson's proposal, must have stumbled on Watson's words indicating that "hundreds of reports have been received and investigated" by TID since the previous October. Cabell knew this wasn't true, and he probably didn't want to read more of Watson's letter, or be subjected to more "lies" from Watson.

Watson was attempting to sell the idea that ADC should be considered a first responder in the receipt of sighting reports. It would provide a measure of efficiency, and provide quicker response for national security reasons. JANAP 146A (mentioned in Chapter 25) is what clued him in to this idea. With this proposal, Watson did not want to suggest that his solution would serve to bypass the Directorate of Intelligence at Air Force Headquarters, but he did indicate that AMC would then become more of a "technical intelligence service organization as envisioned." But Watson also knew that TID would be able to continue ignoring sighting reports, which might possibly be passed on from ADC. He also knew his solution would be an advantage to those in-the-know, and to the Control Group who wanted to remove TID from the flying object investigation business.

Perhaps Watson thought he had a good hand in his game of strategy, but he probably didn't realize he actually lost his battle when his letter initially mentioned that investigative conclusions on flying objects were determined from "facts available," and that there was a "lack of facts." This was the critical information Cabell needed in order to proceed in making his decision about Watson and TID's investigative role with the UFO question. Cabell would not be further swayed by Watson's letter, and Cabell may have already made certain decisions in regard to Watson.

❧❦

The months of January through April 1951 were very interesting and quite revealing in regard to how the Air Force was managing the UFO question. After application of heavy pressure by Watson to get out of the investigation business, and with orders by General Cabell to revive Project GRUDGE, Colonel Watson was not about to cave in. It was not an option for him. Watson's blatant disregard of a general at Air Force Headquarters was extreme, and Watson would continue this by ramping up debunking rhetoric and urging others to do the same, which probably aggravated Cabell greatly. But Watson also played his "Ace Card" in his 23 April letter to Cabell.

The debunking help Watson received from Bob Considine was substantial, but the help received from Dr. Liddell was a bit over the top. Most people didn't recognize this, except for Donald Keyhoe and credible witnesses involved with significant sightings, which included Nick Mariana. When Air Force scientist Anthony Mirarchi stepped in to add his "two cents," it only confused the matter, but the Air Force did not appreciate his revelation that Green Fireballs were activities of a foreign enemy probing defenses. Now it was up to Cabell to make a decision regarding investigation of the UFO question, although he may have already decided.

The twists and turns associated with safeguarding the secret of Roswell, and the UFO question, was a taxing situation for Air Force officials at high levels, and also the Control Group. This, however, was becoming standard procedure, and there was a prospect that it would continue. It would be a matter of contending with issues, while keeping the secret intact. But would the Control Group be able to contend with General Cabell, who wanted to overcome his frustration with Colonel Watson. Or would Watson prevail?

CHAPTER TWENTY-EIGHT

GRUDGE in Transition

Project GRUDGE: May - October 1951

The Science of Mind Control

The rhetoric attributed to Colonel Watson in articles by Bob Considine did not go unnoticed by General Cabell. Articles quoting Navy scientist Dr. Urner Liddell did not go unnoticed either, and Cabell knew that not all flying objects in various sightings were Skyhook balloons. News articles in magazines such as *Look*, *Time*, and *Aviation Week* must have caused Cabell to become quite mystified about what was appearing in the press, although articles mentioning Colonel Watson's name provided a clue.

On the other side of this, many in-the-know were pleased with information coming out, including high Air Force officials and the Control Group. The press was also generally very accepting of debunking views being expressed. This included criticism about secrecy in the Air Force regarding flying object investigations, and the wasting of time and energy over sightings, which were considered to be misperceptions and fantasies of the mind.

It must be noted, however, that great emphasis was being directed by the government at this time on psychological manipulation of the masses, and creation of a "social consensus." This is what the RDB was somewhat involved with via their Human Resources Committee (HRC) run by Dr. Donald Marquis, a well-known psychologist and president of the American Psychological Association. But a much greater effort was being conducted by others, such as the National Psychological Strategy Board (NPSB) within the executive branch of government (State Department), the ONR's Human Resources Division, and the CIA's Office of Scientific Research. Each was involved in the study of mass psychology, and heavily involved in the attempt to change minds through the use of public media. There was also much experimentation taking place with effects of various drugs on the human mind. It was recognized that there were effective ways of changing attitudes and ideas in masses of people, and this was being exploited in different ways.

One would normally assume that most activity in this area was meant to turn around the minds of those in foreign states, including communist countries and North Korea, but a very great effort was also directed toward managing a consensus of thought among the American people. This meant that news organizations would cooperate in this process, especially the heads or presidents of media organizations coming under government influence.

The reporting of news, normally orientated to the communication of unbiased facts and information, was now being exploited to disguise it with innuendo, opinion, and outright distortion in order to sway public attitude in certain directions. At this particular time, mind control of the masses was a particular interest of General Walter Bedall Smith, and it was becoming developed and utilized by certain government agencies with specific agendas. Today, many media organizations and political parties have also recognized the power they now have available to sway or swing public opinion. They use the same techniques, with many different avenues of communication technology.

TID Becomes ATIC Under General Cabell

After many months of pressing Watson for reactivation of Project GRUDGE, and providing instructions to various commands to send sighting reports to TID, Cabell must have been quite perturbed that Colonel Watson was vehemently denying flying objects, and viciously denouncing credible and responsible witnesses. Cabell must have also been taken aback by Navy confirmation that all flying objects were Skyhook balloons. He must have wondered why the report by Liddell appeared to be so oddly contrived. What was going on? Quite likely, that was the question running through Cabell's mind, and he needed to find answers so that the investigation of flying objects could move forward.

It would be interesting to know what Cabell was finding out, or what he was being told by the Navy, but no documents have surfaced in that regard. Perhaps he made an effort to contact ONR to find the reason for Liddell's comments, and it's quite likely he had conversations with General Chidlaw at AMC regarding Watson. In any case, he must have received answers that caused him to take stock of the situation, and he then decided on a course of action. His immediate need to take action was probably confirmed when Nick Mariana filed a libel lawsuit on 4 May against Hearst Publications in regard to their publication of Watson's slanderous remarks toward sighting witnesses. And it was probably understood that Mariana had good reason to bring on his lawsuit.

With orders issued on 21 May, General Cabell finally got what he wanted. On 1 June, TID officially became Air Technical Intelligence Center (ATIC), and it was designated the 1125th Field Activities Group under direct command of General John Schweizer, who was Cabell's assistant chief of staff at AFOIN. At this time, there were about 410 people assigned within three divisions of ATIC, which included Technical Requirements, Technical Analysis, and Technical Services. Within that large organization, Colonel Watson suddenly came under command and control of General Cabell, and not General Chidlaw at AMC.

One might wonder how Cabell pulled this off in taking such a large organization under his "wing" and away from AMC, but he did. One can speculate there was more to the situation going on at TID that forced Cabell to make this change, and one might also consider that Cabell received help from General Twinning, along with coordination from General Chidlaw.

The most important thing that comes to mind in all this is the fact that Cabell may have also inherited wreckage and debris of Roswell and other incidents, which were under original control of T-2 and TID. It is more than likely, however, that evidence isolated at Wright-Patterson AFB remained tightly secured by a highly specialized unit under leadership of Colonel McCoy, with direct ties to Vandenberg, Twining, and the Control Group. McCoy probably remained in control of that group when Colonel Watson originally took over TID.

Robert Ginna Visits ATIC

On 22 June, ATIC was visited by Robert Ginna of *Life* magazine after receiving permission from the Public Information Office at the Pentagon. Ginna's intent was to write an article for the popular national magazine, and he wanted to find out the current status of ATIC's investigation into flying objects. He also wanted to look into specific sighting cases he had in mind. It is quite likely his inspiration for this came from the *Look* magazine article in February, where Dr. Liddell mentioned that all sightings were misidentification of Skyhook balloons. Ginna was determined to research specific Air Force sighting cases, and draw his own conclusions. He knew the Air Force had never specifically addressed Skyhook balloons as being the sole reason for flying object sightings.

Considerable discussion had been taking place in the media over the last few months regarding misidentification of Skyhook balloons, and this was facilitated by encouragement from Colonel Watson, who promoted the fact that the Air Force was investigating "nothing of substance." Quite likely, there was also considerable discussion at Air Force Headquarters by General Cabell and others who knew that flying objects were real, although they didn't have a handle on them.

Because General Cabell wanted to get to the bottom of the situation, it was a perfect time to confront Watson and the people at ATIC with someone like Ginna. After all, Watson was confirming that ATIC was actively involved with many "worthless" investigations, and had investigated "hundreds" since last October. There should be plenty of incidents from previous months that he would have information on.

Exactly how Ginna was granted permission to open ATIC's door is not known, but it was arranged by Air Force Special Assistant to Director of Public Relations John Shea. One can speculate there was inside help to smooth the way, and this is reminiscent and similar to how Sidney Shalett was provided the same access (see Chapter 17). Although Cabell was against Shalett having access at the time, Cabell now seemed quite comfortable with the prospect of Ginna's visit. Ruppelt would later state the following:

> The "maybe they're interplanetary" with the "maybe" bordering on "they are" was the personal opinion of several high-ranking officers in the Pentagon—so high that their personal opinion was almost policy. I knew the men and I knew that one of them, a general, had passed his opinions on to Bob Ginna.

Upon arrival at the Public Information Office at Wright-Patterson, Ginna was escorted to ATIC by Albert Chop, who had replaced Colonel Clare Welch as the new chief of Wright-Patterson's press section. Little did Chop realize he would be located at Air Force Headquarters another year later and find himself directly involved with a major flying object incident in Washington, D.C.

When Ginna opened the door at ATIC, it was a time of panic and feverish action by James Rodgers, Watson, and others. They were totally unprepared. Case files that were previously bundled and stored away, or even thrown away due to no investigative activity taking place, were difficult to locate. Project GRUDGE had been closed down and inactive for a very long time. This, however, was not exactly the impression Watson and his organization wanted Ginna to have, primarily because the Public Information Office and Directorate of Intelligence at Air Force Headquarters was previously given the impression that Project GRUDGE was actively in the business of investigating sighting reports. It didn't take long, however, for Ginna to figure out the situation, especially after being repeatedly told that specific case files were classified, or not readily available.

Word of discovered inactivity at ATIC must have filtered up to General Cabell, because it is known that Cabell again communicated strong displeasure to Watson. It would be the beginning of the "end" for Watson. Also, Major Jere Boggs and his superiors were taken to task for continued efforts to debunk the UFO question. It became apparent to Cabell that they were colluding with Watson, and one is tempted to wonder if Colonel Frank Dunn was able to provide information on this to Cabell.

Colonel Dunn Replaces Watson

It did not take long for Cabell to react to news he received. Colonel Watson was now named chief of the Foreign Affairs Department within the Technical Requirements Division of ATIC. As one might guess, he also received new orders for reassignment, which would be effective on 24 September. That is when he would head to France and become deputy chief of the Intelligence Analyst Division of the 7470th Headquarters Support Squadron for USAFE. Although it would be another few months before Watson would depart, little more would be heard from him again at ATIC.

Colonel Frank Dunn, who previously worked in AFOIR under Cabell, and then with Watson for more than six months, was now chief of ATIC. The Technical Analysis side of ATIC was a large organization, and the small inconspicuous investigative unit dealing with the UFO question was buried within its Aircraft and Propulsion Section. It was a unit Colonel Dunn would pay little attention to, but he would rely on those under him to guide and maintain an active investigation of flying objects.

Lieutenant James Rodgers, who had been Watson's right-hand man in charge of Project GRUDGE, was reassigned to other duties at ATIC. Replacing him was Lieutenant Jerry Cummings, who recently arrived in June. He would work under Colonel Nathan Rosengarten, who was chief of the Performance and Characteristics branch of the Aircraft and Propulsion Section. Captain Roy James would continue to work in the radar section.

A New Charter

With ATIC under his command, and personnel changes made, Cabell was expecting to see much better performance under a new charter. But this was still a difficult situation, because several others in Watson's previous chain of command were still in place, and the GRUDGE investigative unit was understaffed. Lieutenant Cummings, however, wasted little time, and immediately started to review Project GRUDGE while attempting to retrieve, assemble, and look at prior cases. He was not receiving many new sighting reports, however, and it took a while before he realized that many were being intercepted and not forwarded to him. He also quickly learned he had a much larger job than he could handle by himself, and he occasionally received help from First Lieutenant Edward Ruppelt and Second Lieutenant Henry Metscher, who also worked for Rosengarten. Ruppelt became a helpful investigator with his engineering degree, and with an interest in the sightings, although he was also immersed in other important matters, especially with analysis of the recovered Soviet MiG-15.

Because the Korean War was a big concern, and the Soviets were constantly exhibiting a threatening posture and attracting attention, the UFO question was not a top priority. However, the flying object situation would receive much more attention in ATIC than from the previous TID organization.

From the beginning, Cabell had been of the opinion that there was not a satisfactory answer to the UFO question, and he wanted the question answered. This would be accomplished by organizing and gathering "data" and "facts" in a database that would later be made available to Project Stork, which was the focus of a special contract initiated on 1 June when Cabell took over ATIC. Air Force contract AF-33(038)-4044 involved an all-encompassing project serving a number of areas, but one specific goal was to enable a statistical survey to determine modus operandi of the flying objects, and what they were about. There was also consideration given to gathering a group of scientists together to go over results of the project. All of this was the "hand" Colonel Watson unknowingly dealt to Cabell in his 23 April letter, but it was a different kind of hand than his "ace in the hole," which was referred to in the last chapter.

By July, the new ATIC organization was getting its feet on the ground with its new charter, and consolidating plans and objectives. Most important, it was now recognized that a great many important flying object sightings were reported by competent, reliable, and credible people since the time of the final GRUDGE Report. Many of the sightings were by military pilots, and also those in the airline industry. They were sightings that were ignored in Watson's letter to Cabell at the end of April. In consideration of this, it was stated that the specific conclusions in the GRUDGE Report were no longer valid. These were the conclusions stating that all sightings were the result of misinterpretation of conventional objects, mass hysteria, hoaxes, and psychopathological persons. The intention was to positively reverse the attitude promoted by Watson, and to put him in his place. Competent witnesses would be exonerated, and there would be no more meddling by Watson in new goals, efforts, and affairs of Project GRUDGE.

This turnaround in attacking the UFO question head-on by General Cabell is very interesting. There was no instigation, or requests from outside the Air Force to do this. It was an internal decision prompted by many sightings by competent military, who provided indications that real objects were appearing in a harassing and flaunting manner. There was no pressure applied by the press or public on this, although there remained plenty of confusion among them between fact and fiction on the UFO question. This acknowledgement about invalid conclusions in the GRUDGE Report was a direct and pointed repudiation of the report

Cabell's decision to proceed with serious internal investigation of the UFO question was similar to the beginnings of Project SIGN and Project Twinkle. Project GRUDGE was now being transformed because the general Air Force expressed a need to get serious about flying objects, and to find out if they posed a threat. Previously, Watson tried his best to divert attention, and prevent further investigation of flying objects. Those efforts, which were backed by those in-the-know and the Control Group, were now very much weakened, and they found that the actions of unearthly flying objects and sightings by many competent witnesses could not be controlled. It now seemed as if General Cabell would finally have his way in proactive investigation of flying objects, at least for a little while, but that would have to change.

Holloman AFB Report

When Project Twinkle was previously under contract, and centered at Holloman AFB and White Sands, a totally separate independent investigation was taking place on aerial phenomena by personnel at Holloman. Their report, released on 25 July, specifically dealt with sightings by Askania phototheodolite crews witnessing flying objects cavorting with research balloons, rockets, and test aircraft. Without any conclusions made, the report established the fact that some type of flying disk-shaped object did exist, which appeared to be intelligently controlled. This report was backed up with claims made two years previously by Charles Moore and General Mills technicians, and also by Navy Commander Robert McLaughlin.

General Chidlaw Receives Assignment

Normally, advance notice of a new assignment would be received four to six months ahead of time. With this in mind, one cannot help but make a connection that Watson may have been aware of General Chidlaw's pending transfer to become chief of Air Defense Command. It may have prompted Watson to recommend in his April letter to Cabell that ADC should take over intercept and investigation of flying objects, and this idea fell in line with JANUP 146A, which was previously revised and released in September 1950.

On 29 July, with promotion to four-star general, Chidlaw reported to his new assignment at Ent AFB in Colorado as base commander. On 25 August, he then assumed command of ADC upon retirement of Lieutenant General Ennis Whitehead. As the new commander of ADC, Chidlaw was well aware of the

UFO question and the magnificent "secret" housed at Wright-Patterson AFB. After all, he was the designated receiver of Roswell material in July 1947 when he was deputy commander of AMC under General Twinning. On becoming AMC commander, he worked closely with Colonel Watson to maintain the secret. Now he was involved with maintaining defense of North American airspace, and this would include radar tracking and interception of flying objects, with pilots making periodic encounters. It is quite likely he would become intimately aware of such encounters, and subsequent loss of pilots.

Chidlaw was replaced at AMC by Major General Edwin Rawlings, who assumed his new official position on 24 August. Rawlings previously worked as Air Force comptroller for Air Force Secretaries Symington, and then Finletter, and he would continue with his significant role of helping build up a stronger Air Force in his work at AMC. With the Korean War still in progress, and the Cold War intensifying, it required a man of his talent to help manage a quick increase in needed air power with advanced technology, which was a vital concern of General Vandenberg.

It cannot be absolutely certain if Rawlings was in-the-know about Roswell, but he was very close to Air Force Secretary Symington, and it's possible he was aware of expenses involved with the UFO question in his job as Air Force comptroller. Also, for many years, Rawlings had a very long and close relationship with General Clements McMullen. It was previously mentioned (in Chapter 9) that McMullen initially received Roswell material, which was then transferred to General Chidlaw at AMC. And Chidlaw was the person whom Rawlings was now replacing. This situation provides speculation that Rawlings must have been in-the-know, and that the "secret" remained intact at Wright-Patterson.

Air Force Four-Star Generals

After the elapse of three years since the Roswell incident, it's appropriate to note the only Air Force generals obtaining the highest rank of four stars. It is appropriate to note them because they were all involved with the Roswell incident in some way. They included: Henry (Hap) Arnold (deceased), Benjamin Chidlaw, Muir Fairchild (deceased), George Kenney, Joseph McNarney, Carl Spaatz (retired), Nathan Twining, and Hoyt Vandenberg.

Another, to be named before the end of the year, was Curtis LeMay. Others named later were Edwin Rawlings (AMC's new commander), Charles Cabell, and William Blanchard (commander of the 509th during the Roswell incident). The above-named people, along with certain others of lesser rank, inherited ultimate responsibility in the Air Force for keeping and maintaining Truth of the UFO question, and the secret of Roswell. One cannot help but believe that those intimately involved with Roswell were promoted and rewarded for their service in protecting the magnificent secret.

Dr. Walter Whitman - New RDB Chairman

On 1 August, Dr. Walter Whitman took over chairmanship of the RDB from Dr. William Webster. He and the previous three chairmen of the RDB (Bush, Compton, and Webster) had connections at MIT. They, along with Lloyd Berkner, were all connected to the ORSD of WWII, then the JRDB, and then the RDB. These were the civilian and scientific leaders involved with the UFO question. After Berkner organized the CIA's OSI in early 1949, the CIA was intimately involved with the Control Group, and with the exchange of information. The RDB was one of two "associations" within the Department of Defense, and it was now becoming much less effective, which was partly due to interservice rivalry, but also because of problems in prioritizing military objectives and coordinating a large bureaucracy of civilian scientists and engineers. There must have been considerable talk between the CIA and RDB at this time regarding a turn over of responsibility in security and control of the UFO question.

The Famous Lubbock Lights Incident

Many reported sightings crossed the desk of Lieutenant Cummings after he began his job in June. But the Lubbock Lights incident, which began on 25 August, and was reported in newspapers on 26 August, is probably the most unique in the history of Project GRUDGE. It was centered at Lubbock, Texas, and also a wide surrounding area of countryside. Lieutenant Cummings, however, did not immediately learn of the incident, and it would take a little over two months to formalize an on-site investigation, which he would not be part of.

The Lubbock Lights sightings was also an incident the Control Group took a huge interest in, and it provided reason for keeping the Air Force investigative process alive. It was also an incident where certain elements have generally escaped attention of researchers, and these will be developed and referred to later.

It all started when four college professors were meeting in the backyard of one of their homes and observing the night sky. Around 9:10 P.M. in the evening, they spotted a high formation of faintly glowing lights passing quickly overhead, and these lights were unlike anything they had seen before. They appeared as scintillating points of bluish-green color, clearly and plainly visible, but not brilliant. Because the lights were speeding so quickly overhead, a very rough estimate of about twenty or thirty lights was determined, traveling in a semi-circular formation located high in the sky, emitting no sound, and moving northeast to southwest. Within seconds, the lights disappeared, but they reappeared later that same evening in the same manner, and the professors took particular notice by taking and recording specific data measurements.

The professors all possessed doctorate degrees, and they included a petroleum engineer, geologist, physicist, and chemical engineer. They were highly trained technical people, and were participating in a micrometeorite study for their college astronomy department. That was the reason their eyes were directed skyward. During much of the time, one of them took notes while others observed. It was a process of documenting their measurements, and pinpointing the location and travel of the "shooting stars" they were observing. This went on for several nights, and they noted that the uniquely-lighted group of flying objects made regular appearances during the next three weeks. They could almost count on when they might reappear, and they estimated that about twelve "official" sightings were made of the lights during that time.

The professors also remarked, in an Air Force OSI interview, that they had observed two other sightings during this period, although they preferred not to make this widely known. This was because the sightings were particularly disturbing to them, and they were reluctant to discuss them. One sighting, observed on 1 September, came in low overhead, and it was described in an affidavit stating the following:

> … The formation included dark diffuse areas, and the arc itself quivered or pulsated in the direction of its travel…having the appearance of a group of from twelve to fifteen pale objects in the shape of a quadrant of a circle…an irregularly shaped yellow light appeared in the rear… The startling characteristics of this one flight made calm observation difficult to impossible.

This observation was too "fantastic" to mention to the local news, but the "very low" flying object with "yellowish lights" and "wiggling" motion had a stark emotional effect on one professor, and this sighting was then referred to as "Henaman's Horror." Evidently he was the one most frightened by the experience, and the others to a lesser degree.

The other sighting was an observation by the wife of one professor, and the professor commented:

> …she came running into the house one evening just at dusk very excited. Due to her usually calm manner, the excitement was very apparent. She said she had seen a very large flying wing type aircraft, making no sound, go over the house.

These very competent and well-respected professors were certain of what they saw. After the first sighting, they were joined by two other colleagues who also saw and confirmed the objects. One of the professors was quoted saying:

> *Frankly, we were astonished. And if I had not had confirming witnesses at the time, I feel sure I should have said nothing about what I saw, for it is incredible to believe they are of terrestrial origin and even more incredible to believe they are from beyond the earth.*

After the initial sighting, numbers of citizens in several towns in the general area surrounding Lubbock also spotted objects passing in "V" formation, or sometimes grouped together. In total, about a hundred people saw the lights. A few, however, reported that the objects were definitely birds with blue-white underbellies, which were flapping their wings and making noise. Others observed that the objects passed soundlessly high in the sky, were brighter than the stars, and disappeared within seconds. In consideration of this, one might say both groups were probably correct in what they saw. Fortunately, however, a photographer happened to obtain pictures of a formation on 31 August, and this put the Lubbock Lights in the headlines.

Except for a couple photos, the pictures showed the glowing objects quite clearly, with minimal streaking or blurring, which required skillful handling of the camera. Also, the objects appeared bright, and not what one would expect from city lights illuminating bird feathers at a distance. The photographer indicated that the formation was "about brighter than the brightest star in the sky" and compared it to "Venus in the early evening." Four of the photo negatives were obtained by ATIC for analytical study and it was concluded that comparative objects in the formation appeared to be identical. The size of each object in formation was calculated to be 310 feet across, plus or minus 30 feet, as seen by a camera one mile away. Curiously, however, the four professors said the pictures did not resemble the lights they saw.

The sightings became national news, but they also generated a profusion of questions that were never totally resolved. They also provided an example of what Colonel Watson was always claiming in regard "mistaken identity." And this was in reference to the "birds," and the fact that a vast majority of sightings involve misinterpretations.

It is highly likely the professors observed something quite significant and very unique, especially in regard to the two "emotional" observations. They definitely knew this, and they were absolutely convinced that authorities and others needed to become aware of what they had observed. In the end, however, the conclusion by the Air Force was "that the objects of this sighting were migratory birds…it is highly probable that they were ducks or plovers." Later, the professors came to an understanding that it was critical to avoid further controversy, and they changed their "tune" in a reticent and restrained manner. Their sudden turnabout came with help from astronomer Donald Menzel! They were concerned about their reputations, and they were no longer available for discussion about the sightings.

Behind the scenes, great interest was generated by the Lubbock Lights because of the very competent witnesses involved, and also because the sightings by the professors corresponded with certain aspects of other coincidental sightings, such as the one described next.

Flying Wing Object Sighted Over Albuquerque

On 25 August, the same date the professors at Lubbock saw their first formation of lights, there was another formation of lights observed near Albuquerque, New Mexico, which occurred about an hour and fifty minutes later. But this was described as a flying-wing-shaped craft. The object was observed by Sandia Base security guard Mr. Hugh Young and his wife outside their home that evening. The next day when Young reported the incident, he signed an affidavit that described the sighting as follows:

> *This aircraft was unusual in the following ways, there was no sound of motors of jets in fact there was no sound at all that I could hear. I could see no fuselage on this aircraft. The size I judged to be at least one and one half times as large as a B-36 and was shaped like a spread V with the wings sloping back at an angle of about 15 degrees. On the rear edge of the wings soft white lights were located in pairs with not less than six of these on each side of center…From the front edge of the wing strips extended to the rear edge of the wings with the strip ending between the lights of each pair…Each pair of lights were separated by about eight times the distance between the lights of one pair. The wings appeared to retain their size from the center to the end without any taper…The aircraft was in my sight about one half minute.*

The wife of the security guard also signed an affidavit stating:

> *On the night of 25 August 1951, at my home at 2158 hours [MTS] I observed for about 30 seconds an unfamiliar aircraft traveling 10 degrees west of north in the direction of 10 degrees east of south. The rate of speed of the aircraft might have been 350 miles per hour. The*

aircraft was low enough for the neon and car lights of East Central Avenue to be reflected from it...The size of the aircraft was apparently one and one half to two times of that of a B-36...with wings swept back...wings did not appear to taper...On the rear edge of each of the wings at least 6 and possibly eight, soft, white lights were visible. These lights were round, not trailing as in jet aircraft exhausts seen at night. These lights were grouped in twos with a larger distance between the pairs than between each one in a pair. The underneath surface of the wings of the aircraft appeared to have stripes running from the leading edge to the rear edge of the wings between each pair of lights...there was no audible sound from motors or jets...no noise that I could hear.

Mr. Young worked the night shift at Sandia Base, and he was waiting with his wife for a ride to work when they sighted the craft. After notifying Kirtland AFB, an interview was set up the next day with AFOSI Special Agent Major Harold Peterson, who quickly determined that there was no question regarding Mr. Young's reliability, primarily because of his "Q" security clearance. The interview produced concise descriptions of the object, as noted above. In later remarks, the object was described as having pairs of glowing bluish lights. It was flying at an altitude of about 800 to 1000 feet, with a speed of about 300 miles per hour, and it made no noise. It must have been quite awesome to view such an object at relatively low altitude. Later, when they were shown a picture of the Lubbock Lights, the couple identified them as very similar to the object they saw.

It is interesting to note that several months prior to this the Air Force had an eight-jet Northrup YB-49 flying wing in operational testing. The testing ended in April 1951, and it was flown back to Northrop's Ontario, California, airport where it remained abandoned until it was scrapped in December 1953. There is little chance the sighting by the Youngs involved this plane, which had already finished its testing and was not flying.

The OSI accomplished a quick "on spot" investigation of this incident with extensive interviews, including verification of security credentials for Mr. Young, and also a check on military and civilian air traffic at the time of the sighting. A report, dated 27 August, was then sent to OSI Headquarters at the Pentagon. ATIC received notification about two weeks later, but no further investigation took place, and the OSI closed further investigative action on the incident.

JANAP 146A Revised

On 6 September, the Joint Army Navy Air Force Publication for CERVIS was updated to the "B" version. This change gave additional special emphasis to "UFO

Reporting," and it maintained a warning regarding the passing of information to unauthorized persons. Of course, the main intent of the publication was to provide pilots with a procedure for reporting unidentified or "unconventional" enemy "aircraft." It also specified that reports be forwarded to the office of the secretary of defense, to ADC, and to the nearest U.S. military command. This revision still did not list ATIC for notification, although ATIC would subsequently be included on most forwarding lists for the document.

The Fort Monmouth Incident

When the Fort Monmouth incident was reported on 10 September, the long lethargic slump in Project GRUDGE was about to be awakened for the first time, but it would not be without substantial internal commotion and repercussions. This was partly due to a series of ineffective communications, and also because close associates of Colonel Watson still exerted influence. They, however, no longer had responsibility for flying object investigation, except for Chief of Technical Analysis Colonel Brunow Feiling, who was Colonel Rosengarten's boss. For the most part, Colonel Watson was out of the picture and managing his Foreign Affairs Department in ATIC. Little was heard from him before he ended his stint at Wright-Patterson on 23 September.

This particular incident began at a radar station near Fort Monmouth, New Jersey, when Private First Class Eugent Clark noted a very fast moving target on his radar screen. He was involved in a demonstration of his radar equipment with students and a few VIPs when it was noticed that the radar screen was unable to keep up with a particular fast-moving object.

A few minutes later, Lieutenant Wilbert Rogers was piloting his T-33 with passenger Major Ezra Ballard. They spotted an object in the same general area over Sandy Hook, New Jersey, and came to within a mile and a half of it. It was unlike anything they had ever seen before. It was disk-shaped with a raised center, shiny silver in color, and with no visible means of propulsion. When first sighted, it was moving steadily in their direction and below them on the left. Rogers then made a sharp turning dive toward it. In giving chase, their eyes were closely focused on the object as it accelerated to more than 900 miles per hour, and then it quickly disappeared out over the ocean two minutes later. ATIC later associated the sighting with other concurrent visual sightings from the ground, and also with two large silver weather balloons that were launched nearby at the time. It was then concluded that the radar operator was confused by his radar controls, and the pilots were fooled by the balloons.

As a side note, senior physicist Dr. James McDonald of the Institute for Atmospheric Physics, and professor at the University of Arizona, conducted a study of this specific case, and he later expressed the opinion that balloon findings on sightings were "strained beyond the breaking point." He testified before Congress in 1968 with the following:

> *UFOs are entirely real and we do not know what they are, because we have laughed them out of court. The possibility that these are extraterrestrial devices, that we are dealing with surveillance from some advanced technology, is a possibility I take very seriously.*

Lovett Replaces Marshall as Secretary of Defense

On 12 September, George Marshall suddenly resigned as secretary of defense "for very personal reasons." He stated that when he took the job a year previously, he had agreed it would only be until June of 1951. Marshall was 70 years old, and it's quite likely he figured it was time to retire. He was previously proclaimed the "originator of victory" in WWII, and was given the office of secretary of state by President Truman after the war. With a successful turnabout in Korea, and a looming settlement or armistice in store, he probably figured it was time to break loose from government service, although his counsel would continue to be utilized.

Marshall was immediately replaced by his close friend Robert Lovett, who had previously served as Marshall's under secretary of state, and also his deputy secretary of defense. Prior to that, Lovett was assistant secretary of war for Air, and he succeeded in boosting bomber production from three per month to more than 1,000 per month during WWII. There was no doubt that Lovett was the right man to replace Marshall.

Cabell Learns of Monmouth

The Monmouth incident became notable because word of it spread to the press and several newspapers before General Cabell became aware of it, which was about 18 days after the incident occurred. It was on Friday, 28 September, that he was notified of the incident in a TWX from General Chidlaw at ADC, who referenced a CIRVIS notice originating from Eastern Air Defense Force (EADF) on 11 September. Chidlaw's message was an urgent request for evaluation of an "object sighted and tracked" in the vicinity of Monmouth, and he suggested "querying of Army, Navy, Research and Development, Grumman, and any other source which may be able to throw light on phenomenon reported."

This message from General Chidlaw was probably instigated because of two additional messages on 21 and 26 September from EADF Director of Intelligence Lieutenant Colonel Bruce Baumgardner. The messages were directed to both ADC and AMC requesting investigation of the Monmouth incident. Baumgardner had already conducted extensive investigation of the incident, with interviews of the pilots and radar personnel.

When Cabell finally learned about the incident, it was not news to some in AFOIN or ATIC, including Lieutenant Cummings who previously reviewed the messages. Cabell, however, was dumbfounded. He previously gave instructions to his people that he wanted personal notification of any new and significant flying object activity. For Cabell, it was upsetting to learn of this incident from General Chidlaw at ADC, which was many days after the fact. He then called Colonel Dunn at ATIC to learn more about the incident, and if any investigation was accomplished. The answer he received, however, was not to his liking, and he requested that ATIC attend a special meeting with him at Air Force Headquarters the following Monday, on 1 October.

Needless to say, a small ruckus ensued, and Lieutenant Colonel Rosengarten and Lieutenant Cummings immediately flew to New Jersey to investigate and prepare for the meeting with Cabell. It also appears that Rosengarten was periodically in contact with Cabell over the weekend, and that Cabell was learning more about the incident as time progressed, especially after Cabell communicated with Lieutenant James Rodgers at ATIC. But Cabell was also curious about how the media became involved, and he wanted answers right away.

After a sleepless weekend of intensive investigation, and after interviewing the radar operator and pilots, Rosengarten sent a "Routing and Record Sheet" to OSI Headquarters in an attempt to find out who had leaked information to the press. The spreading of such information was strictly against several directives, including JANAP 146B. It was then time to meet with Cabell, and this was not expected to be a fun time.

At the two-hour Pentagon meeting, which included many of Cabell's assistant directors and staff, details of the situation at Monmouth were reviewed. It was agreed, however, that a more thorough investigation was necessary before a final summary or report could be issued on the incident. Cabell then asked for a status review of what was going on within ATIC. Although Cummings hesitated for a second, he didn't mince his words. He already knew he was a short-timer, because he had already received separation orders to return to his previous government project at Cal Tech, and he had nothing to lose in speaking out. He proceeded to tell Cabell and others about the prevailing attitude existing in Watson's ATIC when he first arrived in June. He described how sighting reports were previously handled, and what he was now dealing with in order to do his job. In the current Monmouth sighting, it was explained that Cummings and Rosengarten only became involved about the time Cabell did, and this was made clear to Cabell. In regard to Watson, he had already departed from ATIC and Wright-Patterson at this time, but the briefing by Cummings was a confirming moment for Cabell!

The scenario of what happened at ATIC, in regard to this sighting, is quite interesting. It was determined that a news reporter inadvertently overheard details of the sighting, and a naïve public information officer at Mitchell AFB allowed the reporter to interview the T-33 pilot, Wilbert Rogers. This violated regulations and directives. The sighting then made the news, which was about the same time that the sighting report arrived at ATIC on 11 and 12 September.

The sighting report landed on the desk of Rosengarten's boss, Colonel Feiling, who noted that the sighting involved a radar contact. He then gave the report to radar specialist Captain Roy James in the Electronics Branch at ATIC, and James then discussed the sighting report with Lieutenant Rodgers in order to come up with an answer on the sighting. It was then that both Rodgers and James discussed it with Colonel Watson before Watson departed AMC.

After Cabell contacted Colonel Dunn to find out about the incident, Dunn discussed it with Rosengarten, who then summoned Cummings for information. Cummings informed Rosengarten that he didn't have anything to do with it because Rodgers and James took over the investigation. Rosengarten then confronted Colonel Feiling to ask why he was not given the investigative assignment as required. It was an amazing "collision of coincidence" that General Cabell would make contact with ATIC at this "opportune" time.

After hearing from Cummings at the meeting, and pausing for a short moment to contemplate, Cabell loudly complained in a contemptuous voice of being lied to over and over again, and he was probably referring to some of his own staff people, including Major Boggs, but he was undoubtedly referring to Watson in particular. He then brought up the previous GRUDGE Report, which he promptly and severely ridiculed as being a worthless piece of garbage. No doubt he was thinking of conclusions in the report that he had previously repudiated, and with subsequent exoneration of witnesses to sightings. To Cabell, this was one more confirmation that proved he was more than justified in taking over ATIC and getting rid of Watson. It seemed appropriate that Watson was now gone for good. But would he be? The main concern for Cabell was keeping the press and public away from sighting incidents, but he strongly pointed out to everyone at the meeting that he wanted everything in regard to the UFO question handled with an open mind. He then asked:

> *Why do I have to stir up the action? Anyone can see that we do not have a satisfactory answer to the saucer question.*

This statement must be taken to indicate that Cabell was not in-the-know. He was a believer in reality of the objects, although he didn't understand them. With many sighting reports coming in from Korea and elsewhere, he was determined to find the Truth.

With ATIC under his command, he now realized there were still people in the organization obstructing investigation of flying objects, and they would need to be dealt with. He was determined that things would now be different, and he ordered Rosengarten and Cummings to set up a new and more effective investigative organization to unmask flying objects. Lieutenant Cummings, after initial trepidation, came away from the meeting with much respect from General Cabell, and he became Cabell's hero.

After Rosengarten and Cummings returned from Washington, D.C., it did not take long before Colonel Feiling was replaced by Colonel Sanford Kirkland, and Lieutenant James Rodgers removed. At Air Force Headquarters, Major Boggs was reassigned.

৩৵৵

Again, this chapter reveals the "see-saw" adventures of Air Force attempts to manage the flying object situation, and also a massive cover-up. It was a swinging pendulum, but now it was to the side of attempting to get back to real investigation of flying objects. There was an understanding that active investigation was necessary from a national defense standpoint, but also from a practical standpoint. There were many sightings over the last two years, which were witnessed by reliable and competent civilian and military personnel, and the Air Force could no longer afford to remain comatose on the subject.

In regard to General Cabell, it appears he previously assumed that selective investigative effort at TID and ATIC was continuing all along to some degree, although he slowly came to realize that very little was done, or accomplished. That was the beginning of the end for Colonel Watson, and a few others. And it was probably a tremendous disappointment for Cabell to learn that all of his previous efforts to resolve the UFO question were failures. He had relied on many people in the effort to investigate and expose flying objects, but he now realized that he had been deceived and lied to countless times. As a commanding general, it must have been very difficult to believe he had been so flagrantly manipulated.

There were several times in the past where Cabell may have considered his leadership at fault for negative situations that cropped up in the investigation of flying objects, which prevented proactive investigative procedures. For Cabell, it began when the EOTS was about to expose extraterrestrial reality, which was not acceptable to Vandenberg. The Air Force then bungled its handling of Shalett's *Saturday Evening Post* article. Then Cabell realized that the shoddy GRUDGE

Report and its purposeful, flimsy, and made-up investigative conclusions were intentional, but so was Colonel Watson's blatant disobeying of directives and orders to restart Project GRUDGE. But why? Was it simply because one, or maybe a few individuals, thought that all sightings were fantasies gone wild? Why were they willing to hide their lack of investigative inactivity and risk insubordination? Was there more to it than that?

Cabell knew that flying objects were real, and he wanted the Truth. One could easily speculate he now had plenty of reason to suspect that Truth already resided at Air Force Headquarters. In fact, in retrospect, one could speculate he should have learned the Truth much earlier, perhaps several months previous. But Cabell also had great faith, trust, and respect in his people, and he believed in the honesty of people above him who he coordinated with. In his position, it was particularly difficult to conceive that people above him might not have been forthright.

There is good reason, however, to believe that Cabell did not know the Truth prior to 1 October 1951, when he harshly stated in a "complaining" voice that "anyone can see that we do not have a satisfactory answer to the saucer question."

Would Cabell now proceed to discover the Truth of flying objects?

CHAPTER TWENTY-NINE

Cabell Discovers Truth?

Project GRUDGE: October 1951

Cabell, Watson, and the UFO Question

It's necessary to pause a moment to again review what has been witnessed within the Air Force in regard to the UFO question. Director of Intelligence General Cabell at Air Force Headquarters, and the Director of Technical Intelligence Colonel Watson at Wright-Patterson AFB, played on opposing sides in a battle regarding investigation of the UFO question. Cabell depended on Watson's unattached field investigative unit to actively investigate sighting reports of flying objects, and provide analysis of their origin and modus operandi.

From Cabell's perspective, flying objects were mysterious, and quite real. A great many competent and reliable people in the military were reporting significant sightings and encounters, especially near sensitive military areas, and there was a strong demand from the general military that Air Force Intelligence provide answers to it. But there were often many other sightings by the general public that were considered unreliable, and this necessitated that the Air Force take care in not provoking more sightings that would be considered misinterpretations of natural phenomena or man-made objects. It was also feared that public mass psychosis might develop if frivolous reports of sightings were not suppressed.

General Cabell was very concerned about flying objects. He was so concerned that he knew he had to get to the bottom of the problem and understand the true nature and origin of the objects. He understood that many sightings could easily be explained, especially if given extensive analysis, but there were also flying object sightings that defied reasonable explanation, especially when very experienced and credible military personnel related sightings and encounters backed up by radar and other observers. Those reports gave credence to an advanced technology exceeding anything on earth, or so it appeared. Cabell, however, did not want undue press coverage of sightings that might generate unworthy reports, or waste Air Force time. He was more interested in serious encounters that demanded study, and he was consistent in his endeavor to resolve the flying object mystery. The Korean War was a serious diversion to this, which affected everyone, but Cabell never wavered in his desire to get to the bottom of the UFO question.

From Watson's perspective, he had one major job to do, and that was to debunk the UFO question and refrain from any meaningful analysis of flying objects. To him, there was no other option, and he was not about to take action on Cabell's strong directives. His conduct was blatant, and his mind was very set on making the UFO question go away. He continued to do this while hiding TID and ATIC inactivity from General Cabell, and he apparently was not dissuaded by any guilt or negative consequence in doing so. He was very consistent in his resolve, and he never wavered on this.

One can attempt to make excuses for Colonel Watson in order to justify and support his position. It can easily be concluded he was much too busy managing a very large organization, and working hard in the technical analysis of foreign military advancements, especially with a war in progress. This was much more important than dealing with fanciful flying objects. It can also be said he was simply of the opinion that "flying saucers" were a fantasy of the mind gone wild, and the best way to deal with it was to debunk the idea, and remove it from public consciousness.

There can be no doubt, however, that Watson was very aware of the many significant sightings and encounters by competent military, especially near areas with nuclear facilities. And he was aware of many in the military who were demanding answers to the sightings. In this regard, his motivation for debunking, especially when considering strong concern and demands by General Cabell, was an act of self-serving lunacy, with absolutely no benefit to him. But was he really that dumb?

Many excuses can be used as possibilities for Watson's active disengagement from the UFO question, but it is far more reasonable to give Watson greater latitude or credit in this matter. For someone in his position, with strong pressure from Cabell to get the UFO question resolved, he must have acquired great incentive, motivation, and support from elsewhere for his reckless abandon in bucking high authority. He was much too intelligent to see his career tarnished by a headstrong attitude against fanciful flying objects that "did not exist," and then ignore what he was directed to do by General Cabell. His past history in the military proves he was no dummy.

Watson was a shrewd detective, and an experienced technical investigator of foreign technology. That was his specialty and expertise. But his ruse to deceive Cabell does not fit with what it takes be a leader in that position, unless he had extenuating reasons, other than having a "mind-set" against flying objects. One might think he would want to consider some kind of diplomatic communication with Cabell, and come to an understanding of how to deal with the situation. Instead, he played a secretive and clandestine role in making Cabell think his investigative process was ongoing. Then he proceeded to take the offensive to get the public and media on his side, and have them believe that all flying objects were misinterpretations of the mind. Could the answer to Watson's actions be that he was privy to Roswell, and he knew Cabell wasn't? If that was the case, Watson would be inclined to act more independently by avoiding Cabell and squelching the UFO question, while knowing he was protected by those in-the-know at the highest level.

During Roswell, Watson was an "advisor" reporting to the highest level in the Pentagon, and he maintained a close working relationship with many who were in-the-know. When later assigned to AFOAI, his actual whereabouts was questionable, and when Project SIGN investigators proposed the extraterrestrial answer, he did not seem to be around to exert influence. Because temporary duty assignments do not necessarily become part of official records, it is entirely possible that Watson served part time elsewhere before arriving at TID. And it is quite tempting to speculate just where his expertise was put to use during that time. When he finally came to work as chief of TID, his boss was General Chidlaw, who was in-the-know on Roswell. It is also known that when Watson was a new First Lieutenant, he took flying lessons from First Lieutenant Curtis LeMay, who later became involved with Roswell and was now commander of the Strategic Air Command. More than anyone, Watson had all the proper credentials and appropriate connections to be heavily involved with Roswell. And the Control Group found an opportunity to make very "good" use of him.

Cabell Prepares to Depart for New Assignment

After the meeting on the Monmouth incident at Air Force Headquarters, Cabell prepared to depart his job as chief of Air Force Intelligence. For him, however, there was still unfinished business, and his thoughts were probably running through the sequence of events that took place during his tenure on the UFO question. Things must have been adding up for him, and he possibly surmised that there was a reason his efforts to resolve the UFO question were stifled during his entire term as chief of AFOIN.

It was in May 1948 that Cabell first took charge of AFOIN. A few months after that, TID's Project SIGN came to an extraterrestrial conclusion about flying objects, which started to turn things upside down. Cabell's desire to resolve the UFO question became mired in a horrendous mess that he was unable to control. And this didn't improve when Colonel Watson entered the picture at TID in July 1949. The Air Force's investigation then became hog-tied through a combination of events, and it became a situation that Cabell was not able to deal with. But then the mysterious flying objects became pervasive, and continued to be sighted, encountered, and reported by reputable, competent, and credible people, especially by many in the military. Later, Cabell found that little investigative activity was taking place at TID, and it seemed to be purposely disregarded, even when he specifically directed that investigations of reported sightings be reinstituted. Cabell then realized that Watson was sponsoring a campaign to sway media attention against flying objects. They were proclaimed to be fantasies of the mind, or misidentification of research balloons. Next, Cabell received a letter from Watson suggesting that ADC become involved with interdiction and investigation of flying objects, but the letter also contained lies proclaiming that Project GRUDGE was continuing to actively investigate the objects. It was then that Cabell became convinced that Watson needed to be replaced.

When thinking about the last few years, Cabell probably realized it shouldn't have been that difficult to manage investigation of flying objects. There had to be a reason for the continuing turbulent chain of events, and for the situation that prevented active and effective investigation. When Lieutenant Cummings confirmed to Cabell that he had been lied to over and over again, Cabell knew he needed to find out the Truth.

Before departing at the end of the month, Cabell was determined to insure that investigation of flying objects would continue, and that his replacement, and others, would be in tune with a new perspective on how things should be handled. The prevailing attitude now would be that flying objects were real, and that their nature and origin needed to be identified.

The Minneapolis Sighting

On 11 October, balloon technicians from General Mills who were working on a research project near Minneapolis, Minnesota, sighted several flying objects near a balloon they were tracking. Lieutenant Ruppelt, who was now spending more time on Project GRUDGE, became aware of the incident and sent a TWX to notify Colonel Willis of TCB in AFOIN's Evaluation Division of the sighting. His message included the following statement made by General Mills observer Joseph Kaliszewski:

> *...Dick* [Richard Reilly] *and I* [Joseph Kaliszewski] *were flying at 10,000 ft. observing the Grab Bag Balloon when I saw a brightly glowing object to the S.E. of U. of M.* [University of Minneapolis] *airport...I pointed it out to Dick and we both made the following observation: ...This object was peculiar in that it had what can be described as a halo around it with a dark undersurface. It crossed rapidly and then slowed down and started to climb in lazy circles slowly...It went through these gyrations for a couple minutes and then with a very rapid acceleration disappeared to the east. This object, Dick and I watched for approximately five minutes...Shortly after this we saw another one, but this didn't hang around...When I saw the second one I called our tracking station at the U. of M. airport and the observers there on the theodolite managed to get glimpses of a number of them, but couldn't keep the theodolite going fast enough to keep them in the field of their instruments. Both Doug Smith and Dick Dorian caught glimpses of these objects in the theodolite after I notified them of their presence by radio.*

In a later interview, Dick Dorian (mentioned above) indicated that one object appeared to be smoky grey in color, but with no glow, halo, or reflection. It was "cigar shaped" and "retained a definite shape" in the short time viewed. It "acted exactly as if under definitely controlled flight." Two more of the objects appeared, and they "formed in a straight pattern," and then "all departed at the same time." Evidently, the objects observed by those in the air were different than seen by observers on the ground.

The previous day, on 10 October, Kaliszewski and other observers experienced a similar sighting, and he indicated that "from past experience I know that this object was not a balloon, jet, conventional aircraft, or celestial star."

General William M. Garland Arrives

After returning from his meeting with Cabell, Lieutenant Cummings spent his remaining time completing the Monmouth investigation, and finalizing documentation on it, while also preparing Lieutenant Ruppelt to replace him. On 12 October, he made one last communication regarding the Monmouth incident, and then he departed shortly after. From then on, the flying object investigative unit at ATIC was left with a small group of people that included Ruppelt and a few people originally under him, and also Lieutenant Metscher. They would continue to report to their chain of command that included Colonel Rosengarten, Colonel Kirkland, and Colonel Dunn.

On 16 October, Brigadier General William Garland became Cabell's new assistant for Production, replacing General Ernest Moore. He also took over responsibility for the UFO question from General Schweizer, which meant that Colonel Dunn and ATIC came under Garland's control. Since June, Garland had been chief of AFOIN's Air Targets Division, but he would now have much influence in the pursuit and investigation of flying objects. Reportedly, he once sighted a flying object himself, and this leads one to suspect that he would be motivated in his new job and anxious to learn of additional sightings of aerial phenomena. But there is also another side to this that causes one to consider another possible reason for him to be assigned responsibility for the UFO question in AFOIN. There would now be a more open policy toward flying objects, but with certain ground rules, and this could also be an indicator that Garland was possibly in-the-know, which would help limit and control the investigative process.

"Enlightening" Directives from Cabell

Previously in July, Cabell initiated his transformation of ATIC and Project GRUDGE with a new "charter," but after the Monmouth meeting with Colonel Rosengarten and Lieutenant Cummings, he was determined to have them continue to expand flying object investigation with a more effective and aggressive organization. But strangely, there were now new directives from Cabell that prompts one to contemplate whether he may have finally learned Truth of the UFO question.

It started with a 15 October letter to ATIC from Colonel Walter Glover at Air Force Headquarters stating that flying objects did not originate from a "foreign power" (a foreign country). It declared that the objects were a "technical situation" that ATIC needed to deal with and investigate, and that they were not a responsibility of ADC.

This very important communication prompts one to closely consider the actual purpose behind the letter. It leads one to question the motivation and intent for it. Obviously, it was in response to Watson's previous April letter suggesting that ADC become involved with flying objects, rather than TID, but this was the first time Air Force Headquarters declared, and confirmed, that flying objects were not of foreign origin. It was already accepted that they were not of U.S. origin. This meant that there was only one other possibility for origin of the objects, and they were now being referred to as a "technical situation." Could this be a euphemism, or alternative wording for Unearthly Flying Objects?

On 22 October, Cabell followed up with another very interesting letter to Colonel Dunn at ATIC that provides one with additional reason to suspect a new policy. It stated the following:

The reporting and analysis of sightings of unidentified flying objects continues to be an intelligence production requirement of the Air Technical Intelligence Center. It is therefore not considered desirable to shift the responsibility for this function to the Air Defense Command... This is not to imply that the Air Defense Command will not be kept informed...the problem concerns both technical intelligence and air defense. It is desired that your headquarters energetically conduct a continuing analysis of information and material you receive on this subject, and further that you remain prepared to provide this headquarters with up-to-date information as conditions demand...it is directed that you continue to have at least a small section specifically designated as responsible for this function.

On close examination of this letter by Cabell, it brings to attention a change in rational by Cabell, along with certain implications. Previously, Cabell knew that certain incidents provided a mysterious sense of reality to some of the flying objects encountered, and he was anxious to identify what the objects were and where they came from. Now, he still wanted active investigation, but the goal was to gather and forward "up-to-date information as conditions demand," with only a "small section" of ATIC having this responsibility. Serious investigation would be pursued, but the main objective was the maintenance of national security, while tracking the objects.

This was not what would be expected if General Cabell was prepared to resolve the UFO question once and for all. It was more in line with what the Control Group wanted in regard to keeping track of the objects. One would have expected that Cabell would have preferred to reinstitute an organization similar to Project SIGN, with specific objectives, goals, and additional technical personnel in order to investigate and answer the UFO question.

When Cabell finally realized at the Monmouth meeting that he had been lied to over and over again, and then stated, "Anyone can see that we do not have a satisfactory answer to the saucer question," perhaps that is when he examined the words coming out of his own mouth, and then pondered it a bit further. Did he then make a stark realization, and confront Generals Vandenberg, Twining, or Chidlaw? Did he then press for the Truth?

❧☙

General Cabell was director of Air Force Intelligence for almost two and a half years, and he withstood tremendous frustration and stress with the UFO question. His relentless attempt to find answers to flying objects didn't bode well for him, but blame for that can be placed at the feet of General Vandenberg and the Control Group. Cabell undoubtedly harbored great misgivings about his inability to accomplish his job, and there must have been considerable anguish about this, especially after finally realizing it was no fault of his own. Cabell, however, was well thought of, and nobody blamed him for any of the chaos that resulted from his determined efforts to obtain information, and his attempt to resolve the flying object mystery.

One might wonder how Cabell came to learn of Truth. Did he confront Vandenberg, Twining, or Chidlaw,? Or could it have been possible that one of them, in coordination with the Control Group, finally approached him to offer a conciliatory explanation, and then extend sincere appreciation for his unswerving dedication to his job?

On 1 November, Cabell would report to his new job as director of Joint Staff, and his assignment was to assist the chairman of the Joint Chiefs of Staff. This was a significant upgrade for him, along with promotion to Lieutenant General. It would provide great compensation for him, including a later subsequent promotion to four stars. Like Colonel Watson, this would not be the last heard of Cabell, or of his association with the UFO question.

CHAPTER THIRTY

New GRUDGE

Project GRUDGE: October - December 1951

Ruppelt in Charge

By the middle of October 1951, First Lieutenant Cummings departed ATIC and turned over his job to First Lieutenant Edward Ruppelt. Ruppelt was managing a small staff analyzing Russian technology and the MiG-15, but now he inherited additional work to investigate flying objects. Part of his effort, which Lieutenant Cummings previously started, would be to retrieve and sort through old sighting cases of Project SIGN, and attempt to put order to them for additional review.

Also, both Ruppelt and Second Lieutenant Henry Metscher set out to work on more recent sighting reports. Metscher began to sort through information and wrap up unfinished work by Cummings on the Monmouth incident, and Ruppelt began to look at the Lubbock Lights incident and several others coming in. These included a Long Beach, California, incident of a high flying object, an Albuquerque incident of a flying wing, sightings by General Mills Aeronautical Labs researchers of several objects, and an interesting sighting near Matador, Texas. They were significant sightings by competent people, and some of these sightings were described in the previous two chapters.

On 25 October, a letter was sent by ATIC to AFOIN requesting notification to all commands to send sighting reports directly to ATIC's Project GRUDGE at Wright-Patterson. There was a sense that various commands needed to be reminded of this, due to the strong denial of flying objects previously voiced at TID, and the fact that the last notification was more than a year previous. But Ruppelt and Metscher were also uneasy about how much they should advertise this. They could see plenty of work ahead without advertising for more.

In his 1956 book, *The Report on Unidentified Flying Objects*, Ruppelt stated that New GRUDGE was "officially established" on 27 October, although 22 October was specified on the Project's Initiation Form. This reorganized project had a new outlook and new operating plans, which is indicated in the following from Ruppelt:

> *I'd written the necessary letters and had received the necessary endorsements…and now I had the money to operate…I was given a very flexible operating policy for Project Grudge because no one knew the best way to track down UFO's. I had only one restriction and that was that I wouldn't have my people spending time doing a lot of wild speculating. Our job would be to analyze each and every UFO report and try to find what we believed to be an honest, unbiased answer. If we could not identify the reported object as being a balloon, meteor, planet, or one of half a hundred other common things that are sometimes called UFO's, we would mark the folder "Unknown" and file it in a special file. At some later date, when we built up enough of these "Unknown" reports, we'd study them. As long as I was chief of the UFO project, this was our basic rule. If anyone became anti-saucer and was no longer capable of making an unbiased evaluation of a report, out he went. Conversely anyone who became a believer was through.*

This statement by Ruppelt says much about the rules he would enforce on his project, and it provides great perspective on how his investigation of flying objects would be conducted. There would be no more debunking, but there would also be no more thought given to the extraterrestrial possibility. Conveniently, it was more in line with what the Control Group preferred, and also with what General Cabell related in his instructions to Colonel Dunn on 22 October. It was apparent that even if an alien flying object landed for inspection in ATIC's back yard, it would still be put into the "unknown" category." "Anyone who became a believer was through" were the words of Ruppelt. More than anyone else, however, those words would come to haunt him much sooner than he would expect.

Because of the previous ten months spent in ATIC, Ruppelt was now very cognizant of rules by which he needed to operate, and quite aware of instructions from General Cabell. He also knew of people around him who were absolutely convinced of the extraterrestrial answer, since he had talked with some of them, but he was also aware of those who were constantly debunking flying objects. There was a contrast between them. The non-believers were often without a broad perspective, or a personal investigative experience. He was also aware of former investigators that became deeply convinced after talking with witnesses and analyzing sightings. As a result, he knew he must tread a fine line, but he also needed to be very careful not to let his inner thoughts on the subject surface to show bias, which is what happened to other investigators that didn't fare too well.

Ruppelt must have also understood what he was up against regarding Colonel Glover's communication of 15 October, which declared that flying objects were not of foreign origin. That left one other alternative, especially if the objects were not of U.S. origin, or not balloons, meteors, planets, or other commonly misinterpreted things. Ruppelt could read between the lines that anything else would be considered "unknown" and would remain unknown, especially when he considered the demise of Project SIGN. But those unknown sighting cases would still be reported to higher headquarters and ADC. And it also meant that every conceivable effort would be made to identify flying objects with a "reasonable" label, and within a common "identified" category.

Sightings Catch ATIC Attention

ATIC was beginning to learn of new sightings, and many were quite important, but only the Monmouth incident was partially investigated by ATIC at this time. As early as 7 September, notice of the Lubbock Lights incident was received, which included photos. Also, about 17 September, AMC Commander General Edwin Rawlings became personally involved in requesting a thorough OSI investigation of that incident. As a result, a secret OSI spot investigative report was created and delivered on 8 October by Special Agent Howard Bossert of Reese AFB. He produced a very detailed and comprehensive review of what took place in the area of Lubbock, Texas, on 25 September, and for a period of time afterward.

On 11 October, Ruppelt was aware of the Minneapolis incident, and on 12 October, Lieutenant Metscher became aware of the Lubbock Lights, Albuquerque, and Matador incidents. This is when they both recognized that significant investigative action was in store for ATIC.

On 26 October, another copy of Bossert's Lubbock report, with photos, was forwarded by Wright-Patterson's OSI branch to ATIC. This must have signaled Ruppelt that he must get involved with the investigation, and he began to make plans to visit the Lubbock area. But he also needed to coordinate his trip with a visit to ADC at Ent AFB in Colorado.

One of Ruppelt's initial priorities was to touch base with ADC and coordinate with them on the flying object situation. Previously, on 22 October, General Cabell specified in a letter that ADC and ATIC had equal interest in flying objects, although ATIC had investigative responsibility. ADC would need to be advised of flying object investigative activity by ATIC. Of course, General Chidlaw at ADC was extremely interested in flying objects because he was previously in command over Colonel Watson at TID. This was in addition to the fact that Chidlaw's ADC pilots were now on the front line in dealing with flying objects, and that Chidlaw was probably in touch with Cabell before Cabell departed AFOIN.

General John Samford Takes Over From Cabell

On 1 November, General John Samford assumed command of AFOIN, replacing General Cabell. Several years previous, Samford was attached to AFOIN as deputy to the assistant chief of Air Targets, but his most recent assignment was head of Air War College, and also head of the Air Command and Staff School before that. He was now thrown into a position requiring great leadership ability in dealing with the Korean War, and with other matters that required special attention, including the UFO question. Previously, on 11 October, Samford attended an Intelligence Advisory Committee meeting at the CIA with General Cabell, and also with Colonel Edward Porter who was a standing member of the committee. It is assumed that Samford stepped into his new position properly prepared and adequately briefed by Cabell.

U.S. Detonates Hydrogen Bomb

After the Soviet Union exploded its first atomic bomb in September 1949, which occurred much sooner than the United States expected, President Truman announced on 31 January 1950 that the United States would proceed in developing a thermonuclear device. Twenty-two months later, on 1 November, which was the same day Samford took over from Cabell, the United States detonated the world's first Hydrogen Bomb. It was a 10.4 megaton device named "Mike Shot," and it was the first in a series of blasts called Operation Ivy. This particular blast obliterated Elugelab Island in the Eniwetok Atoll of the Pacific Marshall Islands. It created a fireball about two miles wide, and left a crater more than a mile wide. The tremendous flash of light produced by this device was easily seen from outer space.

Ruppelt Visits ADC and Lubbock

On 5 November, Ruppelt combined his visit to ADC Headquarters with his travel to Lubbock, Texas, for investigation of the Lubbock Lights incident. At ADC, Ruppelt met with General Woodbury Burgess, General Chidlaw's deputy, to "coordinate work on the sightings and investigations of unconventional aircraft," which was instigated by Cabell's previous letter indicating that ADC had equal interest in flying objects. Ruppelt found that ADC had no program set up for investigation of sightings, but ADC was "much interested in investigations conducted by ATIC." ADC was prepared to "cooperate to the fullest extent" and requested that ATIC prepare an outline of steps to be taken in reporting of sightings. It was also requested that ATIC send monthly Status Reports to ADC.

After arriving on 6 November in Lubbock, Texas, Ruppelt was met by OSI Special Agent Howard Bossert at Reese AFB, and then they met with the four professors who initially reported the Lubbock Lights described in Chapter 28. Over the course of the next few days, Ruppelt talked with many people, including the amateur photographer of the "lights," the head of the local newspaper, and others. After gathering a mountain of information, he departed, and eventually proclaimed his conclusion in a pending report that the sightings at Lubbock were simply birds. But it was probably about this time, or shortly after, that he must have had his eyes opened considerably to an alternative reality.

Ruppelt - Playing the Game

Prior to visiting Lubbock, Ruppelt learned of the flying wing incident at Albuquerque, New Mexico, and then noted its similarity to the amateur photo taken of the Lubbock Lights. He commented about the photo as follows:

> …*the similarity to the Albuquerque sighting, both in the description of the object and the time that it was seen, was truly amazing.*

Later, Ruppelt mentioned in his book an even more amazing sequel to the Albuquerque flying wing sighting, which occurred as he departed on the morning of 9 July from Lubbock. He was seated on the plane next to a rancher, and they were talking about the meteor that flashed over Lubbock that night. Then, without Ruppelt mentioning who he was, the rancher told him what his wife had seen on the night of 25 August, the same evening the professors made their first sighting. According to Ruppelt, the following is what the rancher stated:

> He [the rancher] *was inside the house reading the paper. Suddenly his wife rushed into the house…this was about ten minutes before the professors made their first sighting…she had seen a large object glide swiftly and silently over the house…it looked like "an airplane without a body"…on the back edge of the wing were pairs of glowing bluish lights…It hit me* [Ruppelt] *right between the eyes. I knew the rancher and his wife couldn't have possibly heard the Albuquerque couple's story, only they and a few Air Force people knew about it.*

For Ruppelt, it was an amazing coincidence for him to sit next to a stranger on the plane, and then have something almost identical to the Albuquerque incident relayed back to him. But there was much more to this flying wing episode than photos from Lubbock, the Albuquerque incident, and this story from the rancher. There was additional information regarding the Lubbock incident, which was brought out in OSI interviews, documented in files, and previously described in Chapter 28. One incident involved the professors at Lubbock in what they called "Heneman's Horror," and the other was an incident one professor told regarding a sighting by his wife of a flying wing. Altogether, there were five independent sightings of a flying wing during the same general period of time. But neither of these other incidents was mentioned in Ruppelt's book, which leads one to contemplate a certain "aura" about Ruppelt, especially when considering many additional "polarities" in his book.

As previously mentioned, Ruppelt stated that the answer to the Lubbock Lights was "birds." This rationale, however, is based on the following statement he provided regarding the Lubbock Lights, which seems a bit odd and contradictory:

> *The best thing I could do, I decided, was to treat each sighting in the Lubbock Light series as a separate incident. All of them seemed to be dependent upon each other for importance. If the objects that were reported in several of the incidents could be identified, the rest would merely become average UFO reports.*

One must remember that Ruppelt established a caveat that anyone who became a believer was finished, and this meant that flying objects would be evaluated according to "half a hundred other things" that might conceivably be possible. Because of the fact that the sightings by various people, over a period of several weeks, identified the objects as groups of birds, Ruppelt felt he had a good enough answer for the Lubbock Lights, which was preferable to further acknowledging the amazing situation that had hit him "right between the eyes." Continued dwelling on the prospect of what he really believed would spell trouble for him, and for the project he was in charge of. There was no way he could allow that, but in all good conscience, he knew the "hidden dichotomy" he was creating. It's not certain he specifically understood the entire Truth of the UFO question, or of the Lubbock Lights, but he knew much more than one would generally suspect, and he knew how to play the "game!" He would continue to play the game to the end of his days, but it is entirely possible that Ruppelt was sending a cryptic message to anyone reading his book regarding Truth of the UFO question.

There is much more to be said regarding the Lubbock Lights incident. When one probes more deeply, it continues to progress into an ever more amazing story that involves government officials obsessed with extreme interest and concern. It confirms that great attention was paid to the UFO question behind the scenes, and that the Control Group, or CIA-OSI, was involved.

Control Group Involved!

Thanks to Ruppelt, his book reveals distinctly important and very telling information regarding the fact that a high level government entity was extremely interested in flying object files at ATIC. He related this in connection with the Lubbock Lights, and he stated the following:

> The only other people outside Project Blue Book who have studied the complete case of the Lubbock Lights were a group who, due to their associations with the government, had complete access to our files. And these people were not pulp writers or wide-eyed fanatics, they were scientists—rocket experts, nuclear physicists, and intelligence experts. They had banded together to study our UFO reports because they were convinced that some of the UFO's that were being reported were interplanetary spaceships and the Lubbock series was one of these reports. The fact that the formations of lights were in different shapes didn't bother them; in fact, it convinced them all the more that their ideas of how a spaceship might operate were correct... This group of scientists believed that the spaceships, or at least the part of the spaceship that came relatively close to the earth, would have to have a highly swept-back wing configuration... (Three years later, the Canadian Government announced that this was exactly the way they had planned to control the flying saucer that they were trying to build. They had to give up their plans for the development of the saucer-like craft, but now the project has been taken over by the U.S. Air Force.)

Other than the top secret memo by Dr. Wilbert Smith released by the Canadian government, and reviewed in Chapter 26, this is one of the most revealing and significant statements ever published in regard to U.S. involvement with the UFO question. It is presented in Ruppelt's book to a naïve public without a hint that it must have been highly classified information. It stated that a "group" associated with the government was "convinced" that some of the flying objects being reported "were interplanetary spaceships." This statement is buried within Chapter 8 of Ruppelt's original book, on pages 109 and 110, and it cannot be read any other way. It brings back recollection of a meeting Ruppelt had in Los Alamos with a group of scientists (see Chapter 23).

Based on what has been reviewed so far, the group referred to in Ruppelt's book was likely a part of the Control Group, and either associated with the RDB or CIA's Office of Scientific Investigation. The group Ruppelt referred to included "intelligence" experts, and it is tempting to believe that this group was part of the CIA-OSI, although it cannot be confirmed. It is known, however, that the CIA was deeply involved with flying object reports at this time, and it was also connected to U.S. and Canadian government efforts in researching the Unified Field Theory in the application of powering and controlling flying objects.

The truly amazing thing about Ruppelt's statement is the mention of Air Force plans to build a flying object in the future. Just how Ruppelt would come to learn of this, or how he would even have the capability, or find courage enough to publish this is hard to imagine. His book was reviewed by the Air Force prior to publication, but somehow his statement must have been ignored or overlooked. And it would appear that his statement was surreptitiously and intentionally included for "unknown" reasons.

One might say that insertion of Ruppelt's amazing statement was rationalized because it was thought to be "contextually relevant" to the Lubbock Lights discussion in his book, but one might question whether this was the "actual" reason for including it. It did not factually enhance the conclusion that the Lubbock Lights were flying "birds," but just the opposite. It stands out in contrast to surrounding dialog, and the information could easily be considered top secret. It shows that Ruppelt was privy to information far beyond what he might normally have been authorized. Did he know far more than he is given credit for in regard to the UFO question, and was his statement a subsequent "protest" for having to play a continuing "game" with the Air Force after he left active duty and began to write his book? Is that why he gave an additional mysterious answer to the Lubbock Lights by stating "they weren't birds," but an "easily explainable natural phenomenon" that he "can't divulge?" Is this why he later gave an additional absurd conclusion in a later addition to his book stating that the Lubbock Lights were possibly "moths" lighted by streetlights, which then forces one to attempt visualization of a "V formation" of moths! Although it might "appear" Ruppelt attempted to take appropriate care in providing a balanced presentation about his role in New GRUDGE, and attempted to convince readers of his book that he was neutral on the UFO question, there is also a sense that there was another kind of message intended by him.

More Green Fireballs

Dr. Lincoln LaPaz, of the University of New Mexico, made mention that bright Green Fireballs, in "amazing numbers," were sighted over New Mexico on four days in October and eleven days in November, with nothing recovered from the ground. Several, however, were seen to explode in a shattering disintegration of red sparks at the end of their flight. LaPaz indicated that the sightings created quite a stir in New Mexico, and it was possible "that they were either U.S. or foreign guided missiles." A spokesman at White Sands, however, confirmed there were no U.S. rocket launches on the dates reported, and this left only one other possibility.

Green Fireballs were quite unlike celestial meteors, including the fact that they only seemed to target one general location on Earth. This caused LaPaz to believe the fireballs were a man-made phenomenon. He stated that, "Their peculiar trajectories did not permit them to be classed as natural phenomena," and he said that his data on the fireballs was available to the military, but no specific military inquiries were ever made to obtain information. He also referenced fireballs observed in 1946 over Sweden and the Baltic, and suggested they were test firings of Russian guided missiles. He said this circumstance "should not be overlooked."

Project Twinkle, which was initiated to study and make observations of Green Fireballs, had been discontinued on 31 March, although some thought it worthwhile to continue the project longer. Project Twinkle observation posts were never able to sight a Green Fireball, although they did discover disk-shaped objects cavorting near research balloons on many occasions. Perhaps the curtailment of the project had much to do with the fact that origin of fireballs was figured out by the U.S. government, and that periodic sightings of objects near the balloons by Project Twinkle needed to be stopped. These "unofficial" object sightings by Project Twinkle, however, were never an "official" subject of investigative interest to ATIC.

Final Report - Project Twinkle

On 27 November, the "Final Report" for Project Twinkle was published, which was long after the project ended, and after confirmed completion of the contract with Land-Air, Inc. As previously mentioned, the Final Report was a gloss over of a poorly run and ineffective investigation that failed to reveal Green Fireballs. More importantly, however, many sightings were made with Askania instrumentation of disk-shaped flying objects cavorting with research balloons, and these were considered "sufficiently significant," but this was not enough to dissuade the negative result given to the project. The Final Report did suggest that "a time correlation study should be made covering the film and verbal recordings at both Askania stations," and this was an indirect reference to movies and observer comments that were recorded during a sighting of the disk-shaped objects, although those objects were not specifically mentioned or identified in the report. The whole scenario regarding Green Fireballs, and Project Twinkle, indicates cover-up of the UFO question, and also possible Soviet mischief.

Status Report No. 1

On 30 November, "Project Grudge - Report No. 1" was produced by New GRUDGE. This report, and others to follow, was intended to keep specific people informed at the end of each month, and this would include those in Air Force Intelligence, and also General Chidlaw at ADC. The report, like reports to follow, would limit included sightings to those originating from "high grade sources such as pilots, and technically trained people." This also meant that other quality sighting reports were not given the same consideration, or thoroughly investigated. Thirty-six incidents were listed, from 25 August to 7 November, and ten were considered "outstanding" and worthy for summarization in appendices. They included the Lubbock Lights, Albuquerque, Matador, Monmouth, Long Beach, and Minnesota incidents. The Lubbock Lights and Albuquerque incidents continued to remain on hold awaiting results of photo analysis. The Monmouth incident was to be further documented in a "Special Report." Other sighting reports were filed away. This type of Status Report is what the Control Group was waiting for, and there is little doubt that serious consideration was given to incidents summarized in appendices of the report.

Ruppelt Visits Air Force Headquarters

On about 11 December, Ruppelt and General Dunn visited Air Force Headquarters to meet with General Samford, General Garland, and others to review overall status of New GRUDGE. Dunn explained the basic operation of the project, and indicated that sighting cases investigated and classified as "unknown" were being set aside without speculation of flying object origin. Garland revealed that the Air Force had sole responsibility for flying object investigation, and that ATIC "was responsible for all [investigation of] UFO reports made by any branch of the military service." It was also mentioned that there were no U.S. secret projects that could be mistaken for flying objects. Ruppelt reported on the current flying object situation, and he stated:

> ...within the past few months the number of good reports had increased sharply...were seen more frequently around areas vital to the defense of the United States. The Los Alamos-Albuquerque area, Oak Ridge, and White Sands Proving Ground rated high. Port areas, Strategic Air Command bases, and industrial areas ranked next. UFO's had been reported from every state in the Union and from every foreign country.

Ruppelt also stated that the previous final GRUDGE Report was not a solution to the flying object problem, and that there was a "residue" of very good, but unexplainable sightings that were classified as unknown. This, of course, is what Major Keyhoe previously pointed out. Ruppelt said that "all recommendations for the reorganization of Project GRUDGE were based solely upon the fact that there were many incredible reports of UFO's from many very reliable people." A great many of those were from the military. This was in concurrence with what General Cabell previously highlighted in promoting his new charter for ATIC.

Sighting by Donald "Deke" Slayton

On 12 December, Captain Donald Slayton was piloting an F-51 near Hastings, Minnesota, when he spotted a unique and curious object, which was subsequently reported to ATIC. His famous aerospace exploits in followings years confirms his credibility, and he described his incident as follows:

> *As I closed on this thing…it looked like a weather balloon, and that's what I presumed it was…so I decided I would just come back around and make a pass on it…I got around where I should have been coming back on this thing, all of a sudden it didn't look like a balloon anymore. It looked like a saucer sitting on edge, about a forty-five degree angle. I didn't have any gun camera film on board unfortunately, or I would have shot some pictures of it…it took off climbin at about a forty-five degree angle and just accelerated, and disappeared…*

Slayton flew fifty-six combat missions in Europe during WWII, and he would later become one of the original seven astronauts for NASA's Mercury space program. He then served as NASA's director of Flight Crew Operations, and was a docking module pilot for the Apollo-Soyuz Test Project in 1975. This was a man who was considered to be completely competent and reliable.

"Civilian Saucer Investigations" Organized

Public interest in flying objects began in 1947, and it was inevitable that clubs or organizations would be created at some point in time by those having an interest in the objects. A few were formed within the last couple of years, and now in December, the Civilian Saucer Investigations (CSI) of Los Angeles was founded by Edward Sullivan, who was a North American Aviation employee. He gathered together some of his fellow scientists and engineers at the company, including Dr. Walther Reidel, who was a former WWII German rocket scientist brought to America under Operation Paper Clip. Reidel was a chief designer and research director at Peenemunde, Germany, on the "V" rockets with Warner Von Braun, who ultimately put American astronauts on the moon with his NASA rocket.

In referring to flying objects, Reidel claimed he was "completely convinced that they have an out-of-world basis," and he said that "there were reports of them over Peenmunde" during the war. In time, more independent organizations would be created and participate in research and gathering of data on flying object sightings and encounters.

AFOIN Letter on Sighting Reports

On 19 December, AFOIN released a letter to all commands titled "Reporting Information on Unidentified Flying Objects." This letter was similar to the one issued by General Cabell on 8 September 1950, which outlined procedures and instructions in handling and forwarding sighting reports. This letter was in response to ATIC's request sent to AFOIN on 25 October, and it was likely discussed in the meeting Dunn and Ruppelt just previously attended with General Samford. It was determined then that all commands needed an additional reminder by letter regarding how to send sighting reports to ATIC

Ruppelt Visits Battelle Memorial Institute

On 26 December, Ruppelt and his commanding officer, Colonel Kirkland, took a trip to Columbus, Ohio, to meet for a couple of days with experts from Battelle Memorial Institute. Battelle was the organization involved with Project Stork, which New GRUDGE depended on to make a statistical study of the flying object situation. They would undertake two studies. One would come up with an interrogation form that would be used in the questioning of sighting witnesses, and the other would provide a system to gather statistical studies from the form. Ruppelt stated that:

> ...when we received a report we could put the characteristics of the reported UFO on an IBM punch card, put it into the IBM machine, and compare it with the characteristics of other sightings that had known solutions.

Special Report No. 1

On 28 December, the first "Project Grudge - Special Report No. 1" was produced to provide investigation results on the Monmouth incident. The objective of this report, and other Special Reports to follow, was to respond to higher authority with a review, results, and conclusions about a specific outstanding incident. In this case, the report was in response to General Cabell's previous request for a report on the Monmouth incident. The conclusion stated for this incident was that the flying object reported by the T-33 pilots "was probably a balloon," and that the radar operator was likely confused by either "anomalous propagation," his own "excitement," or "inability to use aided tracking" with radar controls.

Status Report No. 2

Released on 31 December, and classified secret, "Project Grudge - Report No. 2" provided the idea that New GRUDGE was back in business with a smoothly running project. It stated that much time had been spent in sorting and filing old sighting cases, which dated back to 1946, except for missing records from Watson's tenure. Summary cards were created for each incident, and it was intended that the cards would be "cross-indexed in the attempt to obtain characteristics or trends in object sightings," which was a small part of the Project Stork contract now ending, but to be extended with a new contract. The new contract was previously discussed on 26 December by Ruppelt and Colonel Kirkland in the two-day meeting with members of the Battelle Memorial Institute.

Within this second status report, new cases were cited, and old cases were either "closed" or "continued" pending further investigation. Some incidents that were closed were either given a "probable" or "reasonable" identity to the flying object, or "no conclusion." In the case of the Minnesota sighting of 11 October, the resolution was stated as follows:

> *The object definitely seemed to be controlled. The sources are all experienced engineers with General Mills Balloon Projects and have been observing all types of balloons for several years. No conclusions can be made. It is significant however, that the sources can be graded as very reliable and that they observed an object with which they were entirely unfamiliar.*

The flying object General Mills engineers saw and reported was definitely not the balloon they were tracking. When one reads the above statement, it causes one to take notice. It sends a signal that this particular case, with "no conclusion," might be very significant. Cases were often put away with "no conclusion," especially if they were considered very important or "outstanding" with no reasonable or rational answer possible. To provide a prosaic answer in this case would raise questions about the ability of investigators to do their job as required. This kind of sighting is what higher authority was most interested in, but these cases would be filed in the "unknown" category, which would seem to equate them to being "unworthy." It brings to mind the "unknowns" in the previous final GRUDGE Report by Colonel Watson, which were actually very important cases reported by credible and reliable people.

꙳꙳

Establishment of New GRUDGE was the final effort by General Cabell to put investigation of flying objects on a good footing before departing for his new assignment. In the end, he finally understood the Truth regarding flying objects, which he signaled by providing "instructions" for the operation of New GRUDGE. And this was confirmed by Lieutenant Edward Ruppelt when he specified how his investigations would be conducted. There would be no speculation regarding origin of flying objects, and if a rational or reasonable identification was not possible in a sighting case, there would be "no conclusion." The case would be filed in the "unknown" category. This was acceptable to the Control Group, and it was the only way Ruppelt could manage his investigative unit without getting into trouble. It gives reason to conclude that Ruppelt understood what the "unknowns" actually were. But this also indicates that the Control Group was also now willing to let ATIC proceed with investigation of flying objects, as long as the "ground rules" were understood.

Ruppelt knew that many flying object cases from the past were very mysterious, and that they demonstrated unearthly characteristics not possible or producible by human technology, and this gave cause for previous T-2 and TID investigators to assign or suggest an extraterrestrial origin for some cases. Ruppelt was also familiar with some of those investigators who were strong "believers," and he knew he must remain neutral on the UFO question. Very quickly, however, he also learned that maintaining outward neutrality would be a challenge, and a tough road to follow, although his escape route was the use of "no conclusion" or the "unknown" category as a convenient refuge for the assignment of cases. Still, where given a chance, it was preferable to assign cases to a "reasonable" identified status where possible.

Without acknowledging the Truth, Ruppelt must have internalized the reality of Truth to a great extent, especially after being exposed to coincidences of the flying wing in the Lubbock Lights case. He was also exposed to a very special group of highly respected and expert government scientists, engineers, and intelligence people visiting ATIC, who were convinced that some of the reported flying objects were "interplanetary spaceships!"

This must have greatly impacted Ruppelt early on. He was suddenly introduced to "playing the game" in New GRUDGE, and he may have relished and disdained it at the same time. He was now at the forefront of a monumental secret, and he was the one at the apex of monitoring and "controlling" the "game." It was kind of scary, but also quite exhilarating!

CHAPTER THIRTY-ONE

End of Project GRUDGE

Project GRUDGE: January - March 1952

Garland's Secret Memo to Samford

The year 1952 would begin with a new outlook on the UFO question for AFOIN and ATIC, and it would prove to be a very big year for the phenomena of flying objects, although Lieutenant Ruppelt, General Garland, and General Samford had no idea what they were in for, and neither did the Control Group.

On 3 January, General Garland wrote a secret memo to his boss General Samford, AFOIN's new director of intelligence, and it was titled "Contemplated Action to Determine the Nature and Origin of the Phenomena Connected with the Reports of Unusual Flying Objects." It stated:

> *...the persistent reports of unusual flying objects over parts of the United States, particularly the east and west coast and in the vicinity of the atomic energy production and testing facilities, it is apparent that positive action must be taken to determine the nature of the objects and, if possible, their origin...it is considered mandatory that the Air Force take positive action at once to definitely determine the nature and, if possible, the origin of the reported unusual flying objects.*

It is not certain just what motivated Garland to write this memo, but he was suddenly concerned with flying objects observed over the East and West coast of the United States, and also over nuclear facilities. He also included in his memo that he was extremely concerned that the Soviets were far ahead of the United

States in ability to deliver nuclear weapons, primarily because of assistance from German scientists since WWII. Army intelligence efforts had already discounted the German/Horton possibility, but it was also stated in a 15 October letter a few months previous, by Colonel Glover in AFOIN, that flying objects did not originate from a foreign power. It is possible, however, that the reason for Garland's memo was due to being warned about Green Fireballs, perhaps by Dr. Joseph Kaplan of the AFSAB, or Dr. Anthony Mirarchi, or maybe Dr. Lincoln LaPaz. All were quite certain that fireballs were of Soviet origin, but most sightings of the objects were centered in the New Mexico area. Another possibility is that Ruppelt may have frightened Garland by stating in the last December meeting with Garland that many of the flying objects "were seen more frequently around areas vital to defense of the United States." Whatever it was that sparked Garland's concern, there was a definite interest about flying objects exhibiting a potential threat, and he needed his confidence restored that Russians were not involved, and that they did not possess advanced technology to deliver nuclear weapons.

Garland's memo went on to outline a plan of action for ATIC and ADC to set up three teams for the purpose of taking "radar scope photographs," and also "visual photographs," in the Seattle, Albuquerque, and New York/Philadelphia areas. The idea was to document object sightings, and also to determine if atmospheric inversions, or other effects, were able to impact radar detection of flying objects. Although Garland's plan made sense to him, nothing ever came of it.

The real significance of Garland's memo is the fact that the Air Force, at least Air Force intelligence, was very concerned about flying objects, which were now considered quite real and possibly threatening. It's also quite possible that high Air Force and government officials were very concerned about potential Soviet involvement with Green Fireballs, and the fact that there was virtually no capability for early detection of a Soviet air invasion. For national security reasons, it was imperative that a detection system be developed. All this was in sharp contrast to public perception that the Air Force was totally unconcerned with flying objects. Of course, the public was more than a little suspicious of Air Force motivations or intentions at this time regarding flying objects, and this memo serves to document that the public had a right to be.

Ruppelt Visits General Mills Engineers

Ruppelt was quite aware of Skyhook balloons used in various research projects, and he was particularly aware of the Minneapolis incident on 11 October 1951, when General Mills engineers witnessed flying objects cavorting with their balloon. He recently investigated and summarized the incident in December's Status Report.

On 14 January, he and physicist Mr. Armstrong from Battelle Memorial

Institute, who was working with Project Stork, took the opportunity to visit the Aeronautical Division of General Mills. It was a meeting with engineers and scientists who were intimately familiar with flying object sightings in connection with their balloons. It will be recalled from Chapter 20 that one of them was Charles Moore, who was disgusted by how his sighting in April 1949 was handled by the Air Force. Ruppelt was now able to talk with Moore directly.

Since the time of Moore's sighting, and until recently, the majority of sighting reports from General Mills people were held in strict confidence and kept away from the Air Force. Now, in somewhat unfriendly territory, Ruppelt was able to talk with many witnesses for almost a full day, and he quickly learned that any attempt to infer a "natural phenomena" explanation for the objects was asking for trouble.

People he talked with were intimately familiar with their research balloons, and also with the antics of the many flying objects sighted near them. Ruppelt could not help but recall, from a year previous, Dr. Liddell's remark that "there was not a single reliable UFO report that couldn't be attributed to a Skyhook balloon." Now he was conversing with those who knew differently. Ruppelt would later state:

> *One man told me that one tracking crew had seen so many that the sight of a UFO no longer even especially interested them. And the things they saw couldn't be explained.*

After departing General Mills, it is quite likely Ruppelt was again presented with another enlightening experience that left his eyes wide open. It confirmed to him that the 11 October case was properly diagnosed with "no conclusion," and properly filed as "unknown." In another sense, it also provided affirmation that flying objects were very real. He was becoming well-indoctrinated on that, and it must have made an impression on him. But many more incidents were yet to come for him to verify and confirm true reality of flying objects.

Aerial Phenomena Research Organization

After formation of the CSI organization last month, another independent organization was now created with the purpose of studying flying objects. It was called the Aerial Phenomena Research Organization (ARPO), started by Jim and Coral Lorensen of Sturgeon Bay, Wisconsin, with Coral its designated leader. The purpose of the organization was to advocate for objective scientific investigation of the mysterious flying objects, with the goal of determining answers to their nature and origin. It would later grow into a very large and respected worldwide organization.

The Mitchel AFB Incident

Previously in September 1951, a T-33 jet piloted by Lieutenant Wilbert Rogers, landed at Mitchel AFB in New York after he and his passenger sighted a flying object near Sandy Hook, New Jersey, which was associated with the Monmouth incident described in Chapter 28. Now, on the morning of 21 January, Navy pilot Lieutenant James Zeitvogel took off from Mitchel AFB in a Grumman TBM Avenger torpedo bomber and he sighted an object between him and the base when he was about three miles out to the southeast. At first, he thought the object was a parachute canopy, and then a balloon, but discovered it was neither when he approached to within 2,000 feet of it. The object then began to circle the base counterclockwise at about 300 miles per hour. As he gave chase, the object straightened out, climbed rapidly, and quickly disappeared about seven miles southwest of the base. But it "did not gradually fade away in the distance, it just disappeared."

Within a short time, Ruppelt was on a train to Mitchel AFB to investigate and interview Zeitvogel. Later, Ruppelt commented on the incident by stating the following:

> *I think the pilot summed up the situation very aptly when he told me, "I don't know what it was, but I've never seen anything like it before or since—maybe it was a spaceship" I went back to Dayton [ATIC] stumped—maybe it was a spaceship… The biggest argument against the object's being a balloon was the fact that the pilot pulled in behind it; it was directly off the nose of his airplane, and although he followed it for more than a minute, it pulled away from him. Once you line up an airplane on a balloon and go straight toward it you will catch it in a matter of seconds, even in the slowest airplane.*

Again, Ruppelt was in personal contact with a very reputable and competent individual. It was another incident for him to think about in regard to reality of flying objects. He was being inundated with Truth, but his education was just beginning.

Progress Report to Garland

On 29 January, a little more than a month after his last visit, Ruppelt again arrived at Air Force Headquarters to present current ATIC status to General Garland. He pointed out that the directive of 19 December 1951 required further updating for reporting of flying objects. He also mentioned that several interviews with pilots revealed that "they would be very reluctant to report any type of unidentified object to the Air Force." This was because of stigma attached to such reporting, and it's quite possible that Ruppelt was told this by Lieutenant Zeitvogel, who was involved with the Mitchel AFB incident. Ruppelt also reported that Project Stork consultants had finished a preliminary survey of ATIC sighting reports, and that a formal proposal to ATIC would be submitted in February. Garland also provided feedback that "nothing definite has been decided" on his plan to set up radar and photo equipment at several locations to detect flying objects (or Green Fireballs). This would eventually disappear from Garland's priority list.

While at Air Force Headquarters, Ruppelt also discovered that TCB and other areas in AFOIN were in possession of many additional sighting files that ATIC didn't have. A future trip would be planned to review and make copies of them so that ATIC files would be more complete, and they would also be made available to Project Stork.

Flying Object Sightings Over Korea

On the night of 29/30 January, crews of two B-29s sighted what appeared to be the same flying object. They were flying over widely separate regions of Korea, about twenty-five minutes flying time from each other, and they individually observed and described a spinning disc-shaped object having a light orange color resembling the sun, and with an occasional bluish tint. They further stated that "the outer edge of the object appeared to be fuzzy and it seemed to have an internal churning movement like flames or fiery gases." One plane reported the object flew parallel to the aircraft on the left, side about 600 feet away, and it remained in sight for about five minutes. The other plane reported the object flew by its side for about a minute. All the witnesses "emphatically" stated the object "bore no resemblance" to anything previously experienced.

Many classified reports on these two sightings were transmitted back and forth in days following, between the Far East Air Force (FEAF) Bomber Command, Air Force Headquarters, ADC, and ATIC. If there was ever a time when the Air Force might choose to loosen up on negative talk about pilots and air crews, this was the opportune time to do so without making them appear to be "kooks." And if the public were to learn of such sightings, it was now the time for the Air Force to be the first in releasing information.

Status Report No. 3

On 31 January, "Project Grudge - Report No. 3" was issued, and its contents were quite revealing. It mentioned the difficulty in obtaining flying object sighting reports from pilots due to the stigma attached, which was the same thing Ruppelt previously stated in his 29 January briefing to Garland. This situation was very likely due to Colonel Watson's previous critical debunking of witnesses to flying objects. It quoted one pilot with the following:

> *If a spaceship flew-wing tip to wing-tip formation with me, I would not report it.*

The report also stated:

> *This feeling among people who are in a position to submit good reports is a great handicap to the objective of getting reliable data. The exact nature of some of the objects reported have not been determined, therefore, there is always the possibility that there exists some type of unconventional vehicle possessing extraordinary performance and characteristics. If such a vehicle should appear, its detection would be hampered by the reluctance to report sightings of unusual aerial objects."*

At this time, and even earlier, pilots and many credible people were learning that a negative label could be attached to them if they spoke out on a flying object incident. This continues to the present day, especially for airline pilots, and also for some missileers I know. They are very reluctant to speak out for fear of being labeled a "kook," or being subjected to derogatory comments by the media, skeptics, and others. This has created plenty of reason to remain quiet. Also, because of projected disrespect, they fear possible loss of their jobs or other negative ramifications for speaking out. Those who previously signed an oath of silence with the Air Force have additional reason for not speaking out, even if they may want to.

The January Status Report also provided status on fifteen previously pending, and current sighting reports. The Albuquerque, Matador, and Minneapolis sightings were listed with "no conclusion," which infers they were very significant sightings. It was also mentioned that the Lubbock Lights incident would be published in a special report. Some of the other sightings were also listed with "no conclusion," and some were awaiting further information, or simply closed with no investigation.

Major Dewey Fournet Joins TCB

At the beginning of February, Major Dewey Fournet entered the picture and replaced Lieutenant Colonel Milton Willis of TCB in the Evaluation Division of AFOIN. He reported to Chief of TCB Colonel John Ericksen, and was assigned to be an interface and "Project Monitor" to ATIC. He quickly became interested in the Korean flying object incident, which was grabbing attention at Air Force

Headquarters at the time. He would become a primary interface to Ruppelt, and his duties allowed him to function as a branch office for Ruppelt at Air Force Headquarters. Fournet would soon become a very significant addition to the Directorate of Intelligence, and he would soon be involved in a major flying object incident with Albert Chop.

Green Fireball Sighted

Very early on the morning of 17 February, a Green Fireball was briefly sighted by B-29 pilot Captain Calvin Parker, and his passenger Lieutenant Colonel William Riggs. They were flying about twenty-five miles southeast of Walker AFB (Roswell, New Mexico), when a "ball of green-blue fire light" passed in front of their aircraft about one mile away. It was approximately three feet in diameter, with a tail of fifteen to twenty feet. Observed at an altitude of about 15,000 feet, the fireball stayed in sight for about two seconds, and followed a flat straight-line trajectory. Even through Project Twinkle was discontinued long ago, fireballs continued to make periodic appearances in the New Mexico area, but ATIC simply filed the reported sightings away.

Reclassification of Project Twinkle Requested

On 19 February, Chief of Research Division Dr. Albert Lombard of Air Force Research and Development sent a request to TCB Chief Colonel Ericksen for reclassification of Project Twinkle. The letter also stated the following:

> *"The Scientific Advisory Board Secretariat has suggested that this project not be declassified for a variety of reasons, chief among which is that no scientific explanation for any of the "fireballs" and other phenomena was revealed by the* [final] *report and that some reputable scientists still believe that the observed phenomena are man-made."*

Lombard's letter was instigated because of another internal letter dated 14 January regarding reclassification of the project. Lombard was now looking for help from TCB at the Air Force Directorate of Intelligence because the AFSAB was suggesting that the project not be declassified. Apparently, there was a high security classification attached to Project Twinkle, and now there was confusion on whether the classification needed to be kept in force.

Previously, on 10 February, Dr. Lincoln LaPaz wrote an article in a publication regarding his investigation of Green Fireballs, which caught the attention of Colonel Joseph Caldera, who was assigned to the office of the Joint Chiefs. Caldera thought he had seen one of the fireballs, and he approached Major Fournet to inform him of his sighting. Fournet, who was new to all of this, then queried Ruppelt on what the fireballs were all about, and Ruppelt stated:

> *ATIC has been attempting to get further info on what conclusions were reached by the AF Cambridge Research Laboratory which made a study of this phenomena. They have submitted a report to the Directorate of Research and Development in Washington, but we have been unable to get a copy of the report.*

It is quite conceivable that much more was involved with Project Twinkle than has been made evident. As with anything classified, only those with a need-to-know were privileged to understand why a report with negative results, "no scientific explanation," and no conclusion would remain classified, or why a possible "man-made" phenomenon was enough to maintain the classification. But it does leave one to speculate that the Soviets were the ones targeting the New Mexico area with Green Fireballs, which resulted in a sighting just two days previous.

Ruppelt Provides Briefings

Previously, after the 1 October 1951 contentious meeting with General Cabell, when he indicated that a new direction would be taken with Project GRUDGE, Ruppelt invited a number of scientists and engineers to ATIC to discuss a new direction for the project, and get feedback. He later remarked on this as follows:

> *For the next two weeks every visitor to ATIC who had a reputation as a scientist, engineer, or scholar got a UFO briefing…Unfortunately the names of these people cannot be revealed because I promised them complete anonymity…I found out that UFO's were being freely and seriously discussed in scientific circles. The majority of the visitors thought that the Air Force had goofed on previous projects and were very happy to find out that the project was being re-established.*

One might wonder if any of those visiting ATIC at that time were in-the-know, part of the CIA, or part of the Control Group. This is quite possible, since it was critically necessary for them to learn about the new direction ATIC was taking.

But now, in the later part of February, Ruppelt was also involved in discussions with General Chidlaw, General Woodbury Burgess, and Major Verne Sadowski at ADC, and he received full cooperation from them with plans for New GRUDGE, and plans for working together. It is recalled that Chidlaw (and possibly Burgess) was in-the-know on Roswell. In regard to these discussions, Ruppelt stated:

This briefing started a long period of close co-operation between Project GRUDGE and ADC, and it was a pleasure to work with these people. None of them were believers in flying saucers, but they recognized the fact that UFO reports were a problem that must be considered. With the technological progress what it is, you can't afford to have anything in the air that you can't identify, be it balloons, meteors, planets or flying saucers.

Korean Sightings Make News

The flying object sightings by B-29 crews over Korea on the night of 29/30 January were now making headlines. Contrary to the past, the Air Force was now speaking out quite openly on this matter with the press, and this would promote renewed public interest in flying objects. A United Press article on 19 February stated:

The Air Force disclosed today that objects resembling "flying discs" have been sighted over Korea by crew members of two U.S. bombers... the objects were described by four eyewitnesses as globe-shaped, bright orange in color and emitting an occasional flash of bluish light... The Air Force, which has thrown cold water on hundreds of previous "flying saucer" stories, apparently was impressed by circumstances under which the new sightings were reported by its own personnel... the open-minded Air Force attitude toward the new reports contrasted with the blunt skepticism it has voiced about previous sightings of mysterious objects in the skies.

Quite significantly, this article commented on the new "open-minded Air Force attitude." On 20 February, the following article with a dateline of 19 February appeared in the *New York Times*:

The latest version of the five-year-old flying saucers—strange looking globes over Korea—is being looked into by the Air Force. Bomber crewmen have reported seeing globe-shaped objects of an orange color on flights over Korea on the nights of Jan. 29 and 30. An Air Force spokesman said today that the incidents were being investigated.

When Ruppelt previously briefed Garland on the fact that pilots were reluctant to report flying object sightings, word of it must have been passed around in high places. It could also have been a subject that Major Keyhoe may have complained to

Ruppelt about on his routine visits to ATIC. A warning must have then circulated that it was not wise to discount credible pilots, and not wise to make them reluctant to file reports. Still, it remains somewhat surprising that someone in the Pentagon would suddenly announce the Korean flying object incident to the public via a press release. Was this decision directed by a consensus of high officials, or did some general think it a good idea to now show new openness by the Air Force? Did it come from Garland, Samford, or perhaps Twinning? Ruppelt commented on the new Air Force attitude in the following:

> *...on February 19, 1952, the calm was broken by the story of how a huge ball of fire paced two B-29's in Korea...it was significant in that it started a slow build-up of publicity that was far to surpass anything in the past. This Korean sighting also added to the growing official interest in Washington...*

Senator Russell's Letter to Finletter

On 21 February, Senator Richard Russell sent a letter to Air Force Secretary Thomas Finletter asking that an official report be provided to the Senate Armed Services Committee regarding "Air Force evaluation of recent articles concerning the observation of 'flying saucers' by combat airmen in the Far East."

News of the Korean sightings was spreading far and fast, and now the Senate Committee needed to be involved. This meant that Ruppelt was also heavily involved with phone calls and messages from Washington, D.C. and collecting investigative information. It will be recalled from Chapter 23 that Senator Russell was seriously concerned about Air Force denial of flying objects, while many airline pilots of outstanding character were providing unsolicited testimony about reality of the objects.

About two weeks later, on 4 March, Senator Russell was provided with a response on the Korean incident, but it contained no substantive answer, except for prosaic possibilities. One possibility was that the flying objects may have been conventional enemy aircraft with searchlights attached.

Status Report No. 4

On 29 February, "Project Grudge - Report No. 4" was issued. It stated that more than 600 sighting reports were filed and cross-indexed under fifteen different categories. It was believed this covered a majority of incidents reported since 1947. The report also stated that Green Fireballs were receiving publicity over the last four months, but the report contained little else of interest except for sighting reviews that included the Mitchel AFB incident with Navy pilot James Zeitvoge, and the Korea incident with the B-29s.

Robert Ginna Revisits ATIC / TID

On 3 March, Robert Ginna of *Life* magazine came to ATIC for a second time. The first was in June 1951 (the previous year) to visit with Colonel Watson's new ATIC organization, but now Ginna was given a different kind of reception. Ginna was warmly welcomed by Ruppelt, and files were opened to him. After previously conducting months of personal research, Ginna understood exactly what he was looking for, and ATIC was anxious to give him all the information available, even by declassifying certain files. This cooperation was also extended to Ginna by Air Force Headquarters, with the offering of close coordination for his article, which would subsequently appear in the 7 April issue of *Life* magazine. This openness to Ginna was in stark contrast to his previous experience, and he must have been very impressed.

This turnabout in policy by Air Force Headquarters was quite unexpected. One wonders what took place for the change to occur, and what adjustments to personnel or attitudes were made. Was there a consensus regarding the turnabout, or was there a combination of agitated views and confusion involved, or did one high ranking individual, maybe Twining, make a decision to alter the status quo? Whatever it was, it must have made the Control Group more than a little nervous.

A point of interest on this is that positive change seemed to begin after General Twining entered the picture at Air Force Headquarters in the summer of 1950. Also, AFOIN now had a new director who may not have fully realized that public and press interest in the UFO question needed to be subdued. Whatever it was, there must have been a realization that the previous Air Force attitude toward the public and media was not working, or that it was not realistic to take a "two-faced" stance between the public and military. The big question, however, was whether Air Force Headquarters, or the Control Group, would ever allow release of the extraterrestrial answer, which was always a point of contention, because of need to protect the magnificent secret.

The Air Force was quite aware that a well-respected and very popular national magazine was about to expound on the UFO question, and the nation would soon acquire a new perspective. The Air Force needed to be prepared. But one also wonders what those in-the-know, and the Control Group, were thinking about this chain of events. There was likely much consternation, and perhaps a desire for Watson to return. But during Watson's tenure at TID, there was little information learned about flying objects. Now, the political arena and military officials in Washington, D.C. were aware of many unique sightings in Korea and elsewhere. It was still bothersome, however, to not know the origin and intent of the objects.

Project Twinkle Reclassification Denied

On 4 March, Major Fournet wrote a letter to Air Force Research and Development, which provided an answer to the request to reclassify Project Twinkle. It was signed by Air Force Directorate of Intelligence Executive Officer Colonel Kieling, and it stated:

> *It is not considered appropriate to declassify the project at this time since no definite conclusions have been reached concerning the origin or nature of the phenomena which have been investigated. It is believed that a release of the information to the public in its present condition would cause undue speculation and give rise to unwarranted fears among the populace such as occurred in previous releases on unidentified flying objects.*

On 7 March, Dr. Joseph Kaplan from AFSAB visited ATIC to discuss the situation regarding Green Fireballs. The fireballs were still of great concern to him because they were still being observed. And some had been observed quite recently. His continued interest in this matter is curious, because it brings to mind General Garland's extreme concern about origin and nature of certain flying objects that seemed to be targeting certain areas of the United States, which was reviewed early in this chapter. He was determined to set up special observation posts for radar and photo analysis of the objects.

This phenomena was still of great interest to some people, including Kaplan's AFSAB, which rejected the attempt to declassify Project Twinkle. Kaplan was now visiting ATIC to recommend several methods to "generate information" on the objects, primarily with sighting outposts using spectrum analysis instruments, which would establish or eliminate possibilities of object identity.

The fireballs were never something that GRUDGE was particularly interested in, since they were previously considered "natural phenomena," but New GRUDGE would now attempt to account for them, and Ruppelt informed Kaplan that action would be taken to determine feasibility of setting up a suggested program to learn more on the objects.

Based on a series of events involving fireballs over the last few years, it appears that scientists looking into the matter became convinced of a foreign effort to target certain areas of the U.S. with the speeding man-made objects. It also appears that high government officials may have learned the source of the fireballs some time ago, and that they constituted a potential form of psychological warfare. But scientists continued to remain in the dark, with high interest about the situation, while government high officials preferred to publicly ignore it for security reasons.

Colonel Doyle Rees of the 17th District OSI responded to Dr. LaPaz on the situation with the following:

> *I enjoyed very much receiving your letter of 3 March 1952 and learning that the fireball phenomena is still continuing...I do not know how much help it would be...to talk with OSI agents on the subject, because as you know, the investigative reports are classified.*

Navy Secretary Dan Kimball Reports Flying Object

Sometime during the week prior to 20 March, Navy Secretary Dan Kimball (a member of President Truman's cabinet) departed with Admiral Arthur Radford to visit several areas in the Far East. On a dark night flying over the Pacific, somewhere after leaving Hawaii and heading to Guam, they experienced a fantastic flying object encounter. Kimball was in a lead plane, and Radford was in a plane about fifty miles behind when Kimball's pilot suddenly came to the cabin to inform him that a flying object had just paced alongside the plane for some distance, and then it quickly shot up out of sight. Both the pilot and co-pilot observed the object, and Kimball suggested to the pilot that the plane behind be informed of the incident. After a couple minutes, Radford's plane also reported that an object appeared just off their wing tip, and then vanished skyward.

Immediately after landing, a report of the incident was relayed through channels to the Air Force. This was the appropriate procedure. When Kimball later returned to the Pentagon, he became very irritated when attempting to find out if the Air Force had processed his incident. Evidently, someone in the Air Force was flippant or unresponsive without releasing information to him. How this happened is not certain, but Kimball took immediate action to activate a Navy investigation into his incident, along other incidents, while keeping the Air Force uninformed. This abrupt response was likely a result of pent-up frustration, and it was a sign of no confidence with Air Force handling of the UFO question.

A few weeks after this incident, Kimball told his story of the flying object incident to a group of Navy officers and cadets at the Naval Air Station in Pensacola, Florida, where a few from the media were present. He remarked that he had total trust in his pilot, and that he had flown with him for a number of years and thousands of miles. The interesting part of this is that there is no indication in GRUDGE files that Kimball's incident was ever investigated, but the media did publish his story.

Little Cooperation Between Military

When one takes a look at military service cooperation on the UFO question, one can reasonably conclude that there was very little, or none. Limited information is available in regard to the Army, but some interaction did take place at times between Air Force OSI and its CIC Army counterpart in New Mexico and Texas. In regard to the Navy, it's apparent that the Navy's ONI and ONR had an interest in flying objects, but interactions with the Air Force tended to have negative overtones.

The thing that really provides an indicator of cooperation is the total lack of flying object reports from the Army or Navy in ATIC files. It might be expected that Air Force witnesses would comprise a majority of flying object sightings, but not to the great extent reflected in ATIC files. The Army and Navy were certainly encountering and sighting flying objects, but their files have never been released, or acknowledged.

❦

Beginning in February 1949, Project GRUDGE endured a long wild ride, and it would seem that any follow-on investigative project on flying objects would endure the same sort of thing. But one is tempted to wonder how long the charade would continue of investigation and denial? How long would the Air Force be able to balance its "see-saw" investigative efforts within a huge cover-up?

In the final months of Project GRUDGE, a new atmosphere of openness became evident, and there was an expectation that investigation of flying objects would be more proactive and productive. New GRUDGE provided a start with that, but another new chapter in the Air Force's investigative efforts on the UFO question (Project BLUE BOOK) was about to begin, and newly promoted Captain Edward Ruppelt would continue with that effort and lead the way.

Project GRUDGE appeared to live up to its name, and even more. The project was an unbelievable foul-up by the Air Force from its beginning. There was no conceivable way that a more chaotic project could have been scripted about "nothing of substance," or of flying objects that "do not exist."

The Control Group and higher levels in the Air Force were desperate to protect and manage their monumental secret, and they relied on lower levels of the Air Force to investigate that secret. One would think that simple intuitive logic would conclude that sponsoring an investigation of your own secret was not very smart. It was a no-win proposition, and it reflected corrupt and inept manipulation by high officials. But Truth of the UFO question and the magnificent secret of Roswell still remains intact and secure behind a wall of Air Force silence and perceived government ambivalence. But Truth is out there for those looking to find it.

Epilogue

Sometimes Truth is elusive, and sometimes it looks you straight in the eye. But even then it can sometimes be very difficult to recognize. And so it is with the UFO question, or an unearthly encounter.

When I started writing this book, I knew the Truth because I was exposed to it and experienced it. But that experience was like standing on the edge of reality, with my reality seriously challenged. It was brought into focus, however, by an Air Force OSI instruction stating, "as far as you are concerned, It Never Happened!" That statement affirmed that I was actually standing on the edge of Truth, peering under its covers, and it involved an unearthly flying object.

It was not until many years later that Truth of my exposure and experience with the UFO question was confirmed. I found that another missileer was involved in an identical experience a few months after mine, but in a different locality. I then realized I was helping to conceal an immense secret that was important to all of humankind, and this became a moral issue requiring serious consideration.

I then discovered other missileers I was associated with long ago that were also involved in their own special experience. They were also instructed by the Air Force to keep silent. This meant that the Air Force was truly involved in a great cover-up involving flying objects. But what was the extent of this, and what was the history behind it? I needed to find out.

When one looks at the history of Air Force investigative action on flying objects, it can be concluded that the government was and is deeply concerned about the objects in regard to national security. But it is desired that the concern not be passed on to the general public. There seems to be a consensus that the public does not have a need-to-know, or that the public would not be able to contend with Truth of the matter!

There has been, however, another side to this, which involves a myriad of questions, speculation, and conjecture about what is going on with the flying objects, and this is exactly what the Air Force, and also a compartmentalized agency of the government, has been critically concerned with. This was reflected by the need of the Air Force to press on with its contorted investigations, and the gathering of information on sighting cases and encounters. Those questions on what the objects were all about, and what they were up to, needed to have answers.

From my perspective, I wondered why and how our Minuteman missiles were taken off-alert by a mysterious flying object. I wondered why the harassing flying objects were playing around in our missile field, where the objects were observed hovering over missile launch facilities. If they were able to disable our missiles by rendering them unlaunchable, could they have also launched them? Could they have been more destructive in their activity, or taken control of the nuclear war heads? Why did they seem to exhibit a mostly harassing, teasing, and flaunting behavior, especially around military or nuclear areas? Why did they seem to be unapproachable, and evasive? One can speculate on it all and make conjecture, but that does not provide answers.

The continuing saga of Air Force investigation into the UFO question is not finished with this Volume 1, which is evidenced by its open-ended closure. But the Air Force's investigative efforts were continued in a revised format, in a project named Project BLUE BOOK, which was initially administered by Captain Edward Ruppelt. That project continued until 1969 when the project was taken underground. There can be no doubt, however, that there has been no less of a concern by Air Force and government agencies regarding the subject of Unearthly Flying Objects.

Because of the open-ended closure of this volume, unanswered questions remain, and the continued role of certain individuals previously mentioned remain for further exposure. Those people include, but are not limited to Lloyd Berkner, Vannevar Bush, Charles Cabell, Benjamin Chidlaw, Albert Chop, Ralph Clark, Edward Condon, Frederick Durant, Dewey Fournet, Miles Goll, Roscoe Hillenkoetter, Allen Hynek, Donald Keyhoe, Robert Low, Donald Menzel, Edward Porter, Roger Ramey, Howard Robertson, John Samford, Walter Bedell Smith, Hoyt Vandenberg, and Harold Watson.

This book documents involvement of many people associated with experiences and activities regarding the UFO question, which helps to reveal the colossal Air Force cover-up. It is observed, however, that there are no chapter endnotes in this book, or any references on where to find or verify information presented. This is intentional, so that people who have an interest will take it upon themselves to verify, confirm, and to also discover new important information that will compliment the scenario presented. Most all information in this book is searchable on the Internet, although great care must be taken to separate out the "garbage," which can easily be misidentified. That includes much information put forth by debunkers and skeptics, and also much disinformation presented by those with an institutional motivation to misdirect and mislead.

It is recognized that some people, because of their skeptical nature, will be inclined to classify much of this book, or a good part of it, as sensational fantasy. That is expected. It is also expected that mainstream media will ignore and stay away from what is presented in this book. They are frightened that it would not serve their "bottom line," or gain favor with the government. But if they were to make a serious effort to confirm and verify material contained herein, they would find what I found in regard to Truth and an Air Force cover-up. And they might find and discover a sensational story of their own regarding the magnificent and monumental secret, which should be made known to all of humankind.

Much of what I learned is documented in this book. However, as previously mentioned, it is only a smidgen of the extensive information available, and only a very small part of what continues to be locked away in government vaults and files. Much of the information is readily available, just by browsing the internet. I have taken some of it and placed it into a historical timeline so that a sequential understanding and bigger picture of Air Force, and government involvement, can be appreciated and make sense.

Some would say that "seeing is believing," especially for those without a sighting incident or encounter, but my situation was unique. I was never presented the opportunity to observe a flying object, but I certainly had to deal with the ramifications of one. I then found other missileers who had their own mysterious encounters. And I also discovered a massive amount of evidence regarding contorted efforts of the Air Force to investigate and cover-up the secret they were protecting. The Truth was presented and confirmed to me.

For me, involvement with the UFO question has been, and continues to be, an amazing chapter in my life. But I have no enthusiasm for receiving personal attention on this, although it comes with the territory. I do feel, however, an obligation to expose the Truth and communicate it to others. During the times I have been able to do that, I have also been amazed at the number of people who have confided to me their own personal experience, or beliefs. One of those in particular is an older gentleman by the name of "Scottie," or more formally named Mr. R. Scott Turner, who had an experience that looked him straight in the eye with the Truth.

Scottie lives with his wife at Harbour Pointe Senior Living in Mukilteo, Washington, where my ninety-six-year-old mother finished her last days about three years ago. I still occasionally visit with Scottie, and he continues to ask me if I've met anyone else "yet" who has seen a flying object. He tells me that he has lived with his experience "every day" for more than fifty years, and he is pretty much convinced that no one else has ever experienced what he did.

After learning of my experience, he confided his story of stopping at an intersection on a lonely country road (which he subsequently took me to) and a huge flying object suddenly appeared above the intersection and hovered in front of his car. It was in the middle of the day, and when he came home and told his wife and children his story, they did not take him seriously, and the children labeled him "Crazy Dad." His wife was more considerate, although it was also difficult for her to deal with the reality. She eventually understood he must have experienced a very startling and shocking incident, which has been reinforced in her because of his vivid memory of the experience after so many years. She recalls his great excitement when he came home directly after the incident. And today, his adult children try to give proper consideration to their father's story, and want to believe.

I've never previously met a person such as Scottie who was so relieved to reveal the experience bottled up inside him. He held back from relating it to others because of the initial reaction he received. At the time, he was a building contractor and housing developer, and he did not want to damage his professional career by relating a story that no one else would believe. His relief in telling me his story reminded me of the relief I once felt when my incident was confirmed to me.

Truth is out there! Hopefully, this book confirms the fact for others!

Index to Acronyms

AAF	Army Air Force
ABC	American Broadcasting Company
ADC	Air Defense Command
AEC	Atomic Energy Commission
AFB	Air Force Base
AFCRL	Air Force Cambridge Research Laboratory
AFIT	Air Force Institute of Technology
AFOAI	Air Force Office of Air Intelligence
AFOIC	Air Force Office of Intelligence Collections
AFOIN	Air Force Office of Intelligence
AFOIR	Air Force Office of Intelligence Requirements
AFOIR-CO	AFOIR Collections Branch
AFOSI/OSI	Air Force Office of Special Investigations
AFSAB/SAB	Air Force Scientific Advisory Board
AISS	Air Intelligence Service Squadron
ALCOM	Alaska Air Command
AMC	Air Material Command
AP	Air Police, Associated Press
ARPO	Aerial Phenomena Research Organization
ATIC	Air Technical Intelligence Division
CIA	Central Intelligence Agency
CIA-OSI	CIA Office of Scientific Investigation
CIC	Counter- Intelligence Corps
CIG	Central Intelligence Group
CIRVIS	Communications Instructions for Reporting Vital Intelligence Sightings
ConAC	Continental Air Command
CSI	Civilian Saucer Investigators

DCI	Director of Central Intelligence
DI	Director/Directorate of Intelligence
DMCCC	Deputy Missile Combat Crew Commander
DRB	Defense Research Board (Canadian)
DST	Daylight Savings Time
EADF	Eastern Air Defense Force
EEI	Essential Elements of Information
EMF	Electromotive Force
EMP	Electromotive Pulse
EOTS	Estimate of the Situation
EST	Eastern Standard Time
ETH	Extraterrestrial Hypothesis
EWO	Emergency War Order
FBI	Federal Bureau of Investigation
FCC	Federal Communications Commission
FEAF	Far East Air Force
FOIA	Freedom Of Information Act
FSC	Flight Security Controller
FTD	Foreign Technology Division
GCI	Ground Control Intercept
GRD	Geophysical Research Division (associated with AFCRL)
GSC	General Security Corp
HF	High Frequency
HRC	Human Resources Committee (associated with RDB)
IAC	Intelligence Advisory Committee
ICBM	Intercontinental Ballistic Missile
IE	Intelligence Estimate
IZ	Inner Zone Alarm
JANAP	Joint Army Navy Air Force Publication
JIC	Joint Intelligence Committee
JRDB	Joint Research Development Board
JWN	Joint Committee on New Weapons and Equipment
LCC	Launch Control Center
LCEB	Launch Control Equipment Bay
LCF	Launch Control Facility
LF	Launch Facility
MCCC	Missile Combat Crew Commander
MIT	Massachusetts Institute of Technology
MOL	Manned Orbiting Laboratory

MW	Missile Wing
NASA	National Aeronautics and Space Administration
NASIC	National Air and Space Intelligence Center
NBS	National Bureau of Standards
NDRC	National Defense Research Committee
NICAP	National Investigations Committee on Aerial Phenomenon
NORAD	North American Air Defense
NPSB	National Psychological Strategy Board
NSC	National Security Council
NSF	National Science Foundation
NSRB	National Security Resources Board
ONI	Office of Naval Intelligence
ONR	Office of Naval Research
OSI	Office of Special Investigations Air Force
OSRD	Office of Scientific Research and Development
OTS	Officer Training School
OZ	Outer Zone Alarm
PAS	Primary Alerting System
PIO	Public Information Office/Officer
PLSS	Portable Life Support System
PRP	Personal Reliability Program
RAND	Research and Development (independent corporate "think tank")
RAAF	Roswell Army Air Field
RDB	Research and Development Board (Department of Defense)
SAC	Strategic Air Command
SCP	Squadron Command Post
SETI	Search for Extraterrestrial Intelligence
SMW	Strategic Missile Wing
TCB	Technical Capabilities Branch (AFOAI)
TWA	Trans World Airlines
TWX	Teletypewriter Exchange (circuit switched messaging)
TID	Technical Intelligence Division
UFO	Unidentified Flying Object
UP	United Press
VIP	Very Important Person
VHF	Very High Frequency
USAFE	United States Air Force Europe
VRSA	Voice Reporting Signal Assembly
WWII	World War Two

Index to Names

Acknowledgements

Writing a book such as this can be fraught with a sense of trepidation, mainly because the U.S. Air Force and other entities of government would not want to acknowledge or confirm their cover-up of the UFO question. On the other hand, there have been many before me who have previously documented some of the same information provided in this book, which was obtained from released secret documents and government files. So, in another sense, I must sincerely thank the U.S. government for release of that information.

Of course, this book could not have been undertaken with out help and support of many individuals. I've received support from my local chapter of the Air Force Association, but I especially thank those missileers who supported me and urged me forward through many years of continued research and investigation. There are many of them, and they know who they are. I also thank my friend Tom Waterman and also missileer Bill DeGroodt who both previewed this book and offered advice.

I must sincerely thank my missileer friends who have allowed me to provide witness to their unworldly experiences, which helped confirm my experience. They include former missileers Daniel Grossman, Paul Johnson, Larry Manross, Dave Schuur, Val Smith, and former Air Force patrolman Wilbur Gunther. I also thank wives Cindy Johnson and Judy Schuur for lending continued support after their husbands passed. I must also thank missileer "Anonymous," because he broke his oath of silence just enough to confide to me his participation in a very significant incident (very possibly related to mine). And most of all, I thank retired Captain Robert Salas for initially exposing and confirming Truth for me. He is the one who helped extinguish a haunting memory, and bring it back into a confirmation of unearthly reality, which was to my great relief.

I also thank former boarder patrolman Don Flickenger for meeting with me and providing information and encouragement. He was the person I found in BLUE BOOK files, who then reconfirmed his own experience from fifty years previous. It was an incident that occurred near the general locality and time frame of my incident.

Thanks goes to James Klotz, who has long been involved with the phenomena of Unearthly Flying Objects. He provided information, counsel, and advice in my meetings with him, and he taught me how to retrieve information through FOIA. He garnered fame as co-founder of the Computer UFO Network (CUFON), and for filing over one-thousand FOIA requests. He is responsible for much information now available regarding government involvement with the UFO question. He also helped review this book and provided his endorsement.

Another who cannot go unmentioned is researcher Robert Hastings who wrote the Foreword to this book. He is a significant authoritative figure when it comes to associating flying objects with nuclear facilities and armaments. He has interviewed more than 150 military personnel, including myself in June of 2010, and has become a trusted friend. He is definitely someone who will maintain confidentiality for anyone desiring to disclose a flying object incident. He has written a book titled *UFOs and Nukes*, and also a second edition to it. He has also produced an award winning documentary film by the same name.

Someone who has contributed greatly in creation of this book is my copy-editor Diane English. With many years of experience in the editing field (and with the perfect name), this book could not have been published without her help. Prior to her, several years of work in content and copy-editing was provided by Juanita Homes, a very talented individual who is also an accomplished defense attorney. She was of great help and encouragement in initial phases of the book, while attending to a very busy life.

Last, and most important, I must give praise and provide loving thanks to my wonderful wife Diana Larson. I've sometimes wondered why authors tend to "gush" with effusive praise on sacrifices that significant others put up with, but now I definitely understand! I've been so lucky to have someone so accommodating and loving, despite the long hours, days, and years that have seemingly put her in second place to this book. She has my everlasting gratitude and love. It's going to be hard to pay her back.

Thank you everyone!

About the Author

At seventy-six years of age, I often find myself thinking about the years in my past and events that have led to where I am today. I guess that is what happens after becoming a "senior citizen" and taking time to reflect on it all. But, actually, I really haven't had time to reflect. Time is rushing by so quickly that I wonder where I'll find time to do all the things I want to do in the years to come. And this book has created an interruption to my life. It has kept me from doing so many other things that are still on my mind.

At this particular age, I used to think that one would be sitting in a rocking chair all day and waiting for time to pass, but instead I have been sitting in front of a computer each day and writing a book, which I never previously dreamed of doing. I want to get back on my bicycle and return to cycling. I want to get into the mountains for some hiking and mountain climbing. I want to get back on my downhill skis when the snow flies. I want to take my sailboat for a long cruise, or tour the United States in my motorhome. And I want to get back into photography. After being retired for almost fifteen years, there is much more to do in my life than sit in front of a computer all day. I need to get back to the passions in my life, and get back to what I really enjoy doing, including more interaction with family and friends.

Don't get me wrong, the writing of this book has been my goal for many years, and Volume 2 has been underway for awhile. This book became a self-imposed obligation to inform the world about Truth of the UFO question, or at least what

I know about it.

In mentioning that, I'm reminded of my fiftieth year reunion at Washington State University that occurred in April of 2013. My wife Diana and I were at a luncheon sponsored by the School of Sciences, and we sat at a table with an astrobiologist and other professors, which included my old physics professor and counselor Dr. Edward Donaldson. It was a great surprise to discover that I was seated next this former mentor, for I never expected to see him after so many years. After a bit of conversation, I was suddenly presented with a unique opportunity to publicly address the UFO question and relate my flying object experience. Everyone around the table took careful notice of what I said, and then they became speechless for a brief moment–perhaps checking out if I was really all there or not. But the response I received was quite positive, and I then became encouraged about relating Truth of my experience to others.

In regard to my previous Air Force life, I was twenty-three when I entered the service, and I was active for about five years. In hindsight, I should have made the Air Force a career, since those years were some of the best in my life. Afterward, I stayed in the Air Force reserve long enough to obtain my retirement papers and retain my officer commission.

After a short stint as an aerospace engineer, I then settled into a long career as a computer systems analyst in Seattle, Washington where I worked for two major companies. I also raised a family, which included a daughter and son, who each provided me with two grandchildren. A later marriage to my wife Diana Larson included her family of two children and four more grandchildren. In between all that, I became a certified data processing professional, and also a part time professor at a community college.

It has been a good life, and it included a unique, amazing, and out-of-this-world experience. Hopefully, that experience and Truth of the UFO question will one day be confirmed to the whole world by high national authority!

Made in the USA
Lexington, KY
21 June 2017